Quantum Mechanics—
What is wrong with it and how to fix it

Proof that Einstein Was Right

NEW EDITION

By Janeen Hunt

Quantum Mechanics—What is wrong with it and how to fix it is published by Lulu Publishing, 860 Aviation Pkwy, Ste 300, Morrisville, NC 27560 USA, www.lulu.com

ISBN: 978-1-4303-0967-3
Fourth Edition

© 2006, 2010, 2017, 2019 by Janeen Hunt

All rights reserved. This publication is protected by federal copyright law. No part of this publication may be reproduced, stored in a retrieval system, or transmitted in any form or by any means, electronic, mechanical, photocopying, recording, or otherwise, or be used to make a derivative work (such as translation or adaptation), without prior permission in writing from the author.

Contact information: Janeen Hunt, California, USA, e-mail: Voyajer17@hotmail.com

New Edition Prologue

It is the fall of 2019 and I have finally gotten acceptance of my paper on a new model of the atom in the online journal entitled American Based Research Journal. That is very exciting news and I will copy the published article into this book.

Since I first wrote this book in 2006, I was finally able to discover an amazing proof of my initial thesis by searching the NIST (National Institute of Standards and Technology) tables of the spectrum of the elements. I was astonished to find that by using my original assumption noted in this book in 2006, that the Balmer-Rydberg equation that predicts the spectral lines of Hydrogen is able to be discovered in the first six elements of the periodic table and their cations including Deuterium. This discovery altered some of my initial assumptions about what the new atomic theory should look like, but it also forced me to some new discoveries and the incredible ability to actually predict the ionization energy of these twenty atoms directly from their spectral data…an unheard-of feat. Other patterns began to emerge with each new finding and I found that I have just discovered merely the tip of the iceberg.

I can now see why the Schrödinger equation and the Heisenberg Matrix Mechanics equation are able to approximate nature although they can only calculate the spectral series of Hydrogen and single electron Helium. It is because the Balmer-Rydberg Hydrogen equation can be found in all the elements that I have evaluated. It is more than gratifying. It is incredibly exciting. I hope that you can share my excitement too as you read through this book to find why a new theory is necessary and how that new theory will open many doors to a whole new world of science.

I've already had many skeptics and outright hostility towards even a thought of trying to change Quantum Mechanics as it stands today, even from the very scientists who say they pray for a simpler, more unified theory. But that is always the way. I think of Copernicus and Galileo and I am content. I'm rather more like Tycho Brahe or a Michael Faraday. I found the data and I need a Kepler or a James Clerk Maxwell to find the mathematical formula that will explain the system.

I trust you will find the history of Quantum Mechanics written herein very interesting, sometimes disheartening, and sometimes entertaining, but definitely showing the necessity of change. I know you will be as surprised as I was at the true history of QM.

So let's start at the beginning, because you will find we have just begun and there is so much more to know. So, ENJOY!

Quote from John Bell, 1989: *"The aim remains to understand the world. To restrict quantum mechanics to be exclusively about piddling laboratory operations is to betray the great enterprise. A serious formulation will not exclude the big world outside the laboratory."*

Excursus

When I wrote this book in 2006, my goal was that some fresh, young scientific mind might take up the idea and run with it. Unfortunately, for most of us, life is about making a living and sometimes just surviving. It is a thrill to be able to take the time to understand the universe in which we live. However, sometimes time is a commodity that most of us don't have. The theory presented in this book from 2006 has undergone revisions as I have studied more, however, I have not been able to completely revise this book which presents some ideas that I no longer hold to. But this book shows the evolution of my ideas and is therefore useful. Unfortunately, I've only had time to think about the subject for a little in 2010, and only just now again I've taken up where I left off and made some progress. I hope you will find this little excursion interesting. I will try to update obvious errors herein when I get the chance in future.

In the words of the cousin of Henri Poincaré in 1909:

"If we now disentangle ourselves from contingencies, it will be understood that in reality physical science progresses by evolution rather than by revolution. Its march is continuous. The facts which our theories enable us to discover, subsist and are linked together long after these theories have disappeared. Out of the materials of former edifices overthrown, new dwellings are constantly being reconstructed.

"The labor of our forerunners never wholly perishes. The ideas of yesterday prepare for those of tomorrow; they contain them, so to speak, in potential. Science is in some sort a living organism, which gives birth to an indefinite series of new beings taking the places of the old, and which evolves according to the nature of its environment, adapting itself to external conditions, and healing at every step the wounds which contact with reality may have occasioned.

"Sometimes this evolution is rapid, sometimes it is slow enough; but it obeys the ordinary laws. The wants imposed by its surroundings creates certain organs in science. The problems set to physicists by the engineer who wishes to facilitate transport or to produce better illumination, or by the doctor who seeks to know how such and such a remedy acts, or, again, by the physiologist desirous of understanding the mechanism of the gaseous and liquid exchanges between the cell and the outer medium, cause new chapters in physics to appear, and suggest researches adapted to the necessities of actual life.

"The evolution of the different parts of physics does not, however, take place with equal speed, because the circumstances in which they are placed are not equally favorable. Sometimes a whole series of questions will appear forgotten, and will live only with a languishing existence; and then some accidental circumstance suddenly brings them new life, and they become the object of manifold labors, engross public attention, and invade nearly the whole domain of science."

—The New Physics by Lucien Poincaré

—March 25, 2017

Original Preface

One begins to study science when one first asks, "Why is the sky blue?" possibly even as a toddler. Not too long ago in history, the answer to questions about the world was that 'the gods willed it to be so'. Winter was a time when the gods turned their back in anger upon humankind and spring was the god's return of good will. Science was the beginning of our search for a rational reason to what we sensed in the world around us. After one studies science, one feels confident that what one experiences happens for a logical reason. If a rock drops, an airplane flies, there is a day and night, and the stars shine, there is a cause other than the will of the gods. There is strength and comfort in this knowledge. It appeals to our logic, our sense of order. ... Then one studies Quantum Mechanics and one is confronted by a science whose underlying theory is one that defies logic, reason, and the rational. Why does it? And does it have to be so?

This is the reason that I have written this book. It is firstly to show that Quantum Mechanics is merely one interpretation of the data. Any person knows that the same incident can be interpreted in different ways. A car accident is one example where each observer will put his or her own construct upon what happened. There is no doubt in anyone's mind that the accident actually happened, but the interpretation is open to debate. In the same way, in the development of Quantum Mechanics, the experiments that led to the theory are no way in question. There can be no doubt that the early Rutherford experiments showed that the atom had a nucleus with an opposite charge to the electrons, but is the interpretation correct? Is there another way to interpret the data that would make more sense, be more logical, and fit into the category of scientific rationale better than the explanation given by Quantum Mechanics?

It is the purpose of this book to explain how Quantum Mechanics developed, its loopholes, its leaps of faith, and its sometimes--faulty reasoning. We will examine the words of the men who constructed the theory through a windy and often disconnected path. We will read quotations and explanations that have been lost through the years. This is important, because as Richard Feyman said of Quantum Mechanics, "You know how it always is, every new idea, it takes a generation or two until it becomes obvious that there's no real problem." But why are later generations more disposed to accept a theory? Is it not because we become accustomed to it? For over 2,000 years humans accepted the theory that the earth was the center of the universe although the data could have been interpreted either way. Aristarchus of Samos (c. 270 BC) proposed that the sun was at the center. The data was the same; the interpretation was different. It wasn't until man made meticulous observations and had the capability of the telescope that science could embrace a theory closer to reality. The same is true of Quantum Mechanics. The data suggests more than one interpretation. This book is meant to expose its flaws and examine the alternatives in an attempt to deter atomic science from stagnation. It is my hope that this book will stimulate lively discussions, important questions, more research, and new ideas. There are some amazing quotes and stories too. Enjoy the ride.

— October 18, 2006

Dedicated to all women and men who have ever thought about the universe and tried to make sense of it.

Special Dedication to Henrietta Swan-Leavitt and Rosalind Elsie Franklin

Quantum Mechanics—
What is wrong with it and how to fix it

Table of Contents

1 Quantum mechanics 1
2 Problem with current model of the atom 3
3 Quantum Regrets 15
4 Theories and Philosophy 20
5 Quantum Mechanics Did Not Explain Chemistry 35
6 New Theory 38
7 A new look at the Davisson-Germer experiment 44
8 The Nucleus of the new model 50
9 A look at Bose-Einstein Condensate 54
10 Binding Energy 57
11 Electron Shells 70
12 Radioactive Radiation 91
13 Molecular Bonding 94
14 The Franck-Hertz experiment 95
15 Suborbitals 96
16 The Strong Force 111
17 The Standard Model 115
18 The Formation of the Wave 119
19 Electromagnetic Waves 127
20 Electromagnetism 136
21 Electromagnetic Fields 140
22 Electricity, Classic Electrodynamics, and the Atom 141
23 Doppler Shift 153
24 Feynman Diagrams 155
25 The Partial Reflection Of Light 160
26 On Liquids and Fluidity 161
27 The Uncertainty Principle 164
28 Real Problems for Quantum Mechanics 177
29 Inertia and motion 185
30 Heat 189
31 Heat and Work 191
32 More on Properties of the Wave 193
34 The Implications for Gravity and the Structure of the Universe. 196
35 Energy 197
36 Time 199
37 In Summary 200
38 Paper Proving New Theory………………………………………………………………………198
References: ii

1 Quantum mechanics

Quantum mechanics is a set of rules governing the atom and the subatomic particles. The rules do not determine a physical model of the atom nor can they predict anything about a single atom. The rules govern the probabilities of an atomic system. Natural laws are governed only by statistical characteristics and statistical interpretation. Niels Bohr, the founding father of quantum mechanics, explains that quantum mechanics is defined by "symbolic" operators of "non-pictorial character" and a "renunciation of detailed space-time coordination of the particles" and that in the treatment of atomic problems in quantum mechanics "we are here dealing with a purely symbolic procedure".[90,p.2-5]

A brief history of quantum mechanics (also called Quantum Physics), which is science's mainstream theory of atomic phenomena, begins with Max Planck's radiation formula in 1900. Max Planck saw that empirically observed radiation does not gain infinite energy and in order to show this mathematically, he introduced a constant represented by the letter h which when multiplied by the frequency showed the energy of radiation, that is, electromagnetic waves. Planck's constant called the "quantum of action" was able to mathematically predict the behavior of radiation at increasing temperatures.

This led to Albert Einstein postulating in 1905 that the quantum was actually a particle of light called the photon. As shall be shown, in 1913 Niels Bohr introduced the first scientifically accepted use of Planck's constant to describe electron orbits in the atom which later in 1925 led Werner Heisenberg to formulate mathematics to describe electron transitions in the atom that would predict the atomic spectrum of hydrogen-like atoms and theoretically should apply to all atoms. The next year Erwin Schrödinger incorporated the wave behavior of electrons into a substantially different mathematical formula to describe electrons in the atom. Max Born, a mathematician, taught Heisenberg to multiply his matrix quantum mechanical formula and re-interpreted Schrödinger's equation to be statistical. As Max Born himself describes the development of quantum mechanics and its meaning:

"From the standpoint of these lectures on cause and chance it is not the formalism of quantum mechanics but its interpretation which is of importance. Yet the formalism came first, and was well secured before it became clear what it really meant: nothing more or less than a complete turning away from the predominance of cause (in the traditional sense, meaning essentially determinism) to the predominance of chance."[97]

The development of quantum mechanics centered around attempts to describe the atom. This became essential when it was found that the atom was not a single particle but consisted of a charged nucleus and oppositely charged electrons. The main two methods in which the atom could be studied were spectroscopy and scattering experiments. Two facts were known about the atom by 1911 from experimentation. One was that the atom had a small particle called the electron with a negative charge and that therefore the atom was not the single smallest particle of matter. The second fact was that the atom had a compact oppositely charged nucleus, yet the electron generally remained stable at a distance. From spectroscopy it was known that the light from atoms was emitted and absorbed at discrete frequencies. Making sense of the data from these two fields was necessary for a model of the atom that focused on explaining why oppositely charged electrons remain stable around the nucleus.

Current theory is based on the assumption that nothing exists between the nucleus of the atom and the first electron shell. Current theory is based on the assumption that the electron orbiting the proton in a hydrogen atom does not fall into the nucleus because it is quantized. What did that mean when Bohr said it? Bohr meant that the mathematical measurement between the nucleus and the electrons was separated by a space that electrons could not exist in. This space was assumed to be a void. Bohr gave no reason for this. It was simply assumed.

However, this book intends to build on a different assumption. That something physical exists in the space between the nucleus and the electrons that creates the appearance of quantization, but is an actual "thing" that forces the electrons away from the nucleus, but is permeable.

The experiments have already been done. Cathode rays showed the atom had a negative particle. Rutherford showed that the nucleus had a positive charge and that most of the atom was permeable. This new assumption simply states that the space that is permeable is not void. The experiments previously done, such as the Davisson-Germer Experiment, can be explained more completely using this new assumption as will be shown.

By reinterpreting each experiment on the atom (not by re-doing the experiments) using this assumption, many experimental results are able to be explained that cannot be and never were explained under current Quantum Mechanical theory. Therefore, this assumption is very useful in explaining real phenomena.

The assumption that something exists between the nucleus and the electron orbits causing the space and separation of the electrons does not negate all of Quantum Mechanics. It merely enhances Quantum Mechanics and gives it a better model to explain the previously unexplainable. Many of the precepts of Quantum Mechanics will be shown to still hold such as Quantum Tunneling. However, tunneling will be seen in a new light and actually have a causation.

The reason this is important is not because Quantum Mechanics has been unsuccessful. To the contrary, Quantum Mechanics when one thinks of Einstein's photon theory is one of the most successful theories of science. However, there are experimental results that Quantum Mechanics does not answer in the matter of the atomic model. And current Quantum Mechanical theory cannot unite with the force of gravity under General Relativity. This book will attempt to show that introducing this new assumption will not only explain reality better, but will open up new vistas in science and make predictions beyond the limitations of the current model.

2 Problem with current model of the atom

It is a well-known fact that in 1913, when Niels Bohr introduced the idea of discrete orbits into his model of the atom, he did so by defying the laws of electrodynamics and introducing what, at the time, was an arbitrary constant into the mathematical formulation of the atom.[6] His formulation of the atom is incorporated into quantum mechanics as it is known today.

Werner Heisenberg, a co-founder of quantum mechanics associated with Bohr, explains: "According to Bohr, the atom consisted of a relatively heavy atomic nucleus, surrounded by electrons, just as the sun is girdled with planets. To this planetary system the same mechanical laws were applied as those used in astronomy, namely, the laws of Newtonian mechanics. At the same time, however, it was claimed that there could only be quite specific electron pathways, marked out by quantum conditions. This statement contradicted Newtonian mechanics, since according to the latter, an external perturbation could easily convert a quantum orbit into one that was not permissible in quantum theory. But in reality, it seemed, for example, that an incoming light beam would lift the electron discontinuously from one quantum orbit to another."[74,p.110]

To create his model of the atom, Niels Bohr introduced Planck's constant, which was used by Max Planck for a completely different purpose, that being, to divide an electromagnetic wave into energy packets.

Before we continue, we should thoroughly consider what h represented when Planck created his constant and why h divides energy into packets. Planck had been taught by Gustav Kirchhoff. In 1859, Kirchoff proved an important theorem "that the amount of energy a blackbody radiates from each square centimeter of its surface hinges on just two factors: the frequency of the radiation and the temperature of the blackbody. He challenged other physicists to figure out the exact nature of this dependency: What formula accurately tells how much energy a blackbody emits at a given temperature and frequency?"[i] Wilhelm Wien created an equation from the observed experimental data. However, it was found that Wien's Law broke down in the cooler far-infrared frequencies. *Id.* This is confirmed by Asimov who explained, "Wien worked up an equation designed to express what was actually observed at high frequencies. Unfortunately, it did not account for the distribution of radiation at low frequencies."[ii] "To put a value on molecular disorder, Planck had to be able to add up the number of ways a given amount of energy can be spread among a set of blackbody oscillators; and it was at this juncture that he had his great insight. He brought in the idea of what he called energy elements – little snippets of energy into which the total energy of the blackbody had to be divided in order to make the formulation work. By late 1900, Planck had built his new radiation law from the ground up, having made the extraordinary assumption that energy comes in tiny, indivisible lumps. In the paper he wrote, presented to the German Physical Society on December 14, he talked about energy 'as made up of a completely determinate number of finite parts' and introduced a new constant of nature, h, with the fantastically small value of about $6.6 \times 10\text{-}27$ erg second. This constant, now known as Planck's constant, connects the size of a particular energy element to the frequency of the oscillators associated with that element."[iii]

So Planck's constant, h, represented the smallest indivisible piece of energy and mathematically describes the frequency of radiation emissions. Since all radiating light is emitted by heat, then all electromagnetic waves should correspond to Planck's radiation analysis. Heat produces light. The sun is hot and radiates light. Both heat and light are kinds of energy because they can produce work. Therefore, all light can be described by Planck's theory of quantization. All electromagnetic waves follow Planck's theory of quantization. Einstein himself showed this in 1905 when he used the Planck constant in describing the Photoelectric Effect which was the energy of light causing electrons to be released from metals and could be described by Planck's constant times the frequency of the light. Therefore, Planck's constant was inextricably intertwined with light and electromagnetic waves.

Given what h represents, it then becomes clear why it is strange to introduce h into the atom and as we shall see, Bohr had to make some very strange assumptions to deal with the frequency problem. Planck's constant h was referred to at the time as the "quantum of action" because it specifically meant that energy did not create a frequency or "action" until the energy level reached the Planck minimum of h. Energy multiplied by time is considered "action" by physicists. Planck's quantum is in erg-seconds because light travels and time is involved in frequency. But the trouble was, in the Bohr model of the atom in 1913, neither the nucleus of the atom nor the electrons around the atom were thought of as energy. The only frequency was the circular electron orbits, but they didn't relate to the color of the spectrum of the element. Rather the particles in the Bohr atom were thought of as mass with opposite electrical charges that should attract each other. And h was known to be a packet of energy that created a frequency.

Nevertheless, Bohr used the constant in order to create the spacing between electron orbitals. However, when Niels Bohr did this, he did not consider the atom to be an electromagnetic wave, nor did he consider the space between electron orbitals to be packets of energy. In fact, wave-particle duality had not yet been proposed and wasn't proposed until 1923 by Louis-Victor de Broglie.[12] Therefore the introduction of Planck's constant into the atom was completely arbitrary at this point.

To match the discrete spectral lines, Bohr decided that each electron orbit should be a whole number "n" where the angular momentum was "n" times Planck's constant "h" times a multiplying factor "K".[6] This was a purely mathematical model and showed no real scientific *physical* cause for the separation of the electrons from the nucleus and, in fact, Bohr's model defied known laws.

Bohr suggested in a memorandum to Ernest Rutherford, famous for discovering the nucleus of the atom, in the summer of 1912 that the introduction of a constant K into the atom along with a constant for proportionality between energy and frequency, i.e. Planck's constant, would fix the orbits of the nuclear atom.[23] The approximate size of the atom had been known for over 40 years since the German chemist Joseph Loschmidt had first calculated it in 1865.[22] However, it must be pointed out that Bohr did not originate the idea of using Planck's constant at the atomic scale since in 1910 it had already been introduced into the model of Arthur Erich Haas. And in 1911, John W. Nicholson had introduced Planck's constant in his atomic model to interpret the lines in the atomic spectra.[85] Bohr himself explains about the development of the use of Planck's constant:

"Further, J. Nicholson had in 1912 made use of quantized angular momenta in his search for the original of certain lines in the spectra of stellar nebulae and the solar corona. Above all, however, it deserves mention that, following early ideas of Nernst about quantized rotations of molecules, N. Bjerrum already in 1912 predicted the band structure of infra-red absorption lines in diatomic gases, and thereby made a first step towards the detailed analysis of molecular spectra eventually achieved on the basis of the subsequent interpretation, by quantum theory, of the general spectral combination law."[90,p.35]

The events that preceded Bohr's theory led Hendrik A. Lorentz to say at the Solvay Congress in 1911 that Planck's constant probably has something to do with atomic size or atomic orbits.[24] At the time, Planck's constant was the talk of the physics world with Einstein's introduction of the photon in 1905 and with the debate raging about the Ultraviolet Catastrophe rather than using Planck's constant. It was known that Planck's constant was comparable to atomic scale sizes. However, the conclusion that Planck's constant should be introduced into the atom is a slippery slope argument i.e. Planck's constant is small so anything small should contain Planck's constant. So with its popularity and with the frenzied enthusiasm over it, it is not surprising that Planck's constant was considered for the atom. At the 1911 Solvay Conference one account describes: "There the idea that Planck's quantum would be the key to all outstanding problems on radiation earned a thorough airing. One of the participants, Sommerfeld, observed in reply to probing by Lorentz that there must be some relation between the magnitude of Planck's constant h and the dimensions of the atom as viewed in Thomson's model."[85,p.154] As Bohr himself explains: "The very theme of the first Solvay conference in 1911, Radiation Theory and Quanta, indicates the background for the discussions in those days….As regards his own attitude, Sommerfeld added that instead of trying from such considerations to deduce Planck's constant, he would rather take the existence of the quantum of action as the fundament for any approach to questions of the constitution of atoms and molecules."[90,p.82]

Therefore, Bohr's use of Planck's constant was anticipated, expected and even stipulated by a quorum of physicists. Bohr initially wanted to justify the use of Planck's constant by considering that a circular periodic orbit of an electron particle on the plane of the ecliptic to the nucleus could be considered a frequency, but that frequency was completely ignored. However, after Bohr had made further calculations he realized that the full Planck's constant could not be used to restrict angular momentum otherwise the atom would be too large, therefore he restricted angular momentum of electron orbits to ½ of Planck's constant citing Planck's use of 1/2h in a 1911 paper.[48] Therefore, Bohr introduces ½ a frequency saying: "Let us now assume that, during the binding of the electron, a homogeneous radiation is emitted of a frequency ν, equal to half the frequency of revolution of the electron in its final orbit…"[6] Already this doesn't agree with Planck's formula. Planck's formula said nothing could be smaller than his constant h.

Bohr understood that a radiating electron would lose energy and fall into the nucleus and therefore introduced a theory of the atom in which the electron does not radiate nor make slight changes or shifts in orbits by here introducing one-half of a quantum into the formula. Bohr in 1913 and Einstein in 1905 both said in their papers that particles radiate when they move. This was due to two known effects of the time. One came from electrons in a cathode ray tube and the other came from the atomic spectrum. In both cases, when the electrons moved through the cathode ray tube they produced light across a continuous spectrum and when the electrons moved in the atomic spectrum light appeared as discrete lines in the spectrum. Neither of them knew of the Auger effect or Bohr might have modeled the atom differently. But Bohr knew that normally an electron moving in a vacuum tube radiated and it appeared from the spectrum that moving electrons radiated light. Also, under the Maxwell equations, acceleration creates radiation, so electrons in a circular orbit are experiencing acceleration and should therefore radiate. So Bohr postulated the electron making one-half of an orbit and then making an orbital shift to a different shell to radiate. That way the one-half orbit of the electron wouldn't radiate and only orbital transitions would radiate, but not at the frequency of the orbit.

The problem with this rationale was that Planck in his 1911 paper stated that light could be absorbed at any amount, however, only emitted in whole quanta. Therefore, $1/2h$ could be left unemitted. (This theory eventually led to the theory of zero point fluctuations.) Bohr's actual 1913 formula states that the kinetic energy $W = th1/2w$, where t is "an entire number" and w is one revolution. So he writes the formula as if the energy emitted comes from one-half a revolution which is still the same as one-half a quantum. He must use a frequency and electron revolution in its orbit to excuse his introduction of Planck's constant into the atom because Planck's constant was explicitly known to be a radiation law constant having to do with heat. Because the majority of heat is in the infrared, Planck's constant also has to do with frequency. But even Einstein pointed out that the frequency of the orbit did not match the frequency of the atomic spectra.

So to excuse introducing Planck's constant, Bohr says a "radiation is emitted" equal to half a frequency and then states that there is no radiation: "According to the above considerations, we are led to assume that these configurations will correspond to states of the system in which there is no radiation of energy states which consequently will be stationary as long as the system is not disturbed from outside."[6] (Bohr explains later that "stationary" means "stable".) [90,p.37] It appears that Bohr is trying to explain that electrons consistently make quantum leaps before they make an entire revolution in their orbit. In this way no radiation would be emitted at one-half a cycle according to Planck's theory of 1911. It seems certain from his paper that this is why he introduced quantized angular momentum at ½hπ, that is, in order to say that the electron has not formed an entire orbit therefore no radiation is yet emitted. He speaks of "the frequency of the energy emitted during the passing of the system from a state in which no energy is yet radiated out to one of the different stationary states…"[6] This was ingenious, that is, the use of Planck's theory to postulate one-half of an orbit in order to say that radiation of the electron had not "yet" occurred. However, Bohr could not carry out this argument to its proper conclusion, that conclusion being that every electron made a transition before it could radiate. Instead this use of orbits of ½ revolution was not kept in modern quantum mechanics so that the whole reason for restricting electron angular momentum to multiples of ½hπ is no longer valid and the solution is therefore arbitrary and has no basis whatsoever in observation of any kind. There is nothing in the atomic spectrum or experimentation that suggests that angular momentum is empirically restricted to ½hπ. Bohr in the final analysis himself shows that electron orbits do not always transition after one-half a revolution because they maintain a natural circulating ground state orbit that he calls "permanent" meaning there is no transition at one-half the revolution so indubitably there must be radiation under Planck's theory. Bohr's 1913 theory says "the permanent state" is t=1 which we call today "ground state" in which there is a circular orbit in which the angular momentum is not zero but **h**/2pi. So, in this case, the electron cannot be making only one-half a revolution if it stays in this normal "permanent state". Therefore at every point in which he tries to incorporate scientific data to fit known physics, it is thereby countered with an assumption that contradicts it.

Not only was the size of the atom known by Bohr, but Bohr says the expected ionization value was already known to science by 1913, as Bohr says in the article, by mathematically manipulating Planck's constant, one can arrive at this explanation: "We see that these values are of the same order of magnitude as the linear dimensions of the atoms, the optical frequencies, and the ionization-potentials." He states that measured absorption of radiation is known "in experiments on ionization by ultra-violet light and by Röntgen rays" (known from radioactive radiation) and in the photo-electric effect.[6] Therefore Bohr is introducing arbitrarily chosen values such as Planck's constant, without good cause since an orbit is not a true Planck radiation frequency, and combining them with known ionization energy values to create a mathematical formula that approximates the known data.

Bohr introduces the principle quantum number "n" describing the angular momentum of each electron shell in this 1913 paper by using the arbitrarily introduced Planck's constant and the arbitrarily introduced one-half revolution to presume a non-radiating electron. This is crucial because the principal quantum number of the Schrödinger and Dirac equations, upon which foundation lie quantum mechanics, retains these same properties introduced by Bohr in 1913. This theory of the atom still holds today that the moving electron does not radiate in its electron shell and has arbitrarily restricted orbits. However, this is the same electron that radiates in a CRT monitor or television tube. This is the same electron that produced the electromagnetic radiation described by Maxwell and found by Hertz.[55] This fact has always been a great unsolved paradox which science tends to be comfortable with ignoring or else is completely unable to explain. Bohr's limiting the emission of electrons to discrete wavelengths does not solve the unlimited wavelengths able to be produced in classic electrodynamics. QED, or quantum electrodynamics, also does not solve this matter as will be discussed.

The introduction of an unrelated constant into the model of the atom began with the gold foil experiment. Ernest Rutherford (1871-1937) and his associates showed from this experiment, in which alpha particles were passed through various metal foils, that "in the passage of *a* [alpha] particles through matter the deflexions are, on the average, small and of the order of a few degrees only" and at other times a small fraction of alpha particles "are sometimes turned through very large angles" from which they deduced that there was a concentrated positive charge at the nucleus.[20] In his 1911 paper, Rutherford concluded: "In comparing the theory outlined in this paper with the experimental results, it has been supposed that the atom consists of a central charge supposed concentrated at a point, and that the large single deflexions of the *a* and *b* particles are mainly due to their passage through the strong central field."[4] This has been taken to mean by the scientific community ever since that the atom "is mostly made of empty space"; it is mostly a "void". This is an unscientific conclusion in that it is so dogmatically black and white without allowance for other possibilities. Certainly the gold foil experiment proves that the positive charge was concentrated into a nucleus in the atom, but it didn't prove that between the electrons and the nucleus there was "empty space". An alpha particle of radium is 4.87 MeV and is moving with a speed at roughly 5% of c, the speed of light.[21] The alpha particle has a positive charge that is absorbed in thick materials but not entirely in thin foils. The alpha particle did not pass straight through the foils without any deflection. Therefore, the space between the nucleus and the electrons in the gold foil experiment could have a certain density as long as it was permeable enough to allow passage of alpha particles traveling at nearly relativistic speeds. For example, this space could be filled with a neutral particulate as long as it was permeable enough for alpha particles to pass through it with ease. Instead, by the assumption that the atom was mostly "empty space" between the nucleus and the electrons, it became necessary for Rutherford to assume that the electrons were circling the nucleus in orbits in order not to fall through the "empty space". This led to the adoption of orbits for the electrons and a planetary model of the atom. However, under classical electrodynamics, a moving charged particle should emit radiation, lose energy and fall into the nucleus, therefore, circling orbits were not a viable solution. Rutherford in a 1914 paper says that Bohr had noticed this problem:

"Bohr* has drawn attention to the difficulties of constructing atoms on the 'nucleus' theory, and has shown that the stable positions of the external electrons cannot be deduced from the classical mechanics. By the introduction of a conception connected with Planck's quantum, he has shown that on a [sic] certain assumptions it is possible to construct simple atoms and molecules out of positive and negative nuclei, e.g. the hydrogen atom and molecule and the helium atom, which behave in many respects like the actual atoms or molecules. While there may be much difference of opinion as to the validity and of the underlying physical meaning of the assumptions made by Bohr, there can be no doubt that the theories of Bohr are of great interest and importance to all physicists as the first definite attempt to construct simple atoms and molecules and to explain their spectra."[4] As Rutherford here points out, Bohr's model of the atom would raise a question of "validity" and there appeared to be missing an "underlying physical meaning" to Bohr's model. Yet, today quantum mechanics still holds firm ground nearly a century later based on Bohr's assumptions even though Bohr's model of the atom was created before it was known whether the nucleus had a positive or negative charge. And this was at a time when it was thought by Rutherford as he proposed in the same paper that helium being four times heavier than hydrogen contained four protons ("positive electrons").[5] Rutherford here himself states that this is a "first definite attempt", yet this idea of introducing Planck's constant at one-half angular momentum has persisted. Certainly we have made enough progress since Bohr's model to improve upon this lack of "physical meaning".

In Bohr's own words in his 1913 paper: "The result of the discussion of these questions seems to be a general acknowledgment of the inadequacy of the classical electrodynamics in describing the behaviour of systems of atomic size.** Whatever the alteration in the laws of motion of the electrons may be, **it seems necessary to introduce in the laws in question a quantity foreign to the classical electrodynamics**, i.e. Planck's constant, or as it often is called the elementary quantum of action. By the introduction of this quantity the question of the stable configuration of the electrons in the atoms is essentially changed as this constant is of such dimensions and magnitude that it, together with the mass and charge of the particles, **can determine a length of the order of magnitude required** ... let us assume that the electron at the beginning of the interaction with the nucleus was at a great distance apart from the nucleus, and bad (sic) [had] no sensible velocity relative to the latter. Let us further assume that the electron after the interaction has taken place has settled down in a stationary orbit around the nucleus. We shall, for reasons referred to later, assume that the orbit in question is circular; … Let us now assume that, during the binding of the electron, a homogeneous radiation is emitted of a frequency n, **equal to half the frequency of revolution of the electron in its final orbit.**"[6] [Bold added for emphasis] And after a few more assumptions, of which one is, "(2) That the latter process is followed by the emission of a homogeneous radiation, for which the relation between the frequency and the amount of energy emitted is the one given by Planck's theory." Bohr adds in the same article about this assumption: "On the other hand, in the calculations of the dynamical equilibrium in a stationary state in which there is no relative displacement of the particles, we need not distinguish between the actual motions and their mean values. **The second assumption is in obvious contrast to the ordinary ideas of electrodynamics** but appears to be necessary in order to account for experimental facts." [Bold added for emphasis] It should be noted that the words assume, assumes, assumed and assumption appear 74 times in this one article. A few assumptions are necessary for any new scientific theory, but as Isaac Asimov points out, "Since we must start somewhere, we must have assumptions, but at least let us have as few assumptions as possible."[iv]

So instead of applying any specific cause at all to the fact that there was distance between the nucleus and the electrons, Bohr introduced Planck's constant which was a measurement having to do with the energy and frequency of electromagnetic radiation and having nothing at all to do with the structure of the atom which Bohr excuses by saying that it is the proper "length". At that time, Bohr believed the atom to be disc-shaped on a plane with the electrons circling in an orbit with no wavelike properties to justify such an introduction. Since the movement of a charged particle causes radiation, this would cause the electron orbit to soon lose energy and plummet into the nucleus. So Bohr threw out the basic known physical laws of the universe i.e. electrodynamics.[6] And in their place, Bohr invented a mathematical formula using Planck's constant as a measurement to restrict the angular momentum of the orbits of his assumed non-radiating electron particles and to keep them apart from the nucleus. Perhaps this shows up the greatest failure of quantum mechanics which is its inability to explain how under classical mechanics a moving charged particle such as the electron can radiate in a cathode ray tube at any frequency dependent only upon the voltage applied without the electron losing mass since this cannot happen in the atom. The point here however is that in Bohr's theory only the mathematical formula, and the arbitrary introduction of Planck's constant into it, bars the electron from approaching the nucleus.

An even more serious problem is that Bohr himself in 1913 did not believe in light quanta (now called photons). In the 2009 book by Louisa Gilder, "The Age of Entanglement" in the chapter entitled "The Arguments" p. 55, Gilder quotes Bohr as saying in 1922: "Well, yes," says Bohr. "But I can hardly imagine it will involve light quanta. Look, even if Einstein had found an unassailable proof of their existence and would want to inform me by telegram, this telegram would only reach me because of the existence and reality of radio waves." Bohr and most of the scientific community did not believe in Light Quanta when Einstein received his Nobel Prize for the photoelectric effect. So Bohr threw in Planck's constant "h" without believing in it. I bet you didn't learn that in school.

This whole theory was so preposterous at the time that Max Von Luau, Nobel laureate, and Otto Stern, both important physicists of their day, who were hiking together in 1913, swore to each other that if Bohr were right, they promised each other to quit physics altogether. They called it "absurd" and "nonsense." Stern said, "It's not physics!" But eventually even these physicists were turned by Bohr's strength of character. Paul Ehrenfest, an important early quantum physicist, wrote: "Bohr's work on the quantum theory, in the Quantum mag, has driven me to despair. If this is the way to reach the goal, I must give up doing physics." Ehrenfest called it "monstrous." But in 1919 when Ehrenfest heard Bohr speak on his model, he changed his mind and was won over by the man himself.[v]

Bohr's hypotheses have not changed in almost a century, but are incorporated into Quantum Mechanics although without the orbits. Current quantum mechanics is merely a variation of the Bohr atom into three dimensions with the addition of several other properties i.e. magnetic moment, spin, etc. Erwin Schrödinger's wave mechanics (1925-26 although the Schrödinger equation is three dimensional only for Hydrogen and is actually in q-space which is multi-dimensional.) and Paul Dirac's equation (1928-30) both use Bohr's original quantum numbers "n" and angular momentum L derived from Planck's constant. The principle quantum number n is still to this day described as a multiple of h-bar or h/2pi as in Bohr's 1913 paper. Quantum jumps are still part of QM.

In order to introduce Planck's constant, Bohr assumed that the electron was circling in a periodic orbit around the nucleus and he likened this to a frequency and radiation. Werner Heisenberg, the inventor of the first full quantum mechanical formula, says of this, "So I now have to point out the difficulties and the errors of this model. The worst difficulty was perhaps the following. The electron described a periodic motion in the model, defined by quantum conditions, and therefore it moved around the nucleus with a certain frequency. However this frequency never turned up in the observations. You could never see it. What you saw were different frequencies, which were determined by the energy differences in the transitions from one stationary state to another."[74] So the periodic orbit of the atom is not empirically observed. Heisenberg says that it had to be abandoned and he in fact did abandon it with matrix mechanics which applied point locations for the position of the electron without applying a periodic orbit to the electron. However, interestingly, the great inconsistency of quantum mechanics is that it constantly reverts back to the abandoned periodic orbit of the electron to describe magnetic dipole moment whenever it pleases quantum mechanics to do so. The science community's description of magnetism itself often resorts to an explanation involving the discarded periodic orbit of the electron. Although the orbit of the electron is still called upon at will to explain quantum mechanical phenomena, Heisenberg explains its rejection just previous to inventing matrix mechanics while working with Hendrik Kramers on BKS theory (Bohr, Kramers, Slater theory): "So it was only a very small step from there to saying: Well, let us abandon this whole idea of the electronic orbit, and let us simply replace the Fourier components of the electronic orbit by the corresponding matrix elements…By this time you see that the idea of an electronic orbit, connected with the discrete stationary state, had been practically abandoned. The concept of the discrete stationary state had, however, survived…. It could not be described by referring to an electronic orbit."[74]

It should be firmly grasped that Heisenberg did not invent matrix mechanics with wave-particle duality in mind, but based it solely on the electron "particle" concept. Heisenberg wrote a formula in which the electron orbit did not exist. Yet he did not describe a wave and as he says above replaced the Fourier components having to do with wave mechanics entirely in favor of specific positions for the electron viewed as a particle without an orbital frequency. Quantum mechanics however invokes the periodic orbit of the electron to this day whenever convenient to describe phenomena that it cannot explain. This is a contradiction since it is abandoned in both matrix mechanics and the Schrödinger and Dirac equations. This is a critical point that cannot be overstated. In modern Quantum Mechanics as described by matrix mechanics, the Schrödinger, and the Dirac equations, the electron does NOT orbit. There is no orbit. The electron transitions between shells, but it doesn't circle the nucleus. That means that the fake frequency that Bohr fabricated to suppose that the orbit was a frequency as an excuse to introduce Planck's constant and to invent 1/2h angular momentum does NOT exist. The inconsistencies of quantum mechanics and the inability for anyone, anywhere, especially textbooks and physicists to be truly consistent in describing the characteristics of the atom through it, arise specifically from the fundamentally flawed way in which it was created. Planck's constant was introduced due to a supposed orbital frequency that had no empirically observed evidence for existing. All of this complicated, mysterious contradiction of physical laws arose from a foundational mistake in the development of a theory of the atom.

Bohr himself admitted to Heisenberg that he had not worked out his model of the atom based upon classical mechanics although he apparently tried to make it appear so in his 1913 paper. According to Heisenberg: "Bohr confirmed to me, what Pauli and I in Munich had long suspected, that he had not worked out the complex atomic models by classical mechanics; they had come to him intuitively, rather, on the basis of experience, as pictures—so far as mere mechanical pictures can be suitable at all—for representing events within the atom."[74,p.40]

The fundamental problem is that the conclusion made by Rutherford, Niels Bohr, and others from the gold foil experiment was unscientific in that it uncategorically stated that the only thing between the nucleus and the electron was empty space, a postulate that still holds today. This conclusion led to the introduction of assumptions contradicting known laws of nature. Even with further development of quantum electrodynamics by Richard Feynman et al., when it was shown that empty space should have a vacuum energy, this was never applied to the empty space inside of the atom which is another contradiction and inconsistency itself.

The mysterious nature of the quantum atom became even more pronounced when Bohr had to explain how the electron could change orbits while still only producing discrete spectral lines that did not produce a continuous frequency i.e. energy change but appeared to discontinuously jump from one energy level to another. Niels Bohr fabricated the notion of quantum leaps in which the electron was never between energy levels but disappeared from one level only to reappear in another in order to account for the atomic spectrum. This was an unnatural and irrational explanation since in the natural world as we know it, matter does not become invisible, transport itself, and reappear elsewhere in this fashion that Niels Bohr has promulgated. Because of this Paul Ehrenfest had called Bohr's model of the atom "utterly cannibalistic" in 1916.[vi] It just threw together the planetary solar system, the Rydberg equation, Planck's constant, and mysterious quantum leaps.

Newton, in combining heavenly motion with earthly motion, and Einstein, both held that the laws of the universe were equal to all observers everywhere. The atom should be no exception. There cannot be different universal laws for a single atom than for a collection of atoms since atoms constitute the basic chemical elements making up the universe. A single atom can have different characteristics when combined into a collection of atoms, but these should not contradict universal laws. One cannot disagree that quantum mechanics has been successful and its success came from contradicting the laws of the universe such as electrodynamics, but the success of quantum mechanics is due to the equations being contrived to predict the empirical evidence. This is because Bohr used the empirically derived Rydberg's constant (1888) in his formula developed from the empirically derived Balmer equation (1885) to devise mathematics to fit the atomic spectrum. By 1890, Rydberg had discovered a formula describing the relation between the wavelengths in lines of alkali metals and found that the Balmer equation was a special case.[68,69] Although the Rydberg formula was later found to be imprecise with heavier atoms, it is still considered accurate for all the hydrogen series and for alkali metal atoms with a single valence electron orbiting well clear of the inner electron core.[68] By 1906, Lyman had begun to analyze the hydrogen series of wavelengths in the ultraviolet spectrum named for him that were already known to fit the Rydberg formula before 1913. In 1908 Friedrich Paschen discovered a series of far-infra-red hydrogen lines, fitting the Rydberg equation.[68] Bohr mentions that his formula includes the known Balmer and Paschen series as well as the Rydberg formula.[6] Therefore, Bohr admittedly incorporated the empirically derived formulae with his arbitrary use of Planck's constant, manipulating the size of Planck's constant to fit the known size of the atom and the known ionization energies. He created a mathematical amalgamation that would incorporate different known elements into a formula that however did not contain a physical cause for the separation of the electrons from the nucleus. This is not the scientific method. Manipulating mathematical formulae to match the evidence where the formula assumes at the outset that the universal laws do not hold is bad science especially where systems can be theorized that explain phenomena within universal law. To say that the atom does not obey the laws of physics as Bohr assumed is to say that the atom is supernatural. Atomic theory becomes magical. Yet Bohr founded quantum mechanics upon the assumption that the atom does not obey universal law.[6] This was tantamount to measuring how high a magician can levitate a person off a table and calling it scientific because one has the mathematical measurement. However, it is still magical until one explains the support that the magician is using. Bohr did not. This was all possibly justifiable as a means to an end but the question still exists to this day: "What is the barrier between the nucleus and the electron?" And equally important, "Why does Planck's constant appear to be successful in describing the atom after such an arbitrary introduction?" As will be demonstrated, these questions can both be explained to arise from a common cause.

Actually, one experiment has proven Quantum Theory to be wrong as far as the Heisenberg/Schroedinger model is concerned. The fact that light is made of photons is shown in nature and makes sense in the universe like F=ma makes sense. For lasers to work, light has to be photons. But that is probably the one "given" under Quantum Theory. Anything beyond the fact that light appears in quanta is something that has to stand the test.

Where Quantum Theory fails is in the vacuum energy predicted by the Heisenberg Uncertainty Principle and the Schroedinger wave equation for ground state of the electron.

This value for the energy in the vacuum predicted by the current theories of the Heisenberg Uncertainty Principle and the Schroedinger wave equation is proven wrong and called the Vacuum Catastrophe.

The value given by the Uncertainty Principle for vacuum energy is off by 120 orders of magnitude to the observed universe.

Neil de Grasse Tyson explains this in The Inexplicable Universe episode five when he explains dark matter and dark energy. He says that science has tried to explain dark matter by the vacuum density caused by virtual particles appearing, but it was off by a number described by one followed by 120 zeros. The ground state energy, and therefore, the vacuum energy, and, therefore, the Uncertainty Principle is off by that much.

Neil de Grasse Tyson says, ""Our most successful theory of the universe Quantum Physics allows you to calculate what you might expect to be the energy contained in the vacuum of space. We can do that calculation. It allows it. ....When you do that calculation, the answer you get does not match reality. The answer you get is off by a factor of 10 to the 120th power. That's one followed by 120 zeros. Now that's just embarrassing…In fact, it is the biggest mismatch between theory and observation that there ever was in the history of science."

This has been verified by the Voyager probes according to another source.[vii]

The density caused by the vacuum particles as described by the Schroedinger equation and the Uncertainty Principle would cause so much gravity in our universe that the universe would collapse.

However, that "something" does occur in the vacuum between solids on earth is experimentally verified. It's called the Casimir effect and the Van der Waal forces. Current Quantum Theory does not measure it correctly though. So all of Quantum Field Theory which is based on current Quantum Mechanics of Heisenberg and Schroedinger is also off by this amount as well.

The Vacuum Catastrophe has been described in detail.[viii]

Because General Relativity does not correlate with Quantum Physics, there is probably another model for both that is closer to reality. I believe this model by introducing causality into the atom is the basis for the new model.

In the theory of General Relativity, the measurements are precise and fit the observations of our solar system exactly. On the other hand, Quantum Theory is based on probabilities, so it doesn't have to be as precise. Also, it is describing the atom at atomic sizes which we can't measure accurately yet, so it might be an approximation of reality without being reality. There is a better model that keeps the photon as proportional to Planck's constant. The 3-dimensional picture of this model is not hard to visualize, but the mathematics are illusive, because we have no experience with measuring the effects of unradiated light. This book gives what the new model should look like and what effects should be expected. The paper in this book gives all the data from the spectrum to back up this theory. This should be a springboard for someone with the vision to calculate what the effects and energies of non-radiating light should be upon particles. In this way, causality is satisfied and the Vacuum Catastrophe disappears.

3 Quantum Regrets

It should be noted that Quantum Mechanics has had many detractors including the noted physicists Albert Einstein, Boris Podolsky, Nathan Rosen, Erwin Schrödinger, Max Planck, Max von Laue, and Alfred Landé among others. The feeling, thoughts, intuitions, and general logic of these physicists inclined them to feel that there was something right and correct about quantum mechanics, but in the end there was something fundamentally wrong as well. Even Richard Feynman, awarded the Nobel prize for advancing the theory of quantum electrodynamics, had his doubts about quantum mechanics, so one shouldn't feel blasphemous in questioning it. Feynman is quoted as saying, "We have always had a great deal of difficulty understanding the world view that quantum mechanics represents. At least I do, because I'm an old enough man that I haven't got to the point that this stuff is obvious to me. Okay, I still get nervous with it… You know how it always is, every new idea, it takes a generation or two until it becomes obvious that there's no real problem. I cannot define the real problem, therefore I suspect there's no real problem, but I'm not sure there's no real problem."[36]

Max Born who gave the statistical interpretation to quantum wave mechanics and was highly instrumental in the foundations of the symbolic view of the atom had this to say of Albert Einstein's letters to him rejecting this quantum mechanical view:

"I have quoted these letters because I think that the opinion of the greatest living physicist, who has done more than anybody else to establish modern ideas, must not be by-passed. Einstein does not share the opinion held by most of us that there is overwhelming evidence for quantum mechanics. Yet he concedes 'initial success' and 'a considerable degree of truth'. He obviously agrees that we have at present nothing better, but he hopes that this will be achieved later, for he rejects the 'dice-playing god'."[97,p.123]

Albert Einstein's opinion should not be taken lightly. He, more than anyone, molded quantum mechanics and he more than anyone loudly rejected its interpretation. Einstein more than all others felt that something was missing, some underlying fundamental truth was yet to be found, or as he liked to put it, quantum mechanics had not yet discovered the secrets of the Old One.[62]

Probably the one who in his way regretted the reception of quantum mechanics the most was Niels Bohr himself. He was disappointed at the manner in which the positivists accepted quantum mechanics without objection to its radical nature and with no heed taken in the least to its philosophy. Bohr wanted the thinkers of the 1950s to be taken aback by quantum mechanics and Bohr wanted his philosophy to be debated because to debate a thing is to take it seriously. Heisenberg says:

"Niels had this to say: 'Some time ago there was a meeting … I was asked to address them on the interpretation of quantum theory. After my lecture, no one raised any objections or asked any embarrassing questions, but I must say this very fact proved a terrible disappointment to me. For those who are not shocked when they first come across quantum theory cannot possibly have understood it.'"

This revealing remark is quite astounding when one considers it. Bohr is not merely disappointed, but "terribly disappointed" personally that quantum mechanics with its magical disappearing and reappearing electrons and its contradicting particle and wave description of nature should be accepted so calmly. He complains that the audience is not at all "shocked". He is even outraged that they are not shocked. Bohr thrives on the shock that quantum mechanics contains: the deviance from the norm, the craziness of the indeterminacy, and the overall symbolic nature of atomic science. That the population has received quantum mechanics and assimilated it without protest is Bohr's "terrible disappointment." That it is taught in classrooms and lecture halls with equanimity and without the need to shout "renunciation" and without the need to convert the disbeliever by reproofs that they should forget cause and effect is intolerable to Bohr. How is it that the crowd has become docile as sheep willingly running to the slaughter?

"Wolfgang [Pauli] objected: '…The positivists have gathered that quantum mechanics describes atomic phenomena correctly, and so they have no cause for complaint. What else we have had to add—complementarity, interference of probabilities, uncertainty relations, separation of subject and object, etc.—strikes them as just so many embellishments, mere relapses into prescientific thought, bits of idle chatter that do not have to be taken seriously.'"[91,p.206]

Niels Bohr was dissatisfied with anyone accepting quantum mechanics as a mathematical explanation of atomic phenomena and then dismissing all the philosophical aspects out of hand. Bohr then continues arguing in favor of metaphysics with Heisenberg and Pauli saying,

"Quantum theory thus provides us with a striking illustration of the fact that we can fully understand a connection though we can only speak of it in images and parables. In this case, the images and parables are by and large the classical concepts, i.e., 'wave' and 'corpuscle.' They do not fully describe the real world and are, moreover, complementary in part, and hence contradictory. For all that, since we can only describe natural phenomena with our everyday language, we can only hope to grasp the real facts by means of these images. This is probably true of all general philosophical problems and particularly of metaphysics. We are forced to speak in images and parables which do not express precisely what we mean."[91,p.210]

Bohr sees atomic phenomena as outside of the realm of "classical concepts" and "everyday language." Bohr argues passionately for the language of religion i.e. the language of images and parables as Heisenberg makes clear when he says: "We know that religions speak in images and parables and that these can never fully correspond to the meanings they are trying to express."[91,p.212]

Bohr was not contented with the way science and philosophy had accepted the mathematics of quantum mechanics and forgotten the philosophical, even religious, implications and the very mysticism of its conception.

Another famous lamenter of facets of quantum mechanics was Einstein who abhorred its lack of causality, locality, and determinism. The apparent abandonment of causality would gnaw at him. He wrote Max Born in 1920:

"The thing about causality plagues me very much too. Is the quantumlike absorption and emission of light ever conceivable in the sense of the condition of complete causality, or is a statistical residue left? I must admit that I lack the courage of conviction here. But only very reluctantly do I give up complete causality."[ix]

But it was originally Einstein who backed Bohr's view of the atom.

"..the first phase of the interpretation of [quantum] theory began with well-known discussions between de Broglie, Schrödinger, Bohr and Einstein provisionally in favor of the Bohr view seemed to lead to a solution."[40]

Einstein is said to have thought of ideas for a model similar to the Bohr atom, but "did not dare publish them."[82,p.84] Einstein endorsed a theory of the atom in which electrons hung in empty space due to a mathematical formula and then spent the rest of his life attempting to overturn the outcome of such a non-physical model of the atom. Einstein strenuously opposed non-locality and, apparently, the Uncertainty Principle. Einstein spent many years debating these points with Bohr through thought experiments.

In Einstein's famous thought experiment that he posed to Bohr in 1930, Einstein proposed that a single photon be let from a theoretical box (or hollowed ball) that was timed and the box later weighed so that both the time and energy of the particle could be known at the same time. Bohr showed that through applying relativistic theory, Einstein's argument collapsed. However, Bohr simply did not understand the point Einstein was making concerning locality. Einstein was not debating the Uncertainty Principle.[x]

Thought experiments are highly unscientific "philosophical devices" with more loopholes than tax law due to the absolute absence of empirical verification. Why not imagine a hypothetically perfect spectroscope that is measuring the transitions of a single hydrogen atom. Here is a thought experiment that will succeed. The temperature is controlled to make the transitions happen at speeds lower than relativistic speeds. The spectral lines appear one at a time because there is only one electron. The transitions are timed, the energy levels of each spectral line as it appears are known through quantum mechanics and the velocity of the transition can be known from the timing of the transitions or appearance of spectral lines, the position is known from the shell that is occupied, and the electron is stationary and has no periodic orbit in the shell under Quantum Mechanical formulae. Even though the energy levels are statistical in quantum mechanics, the exact energy level for a single transitioning electron can be measured by its single frequency. Therefore one has velocity, time and energy all at once. Three out of four isn't bad. But if you want to get the position of the electron in its shell at the same time, then introduce the positive end of a bar magnet or some futuristic precisely accurate magnetic field that aligns each electron transition in the same direction as the positive pole. The reason this is defensible is because under the Bohr-Heisenberg Copenhagen matrix mechanics, the electron does not move in an orbital pathway in its shell and its only movement is the anharmonic oscillation between transitions, therefore, the electron movement would only be in the direction of the magnet during transitions and the velocity is the velocity of the oscillation which is the only movement. Then you have all four variables at once although one only needs two variables according to Bohr's and Einstein's thinking to disprove the Uncertainty Principle. This thought experiment is what Einstein was striving for because it disproves Bohr's theory, but what this proves in the end is ambiguous, so why either Bohr or Einstein expected these thought experiments to be important is strange.

The Copenhagen interpretation of the atom is what we call the modern quantum mechanical theory since Heisenberg coined the term in 1955 as opposed to the old quantum theory before 1925. The founders of modern quantum mechanics and the Copenhagen interpretation are Bohr and Heisenberg. Yet, we find even Heisenberg has his regrets. He hung Bohr's electron in empty space through matrix mechanics, then dissolved its orbital pathways in matrix mechanics leaving no physical reason for the electron to be distanced from the nucleus. He then complains of nuclear particle physics:

"The fact that even in the cosmic radiation no quarks have been found, is a very strong argument in favor of their non-existence. …The same skepticism is justified with respect to other particles, which have been predicted, but not found: W-mesons, partons, gluons, magnetic poles, charmed particles. If they cannot be observed, either in big machines or in the cosmic radiation, it is difficult to argue that they are good concepts in a phenomenological description."[74]

In character, Heisenberg's solution is to do away with the concept of particles in particle physics altogether. But it was Heisenberg's imaginative mathematical view of the Bohr atom that has led particle physics to devise whatever mathematical scheme deemed necessary to explain the atom. It was Paul Dirac who began the journey to Particle Physics when he introduced Special Relativity into Heisenberg's equation. It was Heisenberg who said the electron did not have a pathway in the cloud chamber because it is the "theory which decides what can be observed".[74] Since Heisenberg was 19, he had wanted to get rid of orbits in the atom thereby leaving the atom with no reason whatsoever for the electrons not to fall into the nucleus. So particle physics at least had the good manners to stick an imaginary gluon to the nucleons to hold them together and a strong force in the nucleus. What does Heisenberg have to complain about when he invented the process? The gluon is at least a scientific conclusion—at least where there is an effect, science has invented a cause. This is true science. Science is about assigning causes to effects. When you feel warmth, science calls it heat even if it is invisible. When something falls to the ground, science calls it gravity, even if it is invisible. When a lodestone pulls iron to it, science calls it magnetism, even if it is invisible. That is science. Science says where there is an effect, there is a cause. But what does quantum mechanics say? The electron in the atom hangs upon nothing but empty space because we've assigned the mathematical letter h to it even though it has no frequency. This is not science. This is mysticism, superstition, magic, and religion. It is reverting back to the dark ages of the supernatural. The human race will be lucky if future generations do not speak of the Middle Ages, the Age of Enlightenment, the Age of Reason, the Age of the Industrial Revolution, and the Age of Quantum Mysticism. At least gluons are some scientific attempt to give cause to effects.

Now to introduce a new theory: This attempt to assign a cause can be unified into one force—that force can be explained by the concept of an energy wave bound to each particle. The only other force would be electromagnetism. That one simple idea can unify the atom. It can explain the experiments and give a reason for Planck's constant applying. The concept needs developing and clarifying, but the data from the spectrum is there to prove it as will be shown from the paper included in this book. It is one force, but not uncomplicated. Larger particles react differently to it than smaller particles. It is generated in different energies. It has variations of interference when multiple particles are applied to its system. It involves charges. It creates molecular bonds in ways not imagined. It is a simple concept, yet displays in complex ways. This book will examine such a concept and then turn to the spectrum to prove that this wave exists in the elements.

4 Theories and Philosophy

To understand theoretical physics, one must understand the concept of theory. Theories are simple models of complex systems. The universe is a complex system. In order to analyze the universe as the science of physics does, it must be broken down into simple principles and basic ideas that create a visualization or model. A theory must be useful in that it should approximate reality and make it understandable, measurable, and predictable.

To create a theory, one must begin with assumptions or postulates. All assumptions are unprovable hypotheses and foundations upon which to build a model of reality.[55] If the assumptions lead to a useful model for predicting the real world then the theory is a good theory. In this sense the Greek model of the earth-centered universe described by Ptolemy was a good theory. It accurately depicted the position of the planets. In fact, since Copernicus held to circular orbits, his sun-centered universe was actually less accurate at predicting planetary positions than Ptolemy's. It was only after Kepler corrected for elliptical orbits that the sun-centered model became more useful. The point is that quantum mechanics founded upon the theory that there is only empty space in the atom has been useful, but just as is the case with Ptolemy's earth-centered universe, it is probably not reality. Bohr said this himself:

"There is no quantum world. There is only an abstract physical description. It is wrong to think that the task of physics is to find out how nature is. Physics concerns what we can say about nature."[52]

Quantum mechanics discovered that Planck's constant can be used to describe the atom, but Planck's constant is about the divisions of energy in a wave. Since Planck's constant works in the quantum model to describe the angular momentum of the electron shells, then there is a wave inside the atom in which the electrons move. Erwin Schrödinger, later in life, discovered this himself. Unfortunately, he envisioned the wave so strongly that he put forth a new theory suggesting that the atom was only a wave.[53] Einstein rejected such a theory and rightly so as who knew better than Einstein that waves move overall at the speed of light but mass does not. However, Schrödinger, the person who invented quantum mechanics into a system that is still used today, believed so fully that a wave existed in the atom other than the type of wave-particle duality of today's matter-wave, but a full atomic wave, that he said in the 1950s:

"Let me say at the outset, that in this discourse, I am opposing not a few special statements of quantum mechanics held today, I am opposing as it were the whole of it, I am opposing its basic views that have been shaped 25 years ago, when Max Born put forward his probability interpretation, which was accepted by almost everybody."[54]

Even in 1926, Schrödinger had argued with Bohr saying: "Surely you realize that the whole idea of quantum jumps is bound to end in nonsense. You claim first of all that if an atom is in a stationary state, the electron revolves periodically but does not emit light, when, according to Maxwell's theory, it must. Next, the electron is said to jump from one orbit to the next and to emit radiation. Is this jump supposed to be gradual or sudden?...Why does it not emit a continuous spectrum, as electromagnetic theory demands? And what laws govern its motion during the jump? In other words, the whole idea of quantum jumps is sheer fantasy." Bohr's answer is typically cryptic: "What you say is absolutely correct. But it does not prove that there are no quantum jumps. It only proves that we cannot imagine them, that the representational concepts with which we describe events in daily life and experiments in classical physics are inadequate when it comes to describing quantum jumps. Nor should we be surprised to find it so, seeing that the processes involved are not the objects of direct experience."[91,p.74]

This reminds one of Bohr's words to Heisenberg in their first discussion:

"All this, far from being self-evident, is quite inexplicable in terms of the basic principle of Newtonian physics, according to which all effects have precisely determined causes, and according to which the present state of phenomenon or process is fully determined by the one that immediately preceded it. This fact used to disturb me a great deal when I first began to look into atomic physics."[91,p.39] One wonders if Bohr is referring to the fact that Newtonian mechanics is predictable, or to the fact that quantum mechanics is not, that disturbs Bohr. Bohr later continues: "It follows that there can be no descriptive account of the structure of the atom; all such accounts must necessarily be based on classical concepts which, as we saw, no longer apply."[91,p.40]

Bohr conceives the realm of the atom as completely foreign in nature to our world. Aristotle thought of the realm of the celestial bodies in the same manner. On earth, there were four elements: earth, wind, fire, and water. The sun and stars belonged to the fifth element whose laws did not obey the earthly ones. Yet, one can imagine that even Aristotle would have categorized the atom as belonging to the earthly elements and not as foreign in nature as Bohr would have us believe.

What made Isaac Newton great was that he took Aristotle's division between heavenly physics as a fifth element and earthly physics combining them to show a unity in the universe. Quantum mechanics has divided the physics of the atom from the physics of the universe. The two need to be united once again.

Possibly the most horrifying aspect of science and religion is their mutual propensity to state uncategorically that what they have, possess and own is the absolute truth, the only knowable truth, and it cannot be improved upon. Make no mistake; science has done this in quantum mechanics. The following quotes are unmistakably clear. Let us consider excerpts from an entire chapter in which Heisenberg relates his circa 1930 discussion with Grete Hermann whom he describes as follows:

"In Göttingen, she was an active member of the circle around the philosopher Leonard Nelson, and thus steeped in the neo-Kantian ideas of the early-nineteenth-century philosopher and naturalist Jakob Friedrich Fries. One of the requirements of Fries' school and hence of Nelson's circle was that all philosophical questions must be treated with the rigor normally reserved for modern mathematics. And it was by following this rigorous approach that Grete Hermann believed she could prove that the causal law—in the form Kant had given it was unshakable. Now the new quantum mechanics seemed to be challenging the Kantian conception, and she had accordingly decided to fight the matter out with us."[91,p.117]

In her arguments she said: "It follows that science must presuppose a causal law, that science itself can exist only because there is such a law. The causal law is a mental tool with which we try to incorporate the raw material of our sense impressions into our experience, and only inasmuch as we manage to do so do we grasp the objects of natural science. That being the case, how can quantum mechanics possibly try to relax the causal law and yet hope to remain a branch of science?"[91,p.118]

Part of Heisenberg's answer was: "But we cannot—and this is where the causal law breaks down—explain why a particular atom will decay at one moment and not at the next, or what causes it to emit an electron in precisely this direction rather than that. And we are convinced, for a variety of reasons, that no such cause exists."[91,p.119]

Then Heisenberg continues a little later: "'No, we think that we have found all there is to be found in this field,' I insisted…"[91,p.119]

This kind of religious pomposity of saying one knows all there is that can be known is intolerable in science. The lack of humility in the face of nature and the universe is not so horrendous as the arrogant assumption that one so completely understands the atom and the atomic world that there is nothing else that can be learned.

Even further down Heisenberg argues: "…in other words, he [Kant] could not foresee that atoms are neither things nor objects."[91,p.123]

Isn't that just wonderful!? Humans and planets are made of chemical elements that are made from basic materials that are "neither things nor objects". Does that make scientific, logical sense to anyone on earth? Is this science or religion? Is this a scientific explanation of the natural world or a supernatural explanation of the natural world? Just how far back into the primitive dark ages has science sunk? It is insufferable, unsupportable and unendurable. That we have reverted back—de-evolved—into ideas of magical superstition where atoms are "neither things nor objects" is downright outrageous. Yet science has accepted this interpretation for nearly 100 years now without batting an eyelash.

Perhaps you conclude that Heisenberg made the determination that atoms were "not things" from a thorough study of the atom, after years and years of research, in which there could be no other conclusion. You would be wrong. Heisenberg admits that he made this determination before he began to study the atom, in fact, before he had graduated from school and while he was in his Youth Assembly prior to 1920. Heisenberg quotes discussions among the teenage boys and says, "Robert's references to Malebranche had convinced me that our experiences of atoms can only be indirect: atoms are not things."[91,p.11] Heisenberg was not yet eighteen-years-old and had not yet studied the atom nor even made a decision yet to go into the study of physics which he describes as happening the next year, but here prior to any knowledge of physics, he admits at that young age that he became "convinced" as he says that atoms were "not things". So much for letting nature speak to the physicist and learning from nature. Heisenberg went into physics with a preconceived idea and belief that he passed on to science.

Arnold Sommerfeld, Werner Heisenberg's professor, was Bohr's associate on the Sommerfeld-Bohr model of the atom that preceded modern quantum mechanics. Therefore, Heisenberg was apprised of the Bohr model through Sommerfeld's seminars. Of Sommerfeld's view of the atom circa 1920, Heisenberg quotes Pauli as saying:

"Sommerfeld hopes that experiments will help us to find some of the new laws. He believes in numerical links, almost in a kind of number mysticism of the kind the Pythagoreans applied to the harmony of vibrating strings. That's why many of us have called this side of his science 'atomysticism,' though, as far as I can tell, no one has been able to suggest anything better."[91,p.26]

Heisenberg explains further:

"Ever since Planck had published his famous work in 1900, these additional postulates were known as quantum conditions, and it was they which had introduced into atomic physics that strange element of number mysticism to which I referred earlier. Certain magnitudes that could be computed from an orbit were said to be integral multiples of a basic unit, namely, Planck's quantum of action."[91,p.35]

Heisenberg described the situation as follows:

"Einstein had devoted his life to probing into that objective world of physical processes which runs its course in space and time, independent of us, according to firm laws. The mathematical symbols of theoretical physics were also symbols of this objective world and as such enabled physicists to make statements about its future behavior. And now it was being asserted that, on the atomic scale, this objective world of time and space did not even exist and that the mathematical symbols of theoretical physics referred to possibilities rather than to facts."[91,p.80]

Newton, Einstein and the atomic physicists created laws to describe the universe. A law and a force describe the same thing in physics. They mean that "when something does this, then this happens" therefore that is a law or a force and that's the way it is. Every law and every force is invented under the scientific method to create a rational cause for effects. Quantum mechanics does not take this approach. Quantum mechanics says there is no cause for the random radiation of a single atom and there is only empty space between the nucleus and the electrons. But in the final solution, mankind may even learn that for every law and every force there will undoubtedly be found a cause if only we keep searching.

Science, which was once called philosophy, was the beginning of the use of reason instead of superstition to describe nature. If something unexplained happened, it was the duty of science to find an explanation and science demanded that that explanation be rational. The Greeks were the first recorded society to invent science i.e. the search for the rational explanation of nature.[55] This was the opposite of centuries of superstition where "the gods" decided the nature of the universe, such things as whether to bring the sun back every year after the winter solstice but only if the festivals of humans appeased them. When one takes away a rational cause for a physical phenomenon, one reverts back to "the gods" and away from the "rational". This is what quantum mechanics has done. Quantum mechanics is the result of human desire to express human philosophical preferences under the pretense and in the name of science rather than a determination to let nature speak to those wishing to understand the rational nature of the phenomena. Bohr, Heisenberg, and even Einstein show a predisposition to make nature bend to their theory, if indeed Heisenberg has not misquoted Einstein. Bohr definitely had an agenda and a philosophy that he preached to the world through his quantum mechanics. Heisenberg was less concerned with philosophy as long as the math worked out correctly. Both approaches are unscientific. The nature of the world must be allowed to speak to mankind however what it says may deny our own belief system. Quantum mechanics denies nature's right to speak.

First, the importance of "causality" must be discussed. According to theoretical physicists, Brian Cox and Jeff Forshaw in their book, "What is $e=mc^2$ and Why Should We Care," "To proceed further, we must explain the second of our three keys words: "Causality." Causality is another seemingly obvious concept whose application will have profound consequences. It is simply the requirement that cause and effect are so important that their order cannot be reversed." Applied to Quantum Mechanics, the electron cannot be pushed away from the nucleus without a cause. A cause has to exist prior to the effect that the electron is pushed away from the nucleus of the atom thereby overcoming the electromagnetic force. Our two physicists continue regarding spacetime, "Well, we will soon discover that insisting on a causal universe constrains the structure of spacetime to such an extent that we are left with no choice in the matter. There will be only one way which we can merge space and time together to manufacture spacetime while simultaneously preserving the causal order of things. Any other way would violate causality and allow us to do fantastical things like going back in time to prevent our own birth…" What these physicists are saying is that "causality" is the foundation of science. Any theory without causality is "fantastical" and allows "fantastical things." This is the thorn in Quantum Mechanics. It does not destroy Quantum Mechanics or its usefulness, but this thorn makes Quantum Mechanics fantastical instead of scientific and does not allow for a simpler model that unifies the universal forces. Without causality, a unifying theory is impossible. The reason is because General Relativity is rooted in causality and current Quantum Mechanical theory eliminates causality.

It should be noted that the scientific community has been fighting over causality for the past century, but it is rooted in the scientific method. Science is founded on the fact that when one sees an effect such as the story that an apple hit Isaac Newton on the head that science asks the question, Why did the apple fall on Isaac Newton's head? Without a search for cause, we become complacent, and merely look at effects, and simply accept that the effect exists and search no further. This is not the scientific method, but will lead to stagnation in science.

Einstein did not so much object to the indeterminism of quantum mechanics as Einstein objected to its lack of causality that is fundamentally a lack of sound scientific method.

"In another letter, Einstein told Born, 'You believe in God playing dice and I in perfect laws in the world of things existing as real objects.' There he made it as clear as language can allow that, for Einstein, the opposite of dice throwing was not a predetermined world, but one in which real things happen for real reasons… Einstein's objection to quantum mechanics went a step further, disputing the way the theory had taken reality out of the story altogether and replaced it with a mathematics."[46,p.252]

If Bohr and Heisenberg had simply erred, if they had simply let nature reveal its secrets using the scientific method and simply misinterpreted, then they could be excused and even applauded and praised for their insight piercing to some extent into the unknown. But the facts point unquestioningly otherwise. They had choices before them in which to describe the nature of the atom. They could have followed the method of inferring "a force" to a physical phenomenon that did not display any apparent link to a physical cause. However, they refused the scientific method of force. They refused causality altogether, not due to error on their part, but due to choice. One wishes and prefers to confer upon these men of great genius and intellectual prowess honor and dignity, but one cannot escape their own words of admission of guilt. At every turn, they speak of "many pangs of conscience" as Heisenberg admitted when introducing his Uncertainty Principle simply to abolish the known phenomenon of an electron having a path in a cloud chamber (this is discussed in the chapter on The Uncertainty Principle). Bohr acknowledges to Heisenberg that introducing suborbitals that did not correspond to the atomic spectrum was "an almost intolerable contradiction". One may ask in what way suborbitals do not represent the atomic spectrum. One then can remember that when an electron makes a transition from any place in its orbit, it always releases or absorbs the exact same amount of energy. Therefore, Bohr knew from his first paper in 1913 that the electron orbit must be circular. Obviously, considering the electromagnetic force between the electron and the nucleus, the further an electron is from the nucleus, the easier it would be for the electron to escape the atom, in other words, the ionization energy would be less, and also, the transition energy would be less from one orbital to another depending upon the distance from the nucleus under Coulomb's law. If the electron orbits were not circular, such as in the hypothetical case of a spherical orbit as theorized by Sommerfeld, when the electron transitioned, the amount of energy expended would depend upon where the electron was located in its orbit. The spherical shape would create different distances from the next electron orbit according to where the electron was in its orbit. (Also, the later suborbitals deduced from the Schrödinger equation with their odd shapes create even more pronounced variations in distance between orbits.) This reshaping of the orbit into spherical shapes by the year 1920 in the Bohr-Sommerfeld model of the atom in order to account for the Zeeman Effect was "intolerable" to Bohr for the reason that it was mathematically impossible that an electron transitioning from spherical orbits could always produce the same wavelength in the spectrum. Yet, suborbitals following spherical harmonics persist in Quantum Mechanics to this day. Of course, because Quantum Mechanics believes suborbitals to be spiritual or rather symbolic (Is there a difference?) then one can have make-believe suborbitals of any shape, because when the electron transitions in modern Quantum Mechanics, somehow it is magically in its circular orbit and equidistant from the next shell or rather in its now spherical shell which is still equidistant from the next shell. This is true in modern theory whether or not the electron is supposed to be in a "p" suborbital or "d" suborbital such as p-orbital and d-orbital shaped more like twisted balloons. It's a sacred mystery.

One source says that Bohr despised mathematics because people are inclined to think that the math represents an underlying reality which Bohr did not want to express:

"Bohr's distaste for math was his most distinctive trait. It is common in populations as a whole but unusual among theoretical physicists. He did not like the way mathematics encouraged people to believe abstractions were real."[46,p.92]

But isn't that the purpose of science? Isn't it to explain by mathematics what is real? Isn't the math supposed to explain the velocity of something or the position of something in order to explain the cause of something? If one took Bohr's attitude that the mathematics were unreal abstractions, then one could argue to a police officer that one wasn't actually speeding, it was merely a mathematical abstraction of the police officer's speedometer or radar and that it shouldn't be taken literally. If the mathematics of science does not explain reality, but only abstractions, then is not the mathematics of science explaining something akin to Plato's forms? Plato's forms were flawless abstract objects—perfect, eternal and unchanging abstractions. Bohr consistently describes classical mechanics as only an "asymptotic" expression of physics never really attaining the true form of the atomic world. The classical world with its "largeness" does not characterize the "wholeness" of the quantum world. As Bohr says, "Indeed, it became clear that the pictorial description of classical physical theories represents an idealization valid only for phenomena in the analysis of which all actions involved are sufficiently large to permit the neglect of the quantum."[90,p.2] The classical macro world is "an idealization" according to Bohr because only the atomic processes have "wholeness". He says, "A new epoch in physical science was inaugurated, however, by Planck's discovery of the *elementary quantum of action*, which revealed a feature of *wholeness* inherent in atomic processes, going far beyond the ancient idea of the limited divisibility of matter." [Italics Bohr's] [90,p.2] One gets the sense that the classical world in which we live is incomplete and the quantum world is the true form of "wholeness" in much the same way as the classical world of Plato was incomplete and only the true perfection or "wholeness" belonged to "the forms".

This may seem an exaggerated account of quantum mechanics, but when one researches its history one finds that Heisenberg himself admitted not only being attached to the ideas in Plato's Timeaus long before deciding whether or not to be a physicist, but he himself expresses the view that the science of the atom follows Plato when he says:

"I think that modern physics has definitely decided in favour of Plato. In fact these smallest units of matter are not physical objects in the ordinary sense; they are forms, ideas which can be expressed unambiguously only in mathematical language."[95,p.32]

Heisenberg states in no uncertain terms that atoms are in fact Plato's forms. That is exactly how quantum mechanics has explained them and that is exactly what science believes the atom is. Of course, one doesn't begin a nuclear physics class with the professor stating:

"Class, the science of the atom that you will be studying is not science at all but has reverted back over 2,000 years to Plato's forms, so hopefully you've finished your philosophy and history courses as a prerequisite. Not to worry though as we are presenting the philosophy of the atom as one of Plato's forms in a physics textbook and with esoteric, symbolic mathematics so as to deceive you into thinking you are learning science so that you will feel comfortable learning the philosophy of the atom as one of Plato's forms without questioning its origin." This is never explained to the unfortunate student who thinks he has actually come to learn science and not supernatural philosophy.

Although it is true that Bohr remained unconvinced for over a decade about Einstein's theory of the dualism of light quanta, that is its being both a particle and a wave, yet like a resistant non-believer who becomes the most ardent and outspoken convert, so too did Bohr embrace the dualism of wave-particle duality with fervent religious conviction when it became part of quantum mechanics reminding one of Saul of Tarsus who zealously persecuted Christians until becoming the indefatigable proselyte the Apostle Paul. These men, the founders of quantum mechanics, were playing with science, betraying our trust in science, and using it as a platform for their own philosophical persuasion which offends the human spirit. There is no doubt that the facts were twisted to their own ends. There is no doubt that experimental data was abused and misused. There is no doubt that a physical interpretation of the atom could have been presented but was purposely repudiated in order to create in science the religious philosophical viewpoints of both Werner Heisenberg and especially Niels Bohr. Time and again Bohr stood on his pulpit and preached, "Renunciation!"

Bohr says speaking of magnetic moment without electron motion: "The difference between free electrons and atoms, which we come upon here, is connected with the fact that measurements of the magnetic moment of atoms involve a renunciation, in accordance with the general conditions holding for the application of the concept of stationary states, of all attempts to trace the motion of the elementary particles."[94,p.13] We need "a renunciation" of motion connected with magnetic moment.

And again Bohr preaches: "We must, therefore, be prepared to find that further advance into this region will require a still more extensive renunciation of features which we are accustomed to demand of the space-time mode of description than the quantum theory attack on the atomic problem has required thus far…"[94,p.14] Now Bohr says we need even "more extensive renunciation" of expected scientific views.

Bohr says, "These difficulties seem to require just that renunciation of mechanical models in space and time which is so characteristic a feature in the new quantum mechanics."[94,p.50]

Speaking of Schrödinger's wave equation Bohr preaches: "This entails, however, that in the interpretation of observations a fundamental renunciation regarding the space-time description is unavoidable."[94,p.77]

Again Bohr preaches: "Partly in view of the regret, so widely expressed, with regard to the renunciation of a strictly causal mode of description for atomic phenomena…"[94,p.15] We need to further renounce "a strictly causal mode of description".

And Bohr further advocates renunciation at the 1927 Solvay conference where he says: "A main theme for the discussion was the renunciation of pictorial deterministic description implied in the new methods."[90,p.89]

One author speaks of Bohr's 1927 Como address in this way: "He called instead—the word appears repeatedly in his Como lecture—for 'renunciation,' renunciation of the godlike determinism of classical physics where the intimate scale of the atomic interior was concerned."[82,p.131]

Here is a good time to remind the reader that only 29 physicists in the world attended the 1927 Solvay conference and only 29 physicists created Quantum Mechanics. They did not all agree, but you will find in this book that Bohr bullied those physicists into accepting his religious view of the atom. This coercion will become more apparent as the facts are presented in this book.

Heisenberg quotes Bohr as saying: "That is why I consider those developments in physics during the last decades which have shown how problematical such concepts as 'objective' and 'subjective' are, a great liberation of thought…For all that, we have come a long way from the classical ideal of objective descriptions. In quantum mechanics the departure from this ideal has been even more radical."[91,p.88]

In the next chapter Heisenberg himself says: "Moreover, it seemed very much as if the new physics was in many respects greatly superior to the old even on the philosophical plane…"[91,p.93]

Determinism should be left to the philosophers who ask too many questions and have too few answers. Science deals not with "determinism", which is a philosophical concept, but with "cause and effect", which is a scientific concept. So was the lack of causality in quantum mechanics an error or a purposeful introduction of personal philosophy? Just how hard would it have been to invent some force or explanation, even maybe, invisible gluons to hold the electrons in place in their orbit? But Bohr wouldn't have it. He actually enjoyed, relished, and exulted in the idea of the quantum jumps. "Cause and effect" was consciously repudiated.

"And as Kierkegaard's stages are discontinuous, negotiable only by leaps of faith, so do Bohr's electrons leap discontinuously from orbit to orbit. Bohr insisted as one of the two 'principal assumptions' of his paper that the electron's whereabouts between orbits cannot be calculated or even visualized. Before and after are completely discontinuous."[82,p.76]

"Bohr also liked the kind of thinking that led Heisenberg to write 'the "orbit" [of an electron] comes into being only when we observe it … [it] can be calculated only statistically from the initial conditions.' This kind of antirealism appealed to Bohr very much, and doubtless he also found plenty to enjoy in Heisenberg's remark that although a person might argue 'that behind the perceived statistical world there still hides a "real" world in which causality holds, … such speculation seem to us, to say it explicitly, fruitless and senseless.'"[46,p.262]

Although Bohr had every right to believe in the influence of religion, Kierkegaard and Møller's dualism, doubleness and ambiguity, he had no right to let this affect his science. In his way, he pushed his religious beliefs on the world through his science.[82]

Often when one has no persuasive argument, one must resort to character bashing due to the weakness of one's point of view and it may seem that Bohr's character is here being attacked. In this one must agree with Heisenberg who said of the anti-Semitic attack on Einstein's science in 1922, "The choice of bad means simply proves that those responsible have lost faith in the persuasive force of their original arguments. In this instance, the means applied by a leading physicist in his attempt to refute the theory of relativity were so bad and insubstantial that they could signify only one thing: the man had abandoned all hope of ever refuting the theory with scientific arguments."[91,p.44] Unfortunately, unlike Bohr and Heisenberg, we the common people have no one to look to but our scientific leaders because our political leaders have proven false. We have sought truth from science expecting it without bias. This makes it doubly disconcerting when personal philosophy is promulgated as scientific fact.

Heisenberg and Bohr quarreled over the Uncertainty Principle, because Heisenberg was satisfied with the math and Bohr wanted to propound a philosophy. "Such quarreling over interpretation was one of the things science had seemed to abolish. Philosophers had once argued interminably about the interpretation of the most elementary facts, but science after Galileo and Newton made tremendous strides by somehow limiting that tendency. Galileo, like philosophers before him, wrote dialogues in which people disputed the meaning of facts, but once the science of natural law took hold, mathematics and meaning went together. But with quantum physics, from the start, the meaning was elusive. Its laws predicted experience without explaining it.... To Heisenberg's dismay, Bohr was displeased with the new paper, finding it wrong in some of its physics, often superficial, and insufficiently radical."[46,p.260]

"Insufficiently radical"? What is this a student revolt? A new messiah with a new message? It reminds one of persons who wish to get involved in witchcraft, mysticism, the occult, neo-Gothicism, or simply adopt an outlandish style of dress in order to be different, to be radical, to be anti, to resist authority, to have a cause, to create their own individualism, all in a most juvenile manner.

What we appear to have is a man, or small group of men, trying to preach a new message to the world and bring to the world his own personal, private, religious and philosophical views under the guise of a scientific model of the atom. That science uses statistics and probability is not the problem. These are useful mathematical concepts and sometimes can be the only tools for calculating results. The problem is using spiritualistic, supernatural philosophy and passing it off as science.

This is not an attempt at character assassination in order solely to advocate some new atomic theory, but rather a shock and dismay at the betrayal of the greats that we admired. If it is an unfair portrayal of the character of these men, then why do their own words and the words of those who have biographically recorded their histories show these men having such a disposition toward adhering to their own belief system rather than to empirical evidence? How did this happen? Well once upon a time, science was an open community that allowed open and free expression of new ideas. The small group of physicists in the world were all a closely-knit circle of friends. One did not need proof. One only needed an idea and that idea could be shared in a scientific journal. Once a new idea is accepted by mainstream science, then people like Einstein and Bohr are given power—ultimately too much power. In this way, their science was integrated with their philosophy. And nowhere in this book do I mean to suggest that Quantum Mechanics even as it stands today is not successful. I merely mean to prove that one aspect of it, the very foundation of it, needs to be reconsidered and reevaluated so that it will have causality.

According to Heisenberg, Bohr said upon their first meeting in 1922:

"We must be clear that, when it comes to atoms, language can be used only as in poetry. The poet, too, is not nearly so concerned with describing facts as with creating images and establishing mental connections."[91,p.41]

Then Heisenberg quotes Bohr in 1927 as saying:

"Still, religion is rather a different matter. I feel very much like Dirac: the idea of a personal God is foreign to me. But we ought to remember that religion uses language in quite a different way from science. The language of religion is more closely related to the language of poetry than to the language of science."[91,p.87]

Is Bohr not saying that the language of the atom is the language of religion?

Bohr was no longer a young man but a 42-year-old when he endorsed the Heisenberg Uncertainty Principle upon its weak excuse to eliminate paths in the cloud chamber and the Davisson-Germer experiment as proof of the de Broglie wave in 1927 which we will minutely scrutinize in a coming chapter. Why did these men involved in the development of a theory for the atom embrace "atomysticism"? Perhaps this can all be explained in a simple manner. Quantum mechanics developed in the midst of two world wars. It developed in Europe, the heart of the wars, with a great share of physicists being German and Danish. Perhaps they longed for a return to romanticism, a return to mysticism, a return to a time of magic and belief in the supernatural. Perhaps they longed for something mysterious to have faith in, something greater than themselves, something spiritual, so they invented the quantum mechanical mysticism. We cannot know what the horrors of living in Europe between the end of one catastrophic war and the beginning of another could do to the minds of men. Perhaps there lies the answer.

This led Bohr to describe quantum mechanical methods in this fashion:

"It is emphasized in the article that the symbolical garb of the methods in question closely corresponds to the fundamentally unvisualizable character of the problems concerned."[94,p.13]

Bohr was not unaware of his departure from the foundations of science as he himself continues:

"It may perhaps appear at first sight that such an attitude towards physics would leave room for a mysticism which is contrary to the spirit of natural science."[94,p.15]

Whatever the reason for holding onto and perpetuating the theory of a superstitious symbolic irrational atom that was not a real object, everyone must admit the genius of the mind of Niels Bohr. He did figure out that the spectral sequences pointed out by Balmer, Rydberg, and Ritz were emission energies between transitions and not orbital emissions. He did realize that the transition of electrons required both absorption and emission of energy although this was based on the previous work of spectroscopists and he had consulted an astronomer. He knew the importance of keeping the atomic mathematics in synchronization with the atomic spectrum. He was able to out-maneuver Schrödinger, Heisenberg, and Einstein in a debate (in the case of Einstein it was simply that he did not understand the point Einstein was trying to make as pointed out in the book The Age of Entanglement by Louisa Gilder and the book What is Real? by Adam Becker). To Bohr's credit, like Einstein he was able to take other's work and make connections between phenomena. He kept his hand in nuclear physics all the way from the discovery of the neutron by Chadwick in 1932 through the development of the atomic bomb. His real fault was abandoning the scientific cornerstone that there must be a rational cause for any effect. Without that cornerstone, science is nothing more than another religion. This is where he first let us down. Then by propagandizing his duality creed instead of admitting that one day a causal solution might be found, he let us down again. By saying there was no more to learn, he let us down.

And it is no secret that Heisenberg, Bohr, and Wolfgang Pauli (who is noted for the Pauli Exclusion Principle) knew exactly what they were doing and even joked about the Quantum Mechanical theories and rules and abstruse mathematics calling them "swindles." YES! They actually many times said that Quantum Mechanics was a swindle and yet it is still taught today. This is no secret. If we read the Scientific Correspondence with Bohr, Einstein, Heisenberg[xi],[xii], we see that Bohr and Heisenberg in Copenhagen were the first to read the paper concerning Pauli's Exclusion Principle along with a letter from Pauli in December 1924 in which Pauli asserts unashamedly, "what I do here, is not a bigger nonsense than the hitherto existing perception of the complex structure. My nonsense is conjugate to the hitherto customary one" (translated from Pauli's letter 74, p.188 of above reference). Heisenberg immediately sent a postcard back to Pauli exclaiming and jesting about how their Quantum Mechanics was all a "swindle." He says jokingly in the postcard telling the tongue-in-cheek truth, "Today I have read your new work, and it is certain that I am the one who rejoices most about it, not only because you push the swindle to an unimagined, giddy height (by introducing individual electrons with 4 degrees of freedom) and thereby have broken all hitherto existing records of which you have insulted me, but quite generally, I triumph that you too (et tu, Brute!) have returned with lowered head to the land of the formalism pedants (Formalismusphilister); but don't be sad, there you will be welcomed with open arms. And if yourself you think to have written something against the hitherto existing kinds of swindle then, of course, this is a misunderstanding; for, swindle x swindle does not yield something correct and, therefore, two swindles can never contradict each other. Therefore I congratulate!!!!!!!! Merry Christmas!!" (translated from card 76, pp. 192-3 of the collection referenced).

Bohr himself responded to this letter saying, "I have the impression that we stand at a decisive turning point, now that the extent of the whole swindle has been characterized so exhaustively." (Translated from the above referenced December 22 letter, pp. 194 and 195.) We should care today that even the founders of Quantum Mechanics believed it to be so absurd as to be "a swindle."

Nobel Prize laureate, Murray Gell-mann, who deduced the existence of quarks, said that the Copenhagen interpretation of Quantum Mechanics had to be wrong, writing in 1976: "The fact that an adequate interpretation of Quantum Mechanics has been so long delayed is no doubt caused by the fact that Niels Bohr brainwashed a whole generation of theorists."[xiii]

Niels Bohr's pushy personality is so very evident from all the biographies, letters, correspondence, and eyewitnesses that it is hard to read a biography of him without hearing about how he pushed people around although he often did it in a mild but very persistent way. The book "Quantum Moments" describes Bohr as having a "legendary status" which "derives from the intensity with which he spoke to people, one-on-one." He explains Bohr's bullying like this: "Bohr thought by talking aloud. He would entice a student, friend, or colleague into a long walk—and then instead of beginning a dialogue would begin a relentless exposition of his ideas, stating, reworking, and then restating them yet again. If his companions objected, he would push them repeatedly into rethinking their principles. If he sensed they were not agreeing, his voice grew stronger, and Bohr would not let up until he was sure they had capitulated. He browbeat Erwin Schrödinger into a (temporary) retraction of his ideas, while Werner Heisenberg once broke down in tears under Bohr's relentless questioning."[xiv]

The few men who invented Quantum Mechanics in the 1920s known as the Copenhagen Interpretation pretended to agree, but actually did not agree between themselves. Bohr said, "There is no quantum world." But Heisenberg, Pauli and Jordan thought otherwise. As one physics historian wrote: "Thus the myth that these physicists created a unified Copenhagen Interpretation is just that. A myth.... Quantum Physics in short shouldn't be taken seriously as a way the world actually is....""[xv] He goes on to say Einstein did not agree. "The programmatic aim of all physics is the complete description of any individual real situation as it supposedly exists irrespective of any act of observation or substantiation." Einstein knew that Quantum Mechanics was important, but he felt Quantum Mechanics was incomplete and did not explain the true nature of the world.

On the other hand, the one beauty of that scientific era a century ago was that even a youngster of twenty like Heisenberg could be heard by a renowned physicist like Bohr, but today no one listens to anything but the dogmatic adherence to the canons and tenets set up by the infallible masters. Niels Bohr was 27 when he introduced his model of the atom and Heisenberg was 23 when he introduced matrix mechanics. Once upon a time, science listened. It was open to the novel idea, to the unusual approach, to the youthful naïve suggestion. Not so today. Science is no more than politics and bureaucratic red tape in the guise of peer-reviewed articles which stifle new ideas—sometimes on the grounds of grammar alone. Yes, there is ignorance. Yes, there are obvious errors. But just sometimes there is truth. I say this from the many attempts I have made to even get my final paper peer-reviewed before being rejected out-of-hand in this day and age. I was very surprised to find that once I had actual data and proof of my theory, instead of praise and adulation, I got indifference and even outright hostility. It is so strange to have read so many works on Quantum Physics and to hear the words of every physicist available on documentaries and television network series say that they all wish for a unified field theory, or a theory that would combine quantum theory with gravity, or another interpretation other than the Copenhagen interpretation of atomic physics, or a better cleaner answer than string theory, and yet, when it is provided as will be done herein, along with recently discovered proofs, the very same physicists who have spoken aloud the wish for a more coherent, unified theory are antagonistic to one. But I should not be surprised. Even a small difference in interpretation of the Schrödinger equation as set out by David Bohm in which he introduced a pilot wave theory to show that a hidden variables theory could really work in Quantum Mechanics was met with outright hostility.[xvi] And Bohm was pretty orthodox in his theory of Quantum Mechanics which could be reduced to the Schrödinger equation. Even John Stewart Bell knew that Bohm's 1952 hidden variables pilot wave theory worked. Bell is famous for contradicting the proof of von Neumann that asserted that he had given proof of the impossibility of the existence of "hidden variables."[xvii] According to the book "What is Real?" by Adam Becker, Bohm's doctoral advisor, Robert Oppenheimer, called Bohm's ideas, "juvenile deviationism,," saying that, "if we cannot disprove Bohm, then we must agree to ignore him."[xviii] But according to this source, "Bohm's pilot-wave interpretation clearly worked, as Bell knew quite well."[xix]

Bell, in fact, said when ask why fix a theory that is not broken: "It is not right to tell the public that a central role for conscious mind is integrated into modern atomic physics. Or that 'information' is the real stuff of physical theory. It seems to me irresponsible to suggest that technical features of contemporary theory were anticipated by the saints of ancient religions…by introspection." Bell wanted a theory that didn't have "the measurement problem." In the current Quantum Mechanics theory, an observer causes the measurement to suddenly happen. That is the measurement problem. Bell continues: "Surely, after 6i2 years, we should have an exact formulation of some part of quantum mechanics!" Bell said in 1989. "[Measurement devices] should not be separated off from the rest of the world into black boxes, as if [they] were not made of atoms and not ruled by quantum mechanics." Bell spoke out against the Copenhagen interpretation of Quantum Mechanics weakened it the most since Einstein.[xx]

The difference between Einstein and Bohr was that everyone believed Einstein because his papers proved correct. People believed Bohr because he talked them into it, more like, badgered them into accepting his view. Also, Heisenberg kept saying everyone agreed on the Copenhagen Interpretation since 1927 when it wasn't so. In the book, What is Real?, it shows that Heisenberg had ulterior motives (because he had been a German Nazi sympathizer[xxi]) besides Heisenberg had not even created the name Copenhagen Interpretation until after 1945. That book is worth reading as it shows how anyone who even looked at Quantum Mechanics' origins and worked on foundations was blackballed and their careers ruined. Tell me that isn't like a cult religion that excommunicates members.

In fact, when Bohr first explained complementarity in the 1927 Como lecture, the audience of scientists were not only confused, but unimpressed. A prominent scientist of the time, Eugen Wigner remarked to Belgian Leon Rosenfeld, one of Bohr's close associates: "This lecture will not induce any one of us to change his own [opinion] about quantum mechanics."[xxii] The fact is that most at that time subscribed to Schrödinger's view that quantum jumps should be done away with. This is borne out by the response at the German Physical Society debate in July 1926 in which the audience backed Wien when he asked Heisenberg to shut up and sit down.[xxiii] These same scientists had not changed their opinion the next year at Como when Bohr's lecture just went over their heads and was largely ignored because it was positivist philosophy.[xxiv]

In fact when Bohr's Como talk was published in Nature in April 1928, the editors were so alarmed by the Bohr Copenhagen view that quantum phenomena as abstractions only became concrete, defined, and observable when measured, the editors took a very unusual step of adding a preface to the article stating that the Copenhagen Bohr interpretation was "far from satisfying the requirements of the layman who seeks to clothe his conceptions in figurative language," add that "it is earnestly to be hoped that this is not their [the quantum physicists'] last word on the subject, and that they may yet be successful in expressing the quantum postulate in picturesque form."[xxv]

The idea that the Copenhagen Interpretation of Quantum Mechanics was accepted by a consensus of scientists in 1927 was a myth. It was questioned by Einstein. It was questioned by Schrödinger who wrote the equation. It was questioned by de Broglie. It was questioned by Bohm in 1952. And it was questioned by John Bell, not to mention Everett who came up with the Many Worlds Interpretation. Many careers were ruined because students and professors in untenured positions were unable to get work if they questioned the Copenhagen Interpretation of Niels Bohr. That is not how science is supposed to work and we all know it.

In recent decades, Scientific Realism has become vogue in some circles and Bohm has been more accepted along with Everett's MWI as plausible alternatives to Niels Bohr. However, the point is that even Bohm's theory did not unify the atom, nor did it change Quantum Mechanics as it merely pointed back to the Schrödinger equation which can only predict the lines of Hydrogen and supposedly works for atomic systems despite a wrong foundation because it made a correct guess, but does not say anything about the nuclear forces, nor does it unify the atom, nor can it stand alone as it needs a set of rules outside of the equation such as The Pauli Exclusion Principle and the Uncertainty Principle, nor can it be fused with a theory of gravity. That means that Bohm's theory also accepted the loss of causality and locality in the micro world and the same issues that have been previously mentioned in this book.

There is an excellent book by Lee Smolin, a Scientific Realist greatly admired in the physics community entitled, "Einstein's Unfinished Revolution: The Search for What Lies Beyond Quantum Mechanics," 2019. It is wonderful to read that real physicists want Quantum Mechanics to make sense and relate realistically to the real world. How refreshing! However, the theories that are latched onto in the book are based on the Schrödinger equation and trying to make it make sense. However, the problem with trying to be a Scientific Realist is that Quantum Mechanics began with imaginary shells in which the electron stays away from the nucleus for no reason whatsoever.

But Science began with the study of causality. The first Homo Sapiens who asked the question, "Why?" were looking for "causality" — The reason for an effect.

5 Quantum Mechanics Did Not Explain Chemistry

It is common to hear and is always taught in school and textbooks that Quantum Mechanics explained chemistry and the periodic table. This is a myth. The early chemists explained the periodic table. Schrödinger's equation does not have a limit to the number of electrons in each shell in its equation.

Anyone who has taken chemistry will tell you that Dimitri Mendeleev was responsible for first creating a simple periodic table in 1869. There were several chemists who built upon it and were able to predict new chemicals throughout the late 1800s. Many of them began to notice a pattern that was called "the rule of eight" or "octet rule." This was decades before Quantum Mechanics was invented in 1925.

Since I contributed to the article in Wikipedia about the octet rule, I will quote it here:

In 1864, the English chemist John Newlands classified the sixty-two known elements into eight groups, based on their physical properties. (See Wikipedia for picture of Newlands' Law of Octaves periodic table.)

In the late 19th century it was known that coordination compounds (formerly called "molecular compounds") were formed by the combination of atoms or molecules in such a manner that the valencies of the atoms involved apparently became satisfied. In 1893, Alfred Werner showed that the number of atoms or groups associated with a central atom (the "coordination number") is often 4 or 6; other coordination numbers up to a maximum of 8 were known, but less frequent. In 1904 Richard Abegg was one of the first to extend the concept of coordination number to a concept of valence in which he distinguished atoms as electron donors or acceptors, leading to positive and negative valence states that greatly resemble the modern concept of oxidation states. Abegg noted that the difference between the maximum positive and negative valences of an element under his model is frequently eight. In 1916, Gilbert N. Lewis referred to this insight as Abegg's rule and used it to help formulate his cubical atom model and the "rule of eight", which began to distinguish between valence and valence electrons. In 1919 Irving Langmuir refined these concepts further and renamed them the "cubical octet atom" and "octet theory". The "octet theory" evolved into what is now known as the "octet rule". Langmuir said in his 1919 paper: "These electrons arrange themselves in a series of concentric shells, the first shell containing two electrons, while all the other shells tend to hold eight. The outermost shell however may hold 2, 4, or 6, instead of 8."

Walther Kossel and Gilbert N. Lewis saw that noble gases did not have the tendency of taking part in chemical reactions under ordinary conditions. On the basis of this observation they concluded that atoms of noble gases are stable and on the basis of this conclusion they proposed a theory of valency known as "Electronic Theory of valency" in 1916:

During the formation of a chemical bond, atoms combine together by gaining, losing or sharing electrons in such a way that they acquire nearest noble gas configuration.

References from Wikipedia:

1. Werner, Alfred (1893). "Beitrag zur Konstitution anorganischer Verbindungen" [Contribution to the constitution of inorganic compounds]. Zeitschrift für anorganische und allgemeine Chemie (in German). **3**: 267–330.

    English translation: Werner, Alfred; Kauffman, G.B., trans. & ed. (1968). Classics in Coordination Chemistry, Part I: The selected papers of Alfred Werner. New York City, New York, USA: Dover Publications. pp. 5–88.

2. ^ Abegg, R. (1904). "Die Valenz und das periodische System. Versuch einer Theorie der Molekularverbindungen" [Valency and the periodic system. Attempt at a theory of molecular compounds]. Zeitschrift für Anorganische Chemie. **39** (1): 330–380. doi:10.1002/zaac.19040390125.
3. ^ Lewis, Gilbert N. (1916). "The Atom and the Molecule". Journal of the American Chemical Society. **38** (4): 762–785. doi:10.1021/ja02261a002.
4. ^ Langmuir, Irving (1919). "The Arrangement of Electrons in Atoms and Molecules". Journal of the American Chemical Society. **41** (6): 868–934. doi:10.1021/ja02227a002.
5. ^ Kossel, W. (1916). "Über Molekülbildung als Frage des Atombaus" [On the formation of molecules as a question of atomic structure]. Annalen der Physik (in German). **354** (3): 229–362.
6. ^ "The Atom and the Molecule. April 1916. - Published Papers and Official Documents - Linus Pauling and The Nature of the Chemical Bond: A Documentary History". Osulibrary.oregonstate.edu. Archived from the original on November 25, 2013. Retrieved 2014-01-03.

Before the Heisenberg and Schrödinger equations in 1925 and 1926, an article was written by Edmund Stoner attempting to assign electrons into the Bohr-Sommerfeld atom.[xxvi] He likened the electron orbits as shells nesting in the same way that a Russian matryoshka doll is made. Using Bohr's original formula for n-squared, he arranged the electrons per shell in the sequence of 1, 4, 9, ... electrons per shell. However, this was known not to be in accord with the decades old octet rule, so Stoner knowing the chemistry was contradictory to this system of assignment wrote that the electrons should be paired. This was expanded on, especially Walther Kossel's article on noble gases in 1916, showing that these elements were so stable that the valence electrons needed to be in filled shells, or "closed" shells.[xxvii]

Working from Kossel's article, Bohr wrote an article on electron shells in 1922 using the work previously written, yet today Bohr gets all the credit.[xxviii]

So although it is true that odd shaped suborbitals could be deduced from the Schrödinger equation, the number of electrons in each orbital is not apparent from the equation. Rather, the physicists applied the well-known chemist's empirical knowledge of periodic table patterns to assign electrons to the shapes that the Schrödinger equation solutions created. Quantum Mechanics did not answer the question of chemistry. Rather, chemistry told Quantum Mechanics how to delegate electrons into shells. And yet Niels Bohr took credit for predicting an element later called Hafnium which was easy to predict from the blank spaces in the periodic table. He does get credit from predicting its properties per the periodic table pattern, but none of the above means that somehow the Quantum Mechanics equations deduced the electron patterns for chemistry. The truth is just the opposite.

This brings us to another myth about Quantum Mechanics. It is said that Quantum Mechanics made electronic technology such as computers possible. This is simply not true. It is the chemistry and the experimental evidence that was done in the 1800s that made transistors possible. Even the physicist Brian Greene accidentally shows this when trying to say the opposite. In his three part series "Explaining Quantum History with Brian Greene", episode 3, "Our Quantum Future," he explains about the development of transistors that computers started using cathode tubes. Well, cathode ray tubes were invented in the 1800s and as Wikipedia says: "Hittorf observed that some unknown rays were emitted from the cathode (negative electrode) which could cast shadows on the glowing wall of the tube, indicating the rays were traveling in straight lines. In 1890,…" Electrons were discovered in cathode ray tubes in 1897 by JJ Thompson way before the theory of the atom was even attempted to be modeled and before Planck invented the quantum in his 1900 theory of thermodynamics. Electricity had been known earlier than that since the early 1800s and especially since Michael Faraday made the first motor in 1821 and his experiments that lead to the Maxwell equations on electricity in 1862. Computers are made of transistors which are based on light switches that are turned on and off, or more precisely, electronic circuits that are open or closed. So in the above Brian Greene series, he explains: "The big challenge…we needed to make circuits smaller…In the middle of the twentieth century physicists knew they needed to replace a large problematic vacuum tube with a more reliable switch that used less energy. To get there, they tested the electronic conductivity properties of solids, such as crystals and metals at the atomic level." Does that sound like they used the Schrödinger equation? NO. They were testing metals to see which metals conduct electricity better just as Michael Faraday had done a hundred years before. They knew a little more about certain metals because of how the chemists as shown above had described the periodic table, but this is not Quantum Mechanics. This is chemistry. Brian Greene continues: "This launched the field of solid state physics. In the late 1940s, a team of physicists at Bell Labs made a huge discovery. When two gold-point contacts were applied to a crystal of germanium, it produced an output signal greater than its input. They called it a transistor. The fundamental building block of all modern electronics." This is not Quantum Mechanics. This is use of what was known before Quantum Mechanics was invented. This is use of experiments on metals, not Quantum Mechanical mathematics. This is use of knowledge of valence electrons from chemists and use of knowledge of electricity from James Clerk Maxwell. Yet Brian Greene goes on to say that this couldn't be done without knowledge of Quantum Mechanics. NO. This was done experimentally despite Quantum Mechanics.

6 New Theory

Perhaps all it really demands in order to be so bold as to question the physics giants is that one must have no reputation to ruin and therefore can dare to imagine how one might fit the facts together. Of course, that makes this new theory the work of an amateur and a crank, but one can only say "shame on the physicists" for making this necessary and the absolute necessity has become abundantly apparent. At least, I'm in good company. Everyone who comes up with a new idea from Copernicus to Galileo to Darwin to even Einstein is thought to be a crank until their theory is found to be true.

"Since we must start somewhere, we must have assumptions, but at least let us have as few assumptions as possible."—Isaac Asimov[55]

A theory can be postulated where an actual physical barrier exists between the electron and the nucleus of the atom which has no mass, and where a barrier exists between the electron shells as appears to be shown in the spectrum without merely simply stating a mathematical formula contradicting electrodynamics that "appears to be necessary in order to account for experimental facts".[6] To begin this new theory one must start with the Rutherford model of the atom as it was understood in 1911. This eliminates any theories built upon Bohr's model of the atom which interfere with a new model of the atom. In the Rutherford model of the atom, one has a concentrated nucleus and electrons of opposite charge far apart from the nucleus and with discrete spacing between electrons. The question was then raised: How do the electrons maintain their distance from the nucleus so that the atom is stable?

The theory of a barrier can be based on one assumption alone: There is a permeable, finite energy wave in the atom around any massive particles that is formed by waves emanating from each massive particle in which mass and energy are conserved and the energy density and frequency decrease with distance from the massive particle.

(In my first editions I presumed: A property of this energy wave surrounding the nucleus of the atom is that it retains a fixed coherent waveform. Therefore, it is finite, self-contained, and does not emit radiation nor does its particle lose mass.) I have since come to learn from the Balmer-Rydberg formula that the proton-wave is infinite and that only the electron wave is finite. A copy of the paper in which the data forced this assumption upon me will be included in this new edition of this book. This was hard to accept, but the Balmer-Rydberg formula cannot be interpreted any other way.

(In my first editions I presumed: A property of this energy wave is that it contains some of the mass of the particle in the form of energy and is formed when the particle is created. Therefore a true measure of a particle is the measure of the mass of the particle and the energy wave it creates.) However, it appears from the energy levels existing exterior to the particle, that

the mass is created by the amount of energy surrounding the particle. In fact, the energy wave surrounding the particle is equal in energy to the mass of the particle. I have come to realize that for the proton and the electron are the only truly stable particles of all particles with demonstrably much greater half-lives than other individual particles. Therefore, we measure the mass of the proton directly from the fact that the amount of energy surrounding the proton is equal to its mass. All other particles have very short half-lives because they do not have this equilibrium. This is where particle physics goes wrong. Short-lived particles do not make up the universe for the very reason that they are short-lived. An anti-matter universe cannot exist because so-called anti-matter particles are short-lived. This will be discussed further in the chapter on The Standard Model.

A property of this energy wave is that it is permeable to high speed particles. Taking just the simple case of the electron, the energy wave is permeable to electrons that are highly excited or traveling at sufficient velocity so that they pass through the energy barrier. However, in an unexcited atom, the electrons remain in stable discrete shells. Where the velocity of electrons is not great enough to pass through the energy wave, the particles may be deflected as in LEED experiments.

That a physical barrier within the space between the nucleus and electron can explain the entire system of the atom is not a problem. The problem is that such a physical energy barrier in the shape of a wave was not used in the formulation of quantum mechanics.

Why is such a physical barrier a viable solution? Why is such a barrier not arbitrary? Firstly, we have empirical evidence for the movement of electrons by light energy in the photoelectric effect. Therefore, light (meaning electromagnetic waves) emanating as an energy wave from the nucleus could push back on the electrons when it reached the threshold frequency for the photoelectric effect without creating enough kinetic energy to eject it from the atom. (In my first edition of this book, I imagine the proton wave bending back on itself to keep its energy. I now see that is probably untrue.) A better explanation of why the wave around a particle does not radiate is that it does not reach a full quantum of energy. If we look at Einstein's 1905 paper on light as a particle, we see a restating of Planck's equation where only whole numbers create a full quantum of radiable light. This obviously leaves some light unradiated that does not reach this whole number requirement. In fact, this was Einstein's argument in his paper on specific heat[xxix], where he showed that diamond at lower temperatures did not give off heat when lowered by one degree because of this quantum effect of not being a whole number times Planck's constant. Therefore, the energy between electron shells and surrounding the nucleus is such that it is partial quantum energy and therefore cannot radiate, but remains unradiated surrounding the particle.

For all shells, the energy is strong enough to overcome the Coulomb force, therefore, no electrons appear between the nucleus of the atom and the ground state shell. The electrons in each shell also have an antinode of wave energy that surrounds the particle. This could only be the case if the wave emanating from the nucleus were losing antinode strength with distance.

The size of the atom would mean that the light surrounding nuclear particles would have to be in the ultraviolet region i.e. of very short wavelength. This might appear to indicate that the electrons would all be ejected out of the atom. Because of the short wavelength, a great deal of energy would be needed to cause the wave to radiate. Therefore, the wave surrounding the nucleus does not radiate. It is a partial quantum wave. (Note: all references to light waves are to be understood as a reference to the electromagnetic spectrum and not just to visible light.)

Therefore, the world of matter is not translucent because the atom is not mostly "empty space" but mostly energy.

Quantum mechanics manipulates the mathematics to describe a discrete placement of electron shells to match observation of the atomic spectrum without a physical phenomenon to describe the discrete placement. Here is introduced a physical explanation of this discrete placement explained by energy wave antinode barriers. The nodes of the energy wave emanating from the nucleus describe each electron shell thus making them discrete. Between each electron is a quantum of energy separating it from other electron shells. The amplitudes fall to zero at each of the nodes of the wave and produce natural places for discontinuous electron orbitals to occur.

This new theory is simple. A light wave (electromagnetic wave) emanates from each particle and returns to each particle. And, therefore, a light wave emanates from the nucleus of the atom and returns to the nucleus of the atom. Light has the ability to give motion to electrons which is seen in the photoelectric effect.

This is much simpler than Bohr's acrobatics in trying to make the Planck constant not carry energy and not create a full revolution which he described as an orbit or frequency. Bohr's theory fits the observation that the electron does not fall into the nucleus. I keep reverting to Bohr, because although we are all aware that the Bohr model was replaced by the Heisenberg and Schroedinger models, the quantization by Planck's constant h was still carried over into these two later models and is still in use today.

So what does this new theory of the atom predict?
1. It predicts the strong force naturally.
2. It predicts that when particles are accelerated at an atom, there is a speed at which the particle will not pass through the atom, but be deflected back from the atom and that particles of sufficiently low speed will be reflected back as there are no empty spaces or voids in the atom. There are no regions of zero energy. Note: This is not the same as saying there are no areas of destructive interference. This predicts there is no part of the atom that is void without energy.
3. It predicts that an energy wave extends from each particle so that tunneling is a consequence of the light energy wave reaching out from the atom beyond the electron shells.
4. It predicts that atoms can carry excess energy in the energy wave of the nucleus and that particles can carry excess energy in their nuclear waves. Therefore, particles like the electron can flow fluidly and continuously through a cathode ray tube without radiating in light quanta that follows the spectrum of light. The radiation from free particles simply radiates freely from the energy wave of the particle where the extra energy is temporarily stored.

5. It predicts that when electrons are trapped in the nodes of the nuclear wave, the energy that is radiated during transitions is discrete because the wavelength and energy between nodes is discrete. Therefore, the discrete wavelength seen in the spectra of atoms is because the electrons are transitioning through discrete energies of antinodes in the nuclear wave of the nucleus of the atom.
6. It predicts that matter can have potential energy because that potential energy is locked in or stored in the nuclear wave of the atoms or particles.
7. It predicts that $e=mc^2$ means that a lot of energy is stored in the nuclear light wave of the proton of each atom.
8. It predicts that the separation of the electrons orbiting the nucleus is due to energy between the electrons and that the electrons cannot overcome and make transitions through the next antinode without being infused with energy to overcome the waves.
9. It predicts that a large amount of energy is in the antinode between the nucleus and the first shell of the atom, so that the energy causes the nucleus to bind together eliminating the need for a separate force to bind the nucleus.
10. It predicts that it will take a lot of energy to fuse particles into a single nucleus because of the large amount of energy in the first antinode between the nucleus and the electrons in the first shell.
11. It predicts that as atoms become heavier than hydrogen, their nuclear waves superimpose and combine, and the transition energies of the electrons also changes so that transitions of electrons take more energy to get closer to the nucleus and are more likely to be on outer shells or antinodes of the wave.
12. It predicts that in the nucleus, each particle has its own energy wave and that although the wave becomes coherent outside the great sphere of the nucleus, the wave keeps the particles from touching each other because there is always energy between nucleons and external to the nucleus that overcomes the Coulomb force to some extent.
13. It predicts that binding energy will follow sphere packing when corrections are made for the size of the nucleons and the Coulomb force, so that binding energy is not linear.
14. It predicts that the energy carrying properties of particles are related to the mass of the particle which is related to the size of the particle's light wave surrounding the particle.
15. It predicts that imaging of a particle will show a definite point-like particle with a wave surrounding the particle i.e. there will "not" be wave-particle duality where there is either a particle or a wave, but there will be both a particle with its wave.
16. It predicts that the nucleus of the atom will be protected even though an electron is bound in a covalent bond with another atom to form a molecule, because the energy wave around the particle protects the nucleus even when the location of the electron is known.
17. It predicts that as energy is intrinsic to particles, even if the electrons are not whizzing in circular orbits, the electrons will be bounced around by the energy of the energy waves of the nucleus in an atom.
18. It predicts that motions of the electrons in the nuclear wave affect the nucleus as the electrons are in the nuclear wave of the nucleus.

19. It predicts that motions of the nucleons in the nucleus will affect the electrons as the electrons sit in the wave of the nucleus.
20. It predicts that heat is carried as energy in the waves surrounding particles or atoms.
21. It predicts that the Uncertainty Principle is unnecessary.
22. It predicts that wave properties and particle properties of particles can be observed at the same time.
23. It predicts that electrons do not disappear between transitions to different shells, but that only the necessary energy to push through the antinode is either absorbed or released in discrete amounts so that it appears in the spectrum that there is no in-between energy and that absorption and emission is discrete.
24. It predicts that partial quanta of light can be absorbed and emitted without affecting the light spectrum.
25. It predicts there is no necessity for quantum jumps, but only discrete absorption and emission of light by conservation and equilibrium laws.
26. It predicts that light can be absorbed and emitted for each particle or each atom at any frequency as long as the energy absorbed is not greater than the amount that can be contained in the finite energy wave of each particle or atom.
27. It predicts that the energy in the wave around a electron is conserved and finite while the energy wave around the proton is infinite and only these two particles are stable.
28. It predicts that in Compton Scattering, electrons can absorb a partial quantum of x-ray light.
29. It predicts that the energy of the atom and even the particle never drops to zero without the particle disappearing and reorganizing.
30. It predicts that electrons in a cathode ray tube will emit light continuously and not in discrete quantum spectra.
31. It predicts that the electrons can actually form a real crystal lattice in crystals because the electrons can maintain a position in the atom. There is no probability distribution of the electron. The electron can be bonded to a particular part of an atom and remain there without having to be measured to describe a position in the atom. The atoms can form particular structures and maintain those structures because there is no probability cloud or probability wave distribution.
32. It predicts that electrons can move in their own shells without losing energy as the energy is conserved by the automatic equilibrium of the mass and particle wave. Kinetic energy is energy absorbed from the nuclear wave to the particle and transferred from the particle back to the wave without light emission. There is no breakdown of the theory of conservation of energy
33. It predicts that neutrons outside of the proton nuclear wave have too weak of a wave as free neutrons so that they decay quickly, but while protected by the proton wave in a nuclear wave, they will decay at the rate of a proton. This effect is due to that under this theory, the nucleons share their particle waves after undergoing fusion.
34. It predicts that the electron in ground state can move without producing energy because the energy is conserved in the atomic system. Therefore, no Vacuum Catastrophe.
35. It predicts many ground states per atom.

So here we have two theories, one is the current Quantum Mechanical model of the atom and the new theory described here of the atom. Each describe that the electron does not fall into the nucleus of the atom although the electron is negatively charged and the nucleus positively charged. The reason given in current Quantum Mechanics is that for no apparent "causality," the electron simply must not fall into the nucleus. On the other hand, this new theory says that the causality of the observed effect is that a light wave pushes back the electron or electrons from the nucleus. Why is this a better model?

According to Isaac Asimov, we turn to his interpretation of Ockham's Razor. "Therefore, of two theories that explain equal areas of the universe, the one that begins with fewer assumptions is the more useful. Because William of Ockham (1300?-1349?), a medieval English philosopher, emphasized the point of view, the effort made to whittle away at unnecessary assumptions is referred to as making use of "Ockham's razor." Then Asimov continues, "One might suppose that of two theories one out to accept the simpler, as Ockham's razor (see page 5) recommends. However, Ockham's razor is applied properly only when two or more theories explain all relevant facts with equal ease."[xxx]

In the next chapter, I will show that the new model of the atom not only explains the mechanics of the atom more simply, but explains facts that the current Quantum model does not.

7 A new look at the Davisson-Germer experiment

In 1927 it was reported that the American physicists Clinton Joseph Davisson and Lester Halbert Germer and the British physicist George Paget Thomson had shown that a beam of electrons scattered by a crystal produces a diffraction pattern characteristic of a wave.[14] (The Davisson-Germer experiment has been duplicated consistently over the last century using a low-energy electron diffraction (LEED) gun and the results are generally correlated to the Davisson-Germer experiment so this is not an isolated experiment.) The beam of electrons was sufficiently slow moving that it would not simply pass through the crystal. It was concluded that the experiment showed electrons from the beam were being scattered by other electrons at the same angle of incidence with constructive interference that caused the diffraction pattern. It was assumed that the electron particle itself was displaying a different face or aspect and that it was showing a wave property. It was assumed that electrons at times display wave properties and at times display particle properties so that an electron is at the same time both a particle and a wave. Thus this experiment was said to fully support and prove the theory of Louis de Broglie that particles were both particles and waves at the same time i.e. there was wave-particle duality. Later, Bohr explained this wave-particle duality as part and parcel of his Principle of Complementarity stating that two contradicting descriptions of particles as both waves and particles at one and the same instance are necessary for a complete picture of the electron.

In the Davisson-Germer experiment, it is true that due to electrostatic repulsion, electrons will naturally be deflected and scattered by other electrons; however, this can be shown not to be the cause of the results of the experiment. An electron beam aimed at a crystal will encounter atoms arranged in a regular, repeating pattern where many of the atoms will have the same spatial orientation. However, there are some major problems with the interpretation of this experiment.

The first relatively minor problem is shown by Carl Rod Nave's analysis of the original experiment as presented in "Rohlf, James William, Modern Physics from a to Z0, Wiley, 1994",[18] that the peaks of scattered electron intensity with increasing voltage at the same angle did not fully match de Broglie's equation for wave-particle duality when one restricts the orbits to the whole numbers of Bohr's formula where n=1,2,3... (i.e. the set of positive whole numbers). Unless Carl Rod Nave's analysis is formed from incorrect data, then the fact that this experiment was accepted as proof of de Broglie's and Bohr's formulation of electron orbits being quantized by Planck's constant is dubious and should be reanalyzed.[49] It is not surprising that data produced in the last century using Bohr's introduction of Planck's constant is close to the actual spacing of electron orbitals, because Bohr intentionally manipulated the equations using Planck's constant to be consistent with known spectral phenomenon and reduced Planck's constant by half using classical ideas to manipulate the equations further in an attempt to have the equations approximate reality. Nave shows that the de Broglie equation only agrees with the first, third and fifth peaks. And further, Davisson and Germer point out another flaw themselves in their article: "A complete analogy between the phenomena of electron reflection and x-ray reflection would require that the Bragg formula should hold also in the case of electrons. This condition, however, is not satisfied; the wave-lengths at which the beam of reflected electrons attains its intensity maxima are not given by $\lambda = (2d/n) \cos \theta$."[50] It will be elucidated further that the Bragg law does not hold for reflection or diffraction in this case.

The fact that apparently de Broglie's matter-wave (wave-particle duality) equation which integrates Bohr's use of Planck's constant does not fit the Davisson-Germer results exactly could have been minimized in 1927 as being close enough and the matter was eventually ignored.

The second problem with the conclusion made by this experiment is that this experiment was taken as proof of wave-particle duality when it was contrary to accepted theory to assume that the position of the electrons in the atom, even in a crystal, would ever have the regular, repeating pattern of the crystal itself. Heisenberg's Uncertainty Principle further complicates the conclusion of the Davisson-Germer experiment. A main problem with the alignment of electron matter-waves in a crystal is that supposedly under current theory of quantum mechanics, entire atoms do not fall into each other and only maintain distance from other atoms due to the negative electrostatic force between electrons holding the atoms apart. This means that the negative force exterior to the atom should be equally distributed in all directions to keep atoms from encroaching upon each other. However, the electron cloud is a probability distribution. But as electron matter waves have an actual location despite the probability cloud, then there is no scientific way to explain how the electrostatic force could be equal around the entire exterior of a hydrogen atom. Just because one cannot know without measuring where the electron is, doesn't mean it doesn't have a position within the uncertainty restriction. A single electron cannot cover the entire sphere of its orbital at once while still having a measurable position in the probability cloud. This means there would always be places where atoms could collapse into each other. The probability cloud does not mean, and never did mean, that the actual electron is at every point in the cloud forming a "solid" surface that can deflect an electron beam. The points in the electron cloud are probability points where the electron may be. The electron does not occupy one of the points until a measurement is made. However, when the position and momentum of an electron are measured the electron cloud disappears due to the measurement. This is because it is assumed by Heisenberg's Uncertainty Principle that the position and momentum of the electron can indeed be measured and the electron has a location but there is a standard deviation and the location is not exact.[1] Therefore, one must consider that an electron beam pointed at a crystal is actually making a *measurement* of the location of the electron by use of the electron beam. Then it follows from the Uncertainty Principle that the electron is momentarily not in the electron cloud but has a position. Furthermore, with each electron beam measurement of the position of the electrons in the crystal, the position would change in any direction in three-dimensional space that cannot be predicted under Heisenberg's Uncertainty Principle. Rather measurement itself is creating a position and momentum of the electron at the time of measurement. So the electron is being reduced to a particle by the act of measurement and the particle has a variance in position and momentum with a minimum of ½ of h-bar under the Uncertainty Principle. This variation should increase with each successive measurement meaning with each successive electron from the beam that hits an electron in the crystal.[1] Therefore, one is no longer measuring the standing wave of an electron, but the particle itself by use of an electron beam "microscope". However, Bohr had explained waves and particles as different and "mutually exclusive". The conclusion originally drawn from this experiment, that it proved wave-particle duality, is in direct conflict with Bohr's Principle of Complementarity which states: "To observe the properties of an electron is to conduct some sort of measurement. Experiments designed to measure waves will see the wave aspect of electrons. Those experiments designed to measure particle properties will see electrons as particles. No experiment can ever measure both aspects simultaneously and so we never see a mixture of wave and particle."[30] Heisenberg quotes Bohr as saying in 1929: "Different observational situations—by that I mean the over-all experimental setup, the readings, etc.—are often complementary, i.e., they are mutually exclusive, cannot be obtained simultaneously, and their results cannot be correlated without further ado."[91,p.105] In other words, you cannot simultaneously carry on an experiment that hits an electron particle and comes out with a wave nature. In fact, the experiment was not designed to find the wave nature, but was a completely accidental discovery.[19] If anything, the experiment is designed to find the electron particle.

However, even if we eliminate the Principle of Complementarity and then proceed, then under Schrödinger's wave equation an electron being measured with an electron beam is in an eigenstate of position. There is wavefunction collapse so that the wave cannot be detected. Even in matrix theory, the act of measurement in matrix mechanics is taken to 'collapse' the state of the system to that eigenvector (or eigenstate).[1] This taken together with the added factor under the Uncertainty Principle that the electrons are constantly changing in position and momentum for every atom in an unpredictable manner, it is impossible to describe a regular repeating pattern for the electrons from bombardment with an electron beam at a specific angle unless the electron itself comes to a complete halt at the same location in every atom of the crystal at the same time which would violate the Uncertainty Principle. One should ask themselves: Can the electrons in a crystal hold a pattern during repeated collisions with an electron beam under the Uncertainty Principle? One must conclude that the electrons are not holding position in pattern during electron scattering experiments. Therefore, the scattering angles should never align because the electron beam is measuring the position of the electron in the atoms of a crystal rather than the position of the electron cloud or even the position of a standing wave. Therefore, the angle of scattering for individual electrons will be different for every atom in the crystal. There should be no wave aspects being measured. There should be no diffraction pattern under quantum mechanical rules.

One cannot take the less orthodox interpretation of the "duality condition" described by Englert[31] and try to anachronistically superimpose it on the physics of the 1920s. Science tries to justify the results of the experiment by saying that the electron beam aimed at electrons in the crystal was an attempt to view the wave aspects and therefore did view the wave aspects, but this is a non sequitur argument as the mere act of colliding one electron with another is an act of measurement and a collapse of the wavefunction. De Broglie won the Nobel prize in physics in 1929, and Davisson and Thomson won it in 1937 due to the conclusions from this experiment. However, to conclude that this experiment proves wave-particle duality, one would have to discard the fundamental principles of quantum mechanical laws i.e. the Principle of Complementarity, Schrödinger wavefunction collapse, and the Uncertainty Principle.

No one knew better than Niels Bohr and Werner Heisenberg that this experiment did not show wave-particle duality but they remained silent. It may be true that eventually Bohr changed his mind about wave-particle duality being mutually exclusive in a single experiment due to Complementarity, but in 1927, the year of the Davisson-Germer experiment publication, that is not the case. More pointedly, we are speaking of 1927, the very year in which Niels Bohr actually introduced the Principle of Complementarity at the Como lecture. An analysis of Bohr's change in thinking ten years later shows:

"The attribution of particle and wave properties to an object may, however, occur in a single experiment; for instance, in the double-slit experiment where the interference pattern consists of single dots. So within less than ten years after his Como lecture Bohr tacitly abandoned "wave-particle complementarity" in favor of the exclusivity of "kinematic-dynamic complementarity" (Held 1994)."[89]

Bohr proclaimed wave-particle complementarity in 1927, the year of the acclaimed Davisson-Germer paper, which complementarity meant that no experiment could show both the particle aspect and the wave aspect at the same time. Yet that is in fact what the Davisson-Germer experiment was assumed to show i.e. that one could measure the electron particle by collision with another electron in a beam and yet get a wave aspect to display. This was absolutely an undeniable contradiction to Bohr's scientific convictions. However, Bohr and Heisenberg certainly kept silent on the subject in 1927 at a time when "wave-particle complementarity" was unreservedly introduced and thoroughly believed.

There were other such silences in the 1920s development of quantum mechanics. Another such is here described:

"The taste for nonsense probably lay behind the skepticism that Bohr displayed about electron spin when he arrived to celebrate Lorentz's 50 years as a doctor of philosophy. Spin had first been proposed in Copenhagen by a German-American physicist named Ralph Kronig. Heisenberg and Pauli immediately hated the idea. They had just chased all imaginable actions out of quantum mechanics. Now Kronig was proposing to set the electron rotating in space. Faced with such fierce criticism by Bohr's stars, Kronig grew silent and the idea of electron spin had to wait for a second coming, this time under Ehrenfest's encouraging eye."[46]

Kronig introduced electron spin to Pauli who firmly rejected it in early 1925. Later in 1925 Wolfgang Pauli submitted an article saying that "According to the ideas presented here Bohr's constraint expresses itself not in a violation of the permanence of quantum numbers when the series electron is coupled to the atom core, but only in the peculiar two-valuedness of the quantum theoretical properties of each electron in the stationary states of an atom."[8]

In October of that same year Uhlenbeck and Goudsmit published a paper on particle spin that verified Pauli's paper about the two-valuedness of the electron. Not to disparage Pauli's reputation for refusing credit for much of his work, however, it does seem peculiar that Pauli should get a Nobel prize in 1945 for a paper suggesting a two-valued electron which he already had been apprised of and rejected a few months earlier, and then months later another paper is published about spin which confirms Pauli's paper. Three years later in 1928, Pauli quit working on quantum physics and turned to writing fiction instead.[46, p.287] However, Pauli did not fully abandon quantum mechanics as he communicated and worked with Werner Heisenberg into the 1950s on particle physics.[91]

So it should not be surprising that with the Davisson-Germer experiment, the few quantum physicists in the world of the 1920s looked the other way when they knew it rejected the primary concepts of quantum mechanics.

The new theory here presented holds that the spacing of orbitals is not due to Planck's constant itself but due to an actual barrier with a wave formation emanating from the nucleus. The experimental data from the Davisson article shows repeated peaks of scattered electron intensity with increasing accelerating voltage.[14] This data was collected at a fixed scattering angle. And this new model shows how the Davisson-Germer experiment did produce a diffraction pattern consistent with waves even though the electrons themselves could not produce such a pattern as shown above. This new model of the atom predicts that it is possible for the electron beam to be of low enough velocity that it would prevent passage of electrons by the nuclear energy wave due to the photoelectric effect of the waves. Eventually some electrons would be deflected off of the waves of the energy wave antinodes which present at staggered spacing between electron shells because the energy wave is a natural barrier to electrons at low excitation states. Where the nuclear energy waves of the atoms are all presented in the same orientation as in a crystal, the angle of scattering measured would be consistent and show discrete wave distribution. So rather than deflecting off of individual electrons, the beam would deflect off the angular surface of the energy waves creating the nuclear energy wave. Under the theory presented here this diffraction pattern is naturally caused by the wave pattern of the nuclear energy wave rather than by the electrons themselves. As an electron beam is aimed at the crystal at a fixed scattering angle, the electrons in the electron beam will only be able to penetrate the nuclear energy wave about each atom to an average equal extent at a certain voltage and velocity according to the frequency of each antinode of the wave. When the velocity of the electron beam is accelerated by the increase in voltage, the electrons from the beam slowly begin to force their way from a node by pushing against an antinode of the energy wave until finally the beam is strong enough to penetrate into the next node at which time the electrons in the beam pop through the antinode barrier. The pattern of node and antinode, creates increases and decreases in resistance to the electron beam. It should be noted that in the Davisson-Germer data, the intensity between all peaks does not drop to zero as would be predicted by current quantum mechanics. Consequently the nuclear energy wave introduced in this new model of the atom creates the diffraction pattern. Therefore, the new model of the atom being proposed makes the experimental results quite intelligible.

So the Davisson-Germer Experiment did not prove any aspect of current quantum theory, but proved a wave pattern in the atom. This proof was overlooked and twisted to suit the current theory of quantum mechanics that was then in vogue when in reality it proved undeniably and irrefutably that the atom contained a wave pattern that was able to affect the path of electrons; and that at the correct velocity, this wave pattern was continuous in its ability to deflect every electron. There were no gaps in the atom. Yet, the quantization of the electron shells under the current model of quantum mechanics creates gaps of zero energy in the atom. This is not consistent with the Davisson-Germer experiment. The wave pattern shown in the experiment describes a wave for the entire atom without any gaps and this wave was graphed at the time and is unmistakable, yet this wave pattern has been ignored for almost a century now.

No one discussed why the Rutherford gold foil experiment showed that the atom was mostly empty space and yet the Davisson-Germer Experiment showed that an electron beam hitting an atom could be made to deflect every electron at a certain velocity. In other words, in the Davisson-Germer Experiment there was no point in the atom where the electron would pass through empty space. Every electron at low enough velocity deflected back in the Davisson-Germer Experiment in direct defiance of the Rutherford experiment. This effect cannot be explained under the current quantization state of the atom which shows only discrete energy levels with voids in between. The Davisson-Germer experiment proved beyond a shadow of a doubt that the atom was *not* mostly space, not mostly a void, but in actuality a permeable solid that could stop the path of electrons at lower speeds. This refutes current quantum mechanical theory absolutely.

If however quantization is taken as the various energies that electrons emit or absorb while passing from one node or another of a nuclear wave, then quantization still holds in the new model. The observed effects of quantization make sense, but the observed effects of the atom's ability to deflect every electron in an experiment also makes sense. There exists a fundamental flaw in current quantum mechanics due to the anomaly between the results of the Davisson-Germer Experiment and the current model of the atom. This anomaly is readily fixed by an alteration of the atomic model.

We will pick up more details on the Davisson-Germer Experiment in the chapter on Suborbitals.

8 The Nucleus of the new model

A property of the energy wave is that it causes pressure on all free subatomic particles; therefore, all free subatomic particles have binding energy.

I originally thought of the particle wave extending from the particle as having to radiate, so I thought it must behave like a black hole. My old reasoning was: The light waves released into the energy wave curve back in upon themselves creating a self-contained non-radiating wave that causes an outward photoelectric effect as well as a less-pronounced inward photoelectric effect.

I now see that this thinking is incorrect and complicated. When we read both Planck's and Einstein's original articles on the Planck constant and the photon, we see that the mathematics of both is consistent that energy to radiate must be a whole number hv (Planck's constant times frequency). Obviously some light doesn't radiate as this would leave a remainder. Also in Einstein's original paper on Specific Heat when looking at diamond at low temperatures, the energy of the quantum jump can be so large to make the quantum jump that the light is no longer absorbed when raised by one degree nor radiated when lowered by one degree. The equation $E=nhv$ means that the higher the frequency, the higher the energy of light must be to radiate. And the higher the frequency, the smaller the wavelength. Ergo the smaller the wavelength, the higher the energy of light must be to radiate. Wavelengths at atomic size therefore would need tremendous amounts of energy to radiate. Therefore, they do not. So, for each particle in the universe, there is a standing wave of energy around the particle because the wavelength is so short, it cannot radiate. This energy binds the particle.

I will make further corrections to this book using the above assumption.

The true nature of wave-particle duality is that each particle has an energy wave.

A property of the energy wave is that it creates a shield for free subatomic particles that prevents other subatomic particles from complete approach.

The shield is created by the energy of the waves from each particle barring approach by other particles. The energy waves are created at the atomic scale and are intrinsic to the subatomic particles and are "dark energy" due to the lack of radiation.

With free subatomic particles this energy wave is present around each particle as a shield but does not obstruct the movement of the particle. It does however normally obstruct one particle from fully approaching another particle.

In neutral hydrogen, the energy wave of the proton overcomes the Coulomb force between the electron and the proton because it shields them from each other. The electron is not massless. It in turn has its own energy wave. As I was forced to assume from the Balmer-

Rydberg equation discussed in the appended paper, the energy of the proton-wave extends to an infinite extent while the energy of the electron wave is finite. Each has an energy wave proportional to the mass of the particle.

However, when one particle overcomes the shield created by the energy wave of another as in a high velocity collision, they may form a nucleus. Therefore, when there are multiple nucleons in an atom, the protons and neutrons sit in a pressurized particle container which we call the nucleus. The nucleons share energy waves.

Expel a proton from the nuclear barrier created by the energy wave in a more complex element than hydrogen and it regains its original mass and energy wave. The energy in its energy wave is part of its mass under Einstein's equation.

A law would exist in the pressurized atom thus: An energy wave will have a tendency to maintain nuclear equilibrium. This law governs the mass to energy wave ratio of each particle. This means for each particle at its creation, the amount of light energy released into its self-contained energy wave is proportional to its mass. In multi-nucleon nuclei, when the most uniform pressure from all nucleon waves is adequate, the nucleus is stable. When the most uniform pressure from the nucleon waves is inadequate, the nucleus is unstable. This is especially true in larger atoms where the energy of each nucleon wave is part of the nucleus itself so that less energy is exterior to the nucleus holding it together.

This means that there is a tendency to maintain balanced exterior pressure on the nucleus. That is to say that the external pressure in trying to maintain nuclear equilibrium will tend to a spherical shape. The sphere is the natural shape when equal amounts of pressure are applied to any object from all directions. The nucleons themselves have a spherical shape and the nucleus is not a hard sphere.[10] However, a balancing occurs between the nuclear mass and the energy wave in the distribution of mass and energy to form the most uniform spherical pressure under the law of nuclear equilibrium. Adding more protons or neutrons changes the shape of the nucleus so that the amount of external pressure from the energy wave changes with mass and under the law of equilibrium this causes the most uniform distribution, sometimes creating a stable atom and sometimes expelling particles.

Because at the atomic level this energy wave and its wave would be associated with the mass of the particle, it can be seen why the nucleus is at the center of the atom, because in the atom, the proton, being ~1,836 times heavier than the electron, becomes the main contributor to the energy wave in neutral hydrogen.

Since the energy of the wave is according to mass, each atom creates its own unique energy wave described by its atomic weight. And thus the energy wave for each atom of the same element is identical in energy and size and basic shape, although the atoms themselves cannot be truly identical since the electron and nucleons can be in different states within atoms of the same element. The pressurized atom would more accurately account for solids having the property of solidity. As usual, gases remain permeable due to lying on very weak antinodes of the protons between the pressurized atoms.

As each heavier element forms, the energy wave of each particle combines, but not completely since the particles remain separate in the nucleus. In the formation of elements when the energy waves of two or more nucleons overlap, although they initially interfere, at a certain radius from the nucleus the combined waves become constructive and reshape to form a great sphere around the nucleus. This predicts that when a new element forms it does not form fluently. The addition of nucleons into the nucleus causes the reshaping of the energy wave around nuclei as well as the redistribution of mass to energy under the law of nuclear equilibrium. This disturbs the electron shells until the energy wave around the nuclei describes the proper shape again with coherent antinodes and nodes into which the electrons may again settle into stable shells. However, in the nucleus, these antinodes and nodes in multi-nucleon nuclei are no longer regular due to interference from the overlapping waves of the nucleons. Each nucleon is spaced from other nucleons by a certain number of wavelengths according to the energy wave emanating from each particle between them. When a reorganizing of the energy wave occurs at the formation of new elements or isotopes, a gamma ray may be emitted or absorbed as each nucleon settles into a node of the waves emanating from other nucleons. When decay occurs, each nucleon is affected and this may allow for closer spacing of nucleons so that when the nucleons move into closer nodes in relation to each other, a gamma ray may be emitted representing the unnecessary excess energy.

In my first editions, I tried to calculate the energy between the nucleus and first ground state shell. This became unnecessary when I discovered the Balmer-Rydberg equation multi-nucleon atoms. I originally thought: To calculate the distance between the nucleus of a hydrogen atom and the first shell or node in the nuclear wave, the minimum force of the nuclear wave acting between an electron and a proton would give you the minimum force of the wave using Coulomb's Law. The maximum force of that distance would be the force that has to be exceeded for an electron to pass through the atom between the electron and the nucleus, or the force needed to cause the photoelectric effect or electron ionization.

However, the above is inexact. It can be shown that the curve of the energy wave around a proton actually contains a much larger amount of energy than was first expected. But since each antinode of the proton-wave decreases exponentially, the electron comes as close to the nucleus as it can depending upon how much excess energy the electron wave acquires. The more energy held by the electron, the closer it can come to the nucleus.

Rutherford concluded in the Gold Foil experiment that radiation of alpha particles were passing between the void or empty space in an atom. The Davisson-Germer experiment proved that electrons could pass through that space between the nucleus and the first shell. There have been hundreds of thousands of experiments since the 1920s that prove that electrons do pass through the void or empty space in an atom. Modern QM is founded on the principle and hypothesis that an electron cannot exist in that space even though the very earliest experiments proved otherwise. This is yet another phenomena that cannot be proven under current Quantum Mechanics.

The rest of this book will look at different Quantum Mechanical models in the light of this new theory. There will be a lot of new assumptions. This is not to contradict the theory or its usefulness. Rather it is to try to stretch the current theories, question them, and test their limits. To show that the early Quantum pioneers only glimpsed the beginning of the atom. Whether any of the assumptions further discussed are flawed will depend on what is known that may contradict these views. The following discussion is to prompt further scientific mathematical study in relation to the new model assumption that a wave emanates from the nucleus of an atom and from every particle with mass. The data from the spectra is shown in the paper herein but a better statistical relation needs to be created that will account for all the missed data in the spectrum of elements. Therefore, this effect should influence the interpretation of various scientific observations and theories leading to a reinterpretation and expansion on a very good first guess in 1926. The following is simply an exploration of current theories and how they might be affected by this new model of the atom.

In the interests of further investigation, I'm going to question other scientific precepts. I will also attempt to explain unexplained physical phenomena. I do not want to suggest that everything I try to explain is the correct explanation. In some instances, I may be presenting something from a naïve point of view. However, the one thing in this book that I do assert and believe is that science will catch up with this idea of a wave emanating from particles that creates a standing wave that does not radiate. I fully expect that this model will one day be inevitably accepted by scientists based on the fact that it is a simpler and better explanation of the atom, and is in line with all previous scientific thought and experiment; and agrees with the classical explanation of why the electrostatic force between the proton and the electron is overcome in the atom due to a real physical barrier.

I further examine the hypothetical effects of such a barrier. I also agree that the atom must explain the periodic table as does the current model of electron shells. But whatever disruption this model may cause, if it brings us nearer to the truth, then so be it. The history of the introduction in the periodic table is shown in this book to be an ad hoc invention based on chemistry. So this structure should be able to be seen in the spectral data of this new theory.

Unlike the Schrödinger wave model which only predicts a wave for the electron in a hydrogen atom, which is undifferentiated from the electron itself as a particle, this model assumes that every particle has a wave around it so that every particle is both a particle with a wave. Before the Schrödinger wave model, there was no reason to include the Planck constant into the atom where no wave existed.

The theory of the existence of the quantum was founded on empirical evidence by Planck and Einstein i.e. that it is an indivisible packet of energy and that light cannot radiate until it is a full quantum. I incorporate this idea into this model of the atom as this atom has an actual wave with a frequency (not in the exact sense of kinetic energy, but in the sense of stationary energy in a standing wave) around every particle in the atom; and electrons with their own waves sit in the wave of the nucleus of the atom. Therefore, the Planck constant has its rightful place as part of the stationary wave around the nucleus and around each particle, not just electrons. Therefore, the Schrödinger equation is an approximation of reality because it is a wave equation founded upon the equation for classical standing waves. I didn't figure this out until I unraveled the secrets of the atomic spectrum.

9 A look at Bose-Einstein Condensate

This new model of the atom does predict the wave behavior of atoms that is predicted by current Quantum Mechanics. However, this new model adds the causality feature that there exists a particle and wave at the same time in the atom as each particle is surrounded by its own standing light wave which is conserved.

The wave nature of the energy wave of the pressurized atom in the new model of the atom would predict that it could create constructive and destructive interference when two atoms of the same element overlap or superimpose as each and every proton in the universe has a wave of energy stretching to infinity under the Balmer-Rydberg formula as is shown in the research paper appendixed. This model from the spectra results in several grounds states per atomic element with the electrons moving further from the nucleus during cooling to a lower ground state. Therefore, in Hydrogen, the coolest ground state is not in the Lyman series near the nucleus. This is ground state at room temperature which is very hot when compared to zero Kelvin. That is why room temperature ground state is in the Lyman ultraviolet portion of the spectrum. But as the atom cools, the ground state moves to the Balmer ground state, then lower and lower ground states of each series. Therefore, the atom expands and grows larger as it cools. The energy wave around the proton itself starts to lose energy near zero Kelvin and is unable to prevent nuclei of surrounding atoms from passing right through the antinodes and entering each other's nuclei. This is the very nature of the incontestable evidence of the spectra giving rise to this new theory. This theory predicts the behavior of Bose-Einstein Condensate super atoms which came as quite a surprise to the experimentalists who only knew the current Quantum Mechanics which gives no explanation of this behavior. BEC's are one of the biggest testimonies to the accuracy of viewing the spectral analysis in the attached paper as experimentally correct. Each proton is itself surrounded by a standing energy wave.

This phenomenon is observed in Bose-Einstein Condensate.[13] Because it is a fundamental conclusion from this new model of the atom that the nuclear energy wave is a coherent wave, in phase, and maintains that shape proportional to the mass of the particle, it would account for the fact that superimposition of atoms into a so-called super atom would show a wave interference pattern. Again it would be highly unlikely under the current theory that matter waves of electrons could align in such a way as to form a coherent wave diffraction pattern given the theory of matter waves as understood by its inventor De Broglie which has the matter wave in a circular orbit around the nucleus guiding the electron.[12]

Carl Wieman, the leader of the JILA team that discovered Bose-Einstein condensate said about the implosions called bosenovas:

"Understand that atoms have been very well studied. Essentially all the behavior of isolated atoms in general and BECs in particular we thought were quite well understood, and could be predicted accurately by theoretical calculations," said Wieman. "Even for those features that cannot be accurately predicted, the basic physical processes are still qualitatively well understood.

"But the theoretical calculations of what would happen in this situation predict behaviors that are totally unlike what we've observed, so the basic process responsible for the Bosenova must be something new and different from what has been proposed,' Weiman said."

There are two points to mention here. One is that the behavior of atoms is thought to be "well understood" by science and that the behavior of atoms can be predicted "by theoretical calculations" and, two, an explanation of the collapse of atoms near absolute zero is more likely the result of a nuclear wave energy wave barrier in place around the nucleus causing pressure upon it. When the energy of that wave barrier is reduced by freezing to near absolute zero, the nucleus is no longer protected and the atoms fall into one another causing a super atom. In this new theory, everything is made clear.

That the atom is not well understood is evident by uncountable unexplained phenomena. What is heat at the atomic level? How is a magnet explained at the atomic level by quantum mechanics? (Do not confuse the Standard Model as Quantum Mechanics. Quantum Mechanics is based on Heisenberg's and Schrödinger's wave mathematics. The Standard Model is based on Particle Physics and all the forces described as particles in a field. This is not in the QM equations. Particle Physics is an extension over 50 years of Matrix Mechanics and in Particle Physics, everything is particles.) But modern physics of the atom cannot explain that the electrons align in the same orientation relative to the nucleus, but how can they align in quantum mechanics considering the Uncertainty Principle?

Other questions that are not well understood now are: How does a probability cloud align to produce a magnet? How can a single unpaired electron remain in one orientation in the atom under the Uncertainty Principle and, if it does so, does it not break the symmetry of the probability cloud and cause wave function collapse? And then is not the electrostatic force exterior of the atom open to collision by surrounding atoms because only the electrostatic force keeps solids solid? Even if it does not need to be visualized, still there is no quantum mechanical mathematical explanation that can show the orientation of the unpaired electron in a magnet that does not violate the uncertainty relations. How does the grouping of electrons into suborbitals account for chemical properties such as heat conductance, brittleness, reflective properties, opacity, etc.? For instance, hydrogen is a non-metal yet its electron is in an s-orbital and the other non-metals have varying numbers of electrons in the p-orbitals. How is this related? Why is the ion of an element so different from the element itself? How does this arrangement of electrons account for the chemical properties of the element? For instance, why are sulfur and phosphorus non-metals when they are exactly the same as the halogen chlorine in configuration except that chlorine has 5 electrons in its p-orbital and sulfur and phosphorus have 4 and 3 electrons in their p-orbitals respectively. How does the quantum mechanical arrangement of suborbitals account for the difference in chemical characteristics of these three elements? And the answer isn't because chlorine has 7 electrons in its outer shell like other halogens. This isn't a complete answer about chemical properties. Under quantum mechanics, electrons are "identical particles" therefore how does arranging them differently change the properties of the elements?

There is something that quantum mechanics has missed in its neat little packaging of electrons according to the periodic table as suggested by the chemists. There is a physical reason that non-metals are non-metals. How is it that just by changing the number of electrons in the "p" suborbital one gets non-metals, metalloids, halogens, other metals and noble gases? Why should the number of electrons in the "p" suborbital produce such different chemical properties?

There is missing structure in the atom as defined by quantum mechanics. This new model of the atom does not explain all these things either,…yet, but this new theory has some eye-opening solutions as to where we get the ionization levels that match experiment. And the real point being made here is that it is presumptuous to say that the basic workings of the atom are "quite well understood" by science. Science may understand which elements react chemically in a certain way on the one hand and science may have an inkling of the structure of the atom through quantum mechanics on the other hand, but this picture is far from complete and the structure of the atom is far from explaining the chemical properties in their entirety. Atomic science is still in its infancy and probably will continue so until some time in the future when humans are actually able to image the atom other than merely obtaining images that show the position of the atom as a fuzzy blur as is seen today. The only clear evidence is from the spectra of the elements that we can actually see and this has been ignored except for hydrogen. This new theory is based upon the spectra of all elements and explains them all. That is something current Quantum Mechanics cannot answer.

Secondly, and most importantly, scientists have been trying to explain the phenomenon of BECs in which the waves in the super atoms are seen to become coherent and cause interference. Science has tried to explain the so-called "coherence of matter-waves" referring to the de Broglie matter-wave. It is highly unlikely that de Broglie waves for each electron in these complex atoms could become coherent when they must stay in their particular shells. Science has missed the point completely. This new model does explain this coherence of waves in the atom and the de Broglie/Bohm matter-wave cannot explain the super atom. What science is seeing in its full glory for the first time is the coherent light energy wave emanating from the nucleus that is always there and just amplified in the superimposition of atoms in BECs and proven to exist in the attached paper from the spectrum of elements.

10 Binding Energy

Francis William Aston calculated the masses of the elements. "Atoms do not fall apart, Aston reasoned. Something very powerful holds them together. That glue is now called binding energy. To acquire it hydrogen atoms packed together in a nucleus sacrifice some of their mass. This mass defect is what Aston found when he compared the hydrogen atom to the atoms of other elements following his whole-number rule. The density of their packing requires more or less binding energy, and that in turn requires more or less mass: hence the small variations…. Locked within all the elements, he said, but most unstably so in the case of those with high packing fractions, was mass converted to energy…. 'If we were able to transmute [hydrogen] into [helium] nearly 1 percent of the mass would be annihilated. On the relativity equivalence of mass and energy now experimentally proved [Aston refers here to Einstein's famous equation E=mc2], the quantity of energy liberated would be prodigious.'" [82,p.140] (Words in brackets in original.)

The problem with this explanation is that the energy of the mass defect cannot be both "locked within all the elements" as binding energy *and* "liberated" at the same time. Under this new theory, the mass defect is held to the nucleus as binding energy in the energy wave. If the energy were liberated, the nucleus wouldn't hold together. What is happening is that some of the mass is converted to the standing energy wave surrounding the nucleus, that is, the standing light wave that is emanating from each particle and surrounding each particle and therefore surrounding the nucleus. Originally, I thought of this light as curving back upon itself causing a repulsion as in the photoelectric effect thus holding the nucleus together more firmly, binding it. Since finding the Balmer-Rydberg formula in the first 20 ions and neutral elements of the Periodic Table of Elements, I discovered that light does not have to radiate. Rather, the basis for the idea of Planck and Einstein when discussing the quanta was that the algebra showed that light came in packets whose smallest quantum would radiate if multiplied by a whole number. This obviously leaves a remainder. Not all energy inside a black body radiates. It only does so when it reaches a whole number times Planck's constant time frequency. Therefore, there are steps in the ladder that are too high to reach this whole number threshold. This is explained very clearly in Planck's paper on the quantum and even more specifically in Einstein's 1905 paper, but especially in Einstein's paper on Specific Heat. All of these papers show that there are steps in the quantum ladder that cannot be reached by the electron and therefore there is no radiation. I call this radiation, partial quanta radiation, although it shouldn't be called radiation at all as it does not radiate. It is partial quanta energy in the form of a standing wave. When the nucleons unite in the transition from Hydrogen to Helium, they get stuck in each other's particle standing wave and are surrounding by the proton-waves of two protons with a doubling of energy caused by these two protons. There is also a small amount of added energy from the neutron wave. But binding energy that causes a mass defect is the energy released into the surrounding standing wave of the particles in order to hold the nucleons together. Sometimes mass is necessarily released into the environment in order to maintain a stable nucleus with just the right amount of energy in the multiple nucleon waves to hold the nucleus together.

The pressure or force exerted by the energy wave on the nucleus is equal to the binding energy. However, the binding energy of hydrogen is not zero. Binding energy means the energy it takes to split a nucleus apart. If this is applied to subatomic particles, it means the energy it takes to decay the particle. Even in the Standard Model, there is a type of binding energy that keeps quarks in the proton together and keeps the proton from decaying. In this new model of the atom, these two different types of binding energy described in the Standard Model are united in that the same energy wave of each particle causes each particle to hold together and therefore the binding energy is applied to particles as well as nuclei.

Every subatomic particle has an energy wave and therefore has binding energy and is therefore a nucleus.

Since the binding energy for a single proton would be greater than for any multi-nucleon atom, for the atoms of elements in the periodic table, where BE_P is the binding energy of H-1:

$BE_P > 10$ MeV

In fact, due to the data gleaned from the spectra of elements, I was able to make an analysis of the binding energy of a single hydrogen atom. And from there, the empirical data of the spectra revealed so much more than I could have anticipated.

The energy from fission and fusion need to be reanalyzed. The mass defect does not necessarily all become free energy, but is transferred to the energy wave, therefore, the energy in fission is not caused from the mass defect but rather from millions of years of decay occurring in a few seconds. Decay energy is not the same as the mass defect. Decay energy is from the expulsion of subatomic particles from the nucleus. In fusion and fission, decay products lose mass that is converted to energy but this is not the same mass defect that stays intrinsic to the atom as the mass lost in creating a new element is transferred to the energy wave that surrounds the nucleus of the element. There is always a limit of mass to energy wave for each particle in the nucleus under the law of nuclear equilibrium. The natural balance between these two depends upon the density of the nuclear great sphere of each element. Therefore, in decay each new element formed is governed by the law of nuclear equilibrium and the mass defect of the element is not what is released in decay. Rather it is the difference in atomic weight after decay that is released. Mass defect itself is never released because it holds the nucleus in its great sphere.

Rutherford understood this very clearly. Already in 1903 in a paper with Frederick Soddy, Rutherford had calculated the amount of energy released by radioactive decay. As one author recounts:

"A Cambridge associate writing an article on radioactivity that year, 1903, considered quoting Rutherford's 'playful suggestion that, could a proper detonator be found, it was just conceivable that a wave of atomic disintegration might be started through matter, which would indeed make this old world vanish in smoke.' Rutherford liked to quip that 'some fool in a laboratory might blow up the universe unawares.'"[82,p.44]

Rutherford realized the immense energy of radioactive decay before the mass defect was even known and that radioactive decay itself had enough energy to create a great amount of destructive energy. Where mass defect means the amount of mass that goes into the energy wave of an atom that creates binding energy, this is not the same mass defect as the mass that leaves the atom in radioactive decay. The former is energy that never leaves the atomic particles while they are particles, the latter is energy released from the atom as decay products i.e. two different things. One is bound energy and one is free released energy.

Under the law of nuclear equilibrium it is inferred: The denser, the greater the binding energy. This applies to all subatomic particles and nuclei.

The less dense the particle, the less mass to energy ratio the particle has between its mass and its energy wave. For masses greater than the proton, the energy ratio is greater than the mass, causing some of the mass to be crushed. This is the reason that subatomic particles are not stable at a size greater than the proton. For masses smaller than the proton, the energy ratio of the wave is less than the mass, fragmenting the mass. Both the crushing and fragmentation are referred to as decay. The energy to mass ratio also becomes stable at the size of the electron. This is because the universe has two sizes of particles where the standing wave of energy around the particle becomes stable. The proton has a standing energy wave surrounding it that is infinite in extend as it fizzles out to infinity per the Balmer-Rydberg equation. The electron has a finite standing wave that is small enough that electron waves do not normally touch electron waves of other atoms unless in super conductors where the ground state is so far removed from the nucleus in the cold necessary that the electron waves touch each other from atom to atom.

Where m_P is the mass of a proton and m_{SP} is the mass of any subatomic particle and m_e is the mass of an electron:

m_{SP} is unstable where $m_e < m_{SP} < m_P < m_{SP}$

This is due to binding energy from the energy wave of each subatomic particle. The binding energy has directly to do with mass or density. The mass of a particle determines its lifetime because the mass of a particle determines the force upon the particle from its energy wave that surrounds it. In the case of the electron, the energy wave appears to be different than other particles. The nature of the electron is different because it will become clear that although the standing energy wave is the same type of partial quantum energy as the wave of the proton energy wave, the energy it carries is less due to the size of the particles. Protons necessarily need more energy to hold them together whereas electrons are so small, the energy around them is much smaller and finite. Only these two particles are naturally stable on their own in the whole universe of hundreds of particles. Every other particle has an energy wave to particle size imbalance and so decays quickly.

In multi-nucleon nuclei, the law of nuclear equilibrium operates by constantly tending to maintain a balanced spherical pressure on the nucleus unlike with free particles where the spherical shape is established and the pressure from the energy wave is at a set value in relation to the mass. This can be seen from the research article attached to the end of this book.

Where the density of a particle is given by density = mass/volume, the ratio of density of a proton over the pressure from its energy wave equals ideal stability. Therefore, only particles the size of the proton are able to persist in this universe. Again with the electron mass, the universe can maintain nuclear equilibrium and the electron sized particles are stable but finite. However, the amount of energy in the waves surrounding these two particles of different mass would necessarily be different.

In the multi-nucleon nucleus, mass is converted into the energy wave under the law of equilibrium which tends toward uniform spherical pressure. As a result, a free particle such as the neutron that would soon decay, when combined into a nucleus with other nucleons contributes some of its mass into the combined nuclear energy wave and does not decay. It is protected by the energy waves of the electrons in the nucleus. Some mass is converted to energy because the nucleons in the great sphere of the nucleus are perceived as one body that exists from the surface of the great sphere. The bulk of the energy is near the particle in its energy wave. Therefore, the energy wave exterior to the great sphere is missing energy, and mass is therefore converted to energy to keep the ratio correct under the law of nuclear equilibrium. Binding energy in the form of the mass defect can be pictured as being contributed to forming a great sphere around the nucleus of the multi-nucleon atom.

In the great sphere of a multi-nucleon nucleus, where the nucleons are more closely packed so that the great sphere is smaller in volume, the nucleons sit in a stronger portion of their energy waves because a greater portion of the more energetic waves near each particle are allowed to be outside of the great sphere. Therefore, the larger number of nucleons in the multi-nucleon nucleus creates a situation where many of the higher energetic waves of the particles are trapped inside the great sphere of the nucleus and the binding energy per nucleon is less. This causes the binding energy of the heaviest atoms to be unable to hold the nucleus together.

First of all it should be said that science has decided that the nuclei of all atoms are equally dense. This means although the radius for deuterium should include nucleon separation of 3.8 fm, Nuclear physics puts the radius at approximately 1.5 fm to even out the density as follows:

"Various types of scattering experiments suggest that nuclei are roughly spherical and appear to have essentially the same density. The data are summarized in the expression called the Fermi model:

$$r = r_0 A^{1/3} \text{ where } r_0 = 1.2 \times 10^{-15} m = 1.2 \, fm$$

where r is the radius of the nucleus of mass number A. The assumption of constant density leads to a nuclear density

$$\rho_n = 2.3 \times 10^{17} kg/m^3$$

For mass number
A = 2
this gives
r = 1.5119052598738476 fermi."[47]

Scattering experiments appear to show the same density because physicists do not take into account that there is a nuclear energy wave as described in this new model of the atom that affects scattering experiments. The water-drop model of the atomic nucleus with its skin was a useful model because the energy wave actually creates a skin of energy making a great sphere around the nucleus. However, under the above equal density theory, calculating the radius of an atom with mass number 2 at 1.5 fm is an unscientific picture and based upon assumption alone, yet this theory currently stands and that is why this new model of the nucleus may not be taken seriously until one looks at the spectral data which is empirical and cannot be denied. To say that the nucleus of deuterium h as the same density as helium-4 or carbon-12 seems questionable. When one takes into consideration that the nucleus does have nucleon separation and then one calculates the volume of the nucleus according to a great sphere placed around the nucleus, the results are very different. As is known the nucleus is roughly spherical and "*appears*" to have the same density due to the skin of energy forming a great sphere around the nucleus, but it can be shown that it does not.

The nuclear binding energy of the periodic table elements under this new model of the atom resembles sphere packing in a great sphere. To give an approximate representation, if hydrogen-1 is given the radius value of 1 fm, then the volume would be ~4.189 cubic fm.[31] Then in hydrogen-2, the radius of the great sphere is 2 making the volume of the sphere ~33.510 cubic fm. Yet, the mass of the sphere is only about twice that of hydrogen-1. Because the binding energy of the proton is so high, there only needs to be an increase of ~1 MeV in binding energy per nucleon since the binding energy is spread out over the great sphere and because there is so much empty volume in the great sphere. Empty clearly does not mean "empty" since the energy wave radiates from each nucleon. Empty volume is here simply referring to massless volume. Empty and "void" imply the presence of nothing. However, light is an energy that pierces the near vacuum of space and since E=mc2 then where there is energy, there is no void, no vacuum, but a type of mass is present since energy and mass are equivalent and interchangeable. Since radiation, especially gamma radiation, penetrates everything, and radiation is energy then there may not exist a void in the universe. In fact, every proton in the universe has a wave radiating to infinity under the Balmer-Rydberg formula and the new data from the spectra in the article at the end of this book. The Rutherford nuclear atom with its void or emptiness is actually filled with energy.

The great sphere volume itself only enlarges slowly after hydrogen-2. After carbon-12, the mass in the great sphere would be distributed almost spherically with very little extra unfilled volume. Therefore binding energy is higher. This would continue up through the table of elements. However, after the great sphere becomes particle-like meaning dense and spherical, the larger and denser the great sphere gets, the greater becomes the binding energy on the great sphere and crushes larger atoms into decay. If one calculates the radius of a proton as 1 fm and assumes the proton and neutron are spheres with a mass of 1 amu each, then sphere packing in a great sphere results in:

Volume Great Sphere[33]	Mass Number	Density
cubic fm	amu	amu/cubic fm
4.19	1	0.2387
33.51	2	0.0597
41.90	3	0.0716
45.83	4	0.0873

58.63	5	0.0853
58.63	6	0.1023
72.78	7	0.0962
77.95	8	0.1026
85.23	9	0.1056
94.94	10	0.1053
102.16	11	0.1077
102.16	12	0.1175
113.10	13	0.1149
123.59	14	0.1133
129.68	15	0.1157
139.85	16	0.1144
146.46	17	0.1161
153.29	18	0.1174
163.19	19	0.1164
175.55	20	0.1139
177.50	21	0.1183
192.17	22	0.1145
200.36	23	0.1148
208.75	24	0.1150
210.46	25	0.1188
220.89	26	0.1177
231.67	27	0.1165
229.85	28	0.1218
244.12	29	0.1188
251.64	30	0.1192
258.16	31	0.1201
265.57	32	0.1205
272.10	33	0.1213
277.79	34	0.1224
284.49	35	0.1230
291.45	36	0.1235
300.42	37	0.1232
301.06	38	0.1262
315.68	39	0.1235
323.83	40	0.1235
332.19	41	0.1234
334.93	42	0.1254
345.48	43	0.1245
352.66	44	0.1248
358.52	45	0.1255
366.92	46	0.1254
443.27	56	0.1263
451.76	58	0.1284
489.80	63	0.1286

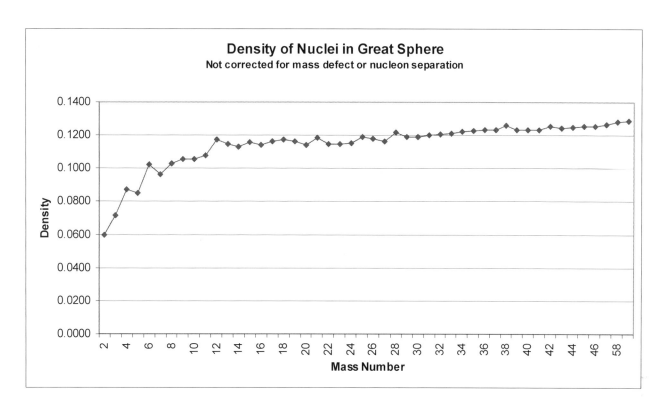

What this shows, even without looking at binding energy, is to expect that a nucleus described as a great sphere will have higher peaks of binding energy with four nucleons in the sphere, six nucleons in the sphere, twelve nucleons in the sphere, twenty-eight nucleons in the sphere, thirty-eight nucleons in the sphere. However, it must be remembered that denser creates more binding energy so that a four nucleon great sphere that is almost as dense as a twelve nucleon great sphere will have almost as much binding energy, that is, pressure from its energy wave. This look at nucleons in a great sphere also leads one to expect that the most significant drop in binding energy is from a single nucleon nucleus to a double nucleon nucleus as from hydrogen to deuterium. The binding energy then is still low in nuclei containing fewer than twelve nucleons. We then can compare this to actual calculated binding energy.

Graph of binding energy per nucleon:

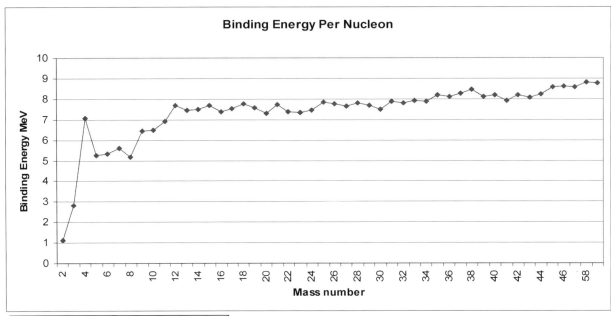

Element	BE per nucleon[29]
H	
2H	1.112
3H	2.827
4He	7.074
5Li	5.266
6Li	5.332
7Li	5.606
8Li	5.16
9Be	6.463
10Be	6.498
11B	6.928
12C	7.68
13C	7.47
14N	7.476
15N	7.699
16N	7.374
17F	7.542
18O	7.767
19Ne	7.567
20Na	7.299
21F	7.738
22O	7.365
23Al	7.335
24F	7.463
25Ne	7.843
26Ne	7.754
27P	7.663
28Na	7.799
29Na	7.683
30Na	7.506

	31Cl	7.87
	32Mg	7.808
	33Ar	7.929
	34Al	7.858
	35Si	8.169
	36Si	8.114
	37P	8.267
	38S	8.449
	39P	8.1
	40Sc	8.174
	41P	7.914
	42S	8.194
	43S	8.059
	44Cl	8.229
	45K	8.555
	46Sc	8.622
	56V	8.573
	58Fe	8.792
	63Ni	8.763

The correlation is significant. It would appear that density of spheres in a great sphere closely follows actual binding energy in the atom. When one calculates density in this way, it is easy to see why binding energy is not absolutely linear as was expected in early modern quantum mechanics. As Yukawa explained, "The binding energies of nuclei heavier than the alpha-particle do not increase as rapidly as if they were proportional to the square of the mass number A, i.e. the number of nucleons in each nucleus, but they are in fact approximately proportional to A. This indicates that nuclear forces are saturated for some reason"[xxxi] The meson field was invented to correct for the non-linear binding energy, but was unnecessary since binding energy actually does follow density. Corrections to the great sphere surrounding the nucleus would need to be made according to nucleon separation, mass defect, as well as the Coulomb force between particles. If a neutron carries an equal amount of positive and negative charge, then the Coulomb force can hold a single proton closer to a single neutron in the nucleus so that helium-4 is dense but helium-6 is not so that an adjustment must be made to the density of an atom with mass number 6. However, even without these adjustments, the binding energy appears to closely mirror sphere packing. While the density is low and the nuclear great sphere is small, the binding energy per nucleon does not need to be great. As the great sphere becomes more spherical and dense, binding energy should increase, however, it does not. The reason is because each nucleon particle has its own energy wave and each wave has the higher energy waves concentrated near the particle. In large multi-nucleon nuclei, the great sphere begins to enclose all of the higher energy waves of the particles so that the skin around the nucleus is weak being made from only the more remote weaker waves of the nucleons near the center of the sphere. To compensate for this loss of the higher energy waves, the larger atoms can only form with many neutrons so that the density is great enough to expose the higher energy waves and to allow the greater mass of the neutrons to be transferred to the great sphere that binds the atom. When the great sphere becomes too large, much of the higher density energy is within the nucleus and fragmentation decay becomes the norm. Bottom line: The strong force, as described by the force of the standing light energy wave surrounding the particle or nucleus, does increase linearly with density just as in Newtonian gravity. There is no need for particle-particle interaction theory. This means there is no need for the Standard Model which is based on particle-particle interaction.

If the binding energy between hydrogen-1 and deuterium is 1 MeV per nucleon and the difference in the sphere volume is ~4.189 for the proton. The great sphere volume for deuterium is radii of two nucleons equals 2 fm corrected for nucleon separation equals radius 5.8 fm so that sphere volume for the great sphere is ~817.283 cubic fm. The nuclear great sphere of deuterium is ~195 greater than hydrogen. If only 2 MeV of binding energy are needed for the great sphere of deuterium, then dividing the binding energy between the two nucleons: ~197 MeV/2 = ~98.5 MeV gives a possible binding energy of the proton, but this is purely conjecture. In fact, I show in my research article attached that under Einstein's famous formula $e=mc^2$ that the amount of energy surrounding each proton is equal to the calculated energy. This assumption obeys gravitational relativistic theory at a quantum level and would allow for a grand theory of quantum gravity. The curve of this energy as it recedes in distance from the proton creates an exponential decay curve that aptly describes both the relativistic change factor of energy closely surrounding the particle and the smooth transition into Newtonian gravitational energy of the inverse square law again at the quantum level.

Considering the energy collisions necessary to study the proton structure such as the ZEUS experiment at HERA in DESY Germany, the binding energy of the proton must be very high.[33] What can be concluded is that the atom has a nucleus with a "skin" much like a water drop, however, this skin is formed by energy in waves emanating from each particle that becomes coherent forming the skin of the nucleus. The skin places pressure upon the nucleus according to density. However, once the waves become coherent from the nuclear skin outwards, the frequency causes a barrier to other particles just as the frequency of the energy of light can deflect particles in the photoelectric effect. However, this wave is not moving like light, but is a standing wave in which each antinode contains energy that creates barriers to other particles.

Science has been well aware since the 1800s that particles are created in cosmic rays. Matter is constantly being created all around us. Very short-lived particles appear in cloud chambers as mass. Matter is being created before our eyes. The only thing that creates these particles is energy transformed into a particle $e=mc^2$.

Because physicists know the radius of a proton and the mass of a proton, they can calculate the density of a proton. Yet, particle physics claims that there is no force exterior to the proton so what is creating its density? In the macro world, density is created mainly by the force of gravity under Newtonian mechanics and presumably by the curvature of space-time under relativity. Planets become spherical when gravity is great enough. In particle physics, the only thing holding protons and neutrons together in the nucleus are gluons which do not surround the particles, but are particles themselves. This is a great flaw of particle physics. The proton is not infinitely dense as it can be fragmented. The fact that it is spherical as are all other subatomic particles (except perhaps those at electron masses which is unknown) strongly suggests a uniform pressure is applied to the exterior of the particle. Even quarks cannot account for uniform pressure causing the proton to be spherical, because there are assumed an odd number of quarks for the proton.

The chemical elements are formed from the most stable of particles, the proton and the electron. That is not to say that other particles have not formed into atoms of very strange elements in supernovae or cosmic rays, but that this universe, in its present state, according to the laws of nuclear equilibrium, decays them quickly. Although these elements may not be stable, protons and neutrons should not be the only particles able to create nuclei and form elements. It is possible that the particles originally termed "the V-particles"[26] such as the kaon and the rho mesons are examples of these fast-decaying types of elements. Rather than being one single particle, the neutral kaon under this new model of the atom presented here may very well be the nucleus of an atom containing two pions. As the atom of the kaon-0s decays the energy wave around each pion becomes unbound from the nucleus and their individual energy waves become separated shooting off in separate directions creating a V path. Instead of a mass defect as in the atoms of the known chemical elements, there appears to be a mass surplus in kaons and rhos because they decay to particles much smaller than their mass. Comparing the case of the neutron, it becomes stable within a nucleus, because it loses mass to the binding energy wave and becomes proton-like while in the nucleus. However with the kaon, if it is a two-pion nucleus, then because individual pions have intrinsic excessive binding energy, then when their energy waves overlap in the nucleus, under the law of nuclear equilibrium, instead of mass being transferred to binding energy, some of the binding energy is transferred to the mass of the nucleus which is the kaon. This would explain why when they decay, a significant amount of mass is unaccounted for.

It is highly likely that the neutron is the main contributor of mass to the nuclear binding energy which is the probable cause of its stability in the atom. However, as the research article I did attached shows, the neutron only adds a small amount of energy to the overall proton-wave, therefore, the reason the neutron decays as a free particle is because it has too little exterior energy in its neutron-wave. It only holds together in the nucleus by being enclosed in proton-waves. Free neutrons have too great or too little of an energy wave to mass ratio and are crushed into decay or cannot hold together. Neutrons are the wrong size for this universe where the existing energy in the universe likes to make protons to become stable and have an equilibrium between particle and proton-wave. I need more accurate data from the NIST tables in my research article to evaluate whether the free neutron has a neutron-wave that has too great or too little energy to hold it together for very long outside the proton-waves.

Because a proton has a lower or higher binding energy than a neutron, deuterium is an isotope that takes mass from the neutron and energy from its neutron-wave and transfers it to the great sphere. But the fact that dineutron nuclei are unstable demonstrates that neutrons do not become completely proton-like in the nucleus, meaning, having exactly the same ratio of mass to energy wave as the proton. In fact, the data for deuterium shows that the neutron is unable to contribute much energy to the energy wave surrounding the nucleus.

In this new theory, the Pauli Exclusion Principle is not the reason for the instability of dineutron nuclei, but rather it is the inability of the combined energy waves of the nucleon to provide stability by the law of nuclear equilibrium. If this is the case, it might follow that certain nuclei made completely of neutrons should be able to exist if nuclear equilibrium can be achieved, however, the evidence from my research article shows this is unlikely.

The strange explanation of an attractive force between electrons in Cooper pairs can be explained under this theory.[17] Electrons are not attractive, but follow Coulomb's law, however, they each have energy waves in equal size to each other, so that they may become trapped inside the nucleus of their combined energy waves. Cooper pairs can be thought of as nuclear anti-ions and the law of nuclear equilibrium will, when possible, bind the electrons. Anti-ion may not be a good term as what I really mean is an electron atom since the electrons bind in their own energy waves.

After having said that the binding energy is not released from the atom but stored in the nuclear energy wave, it must be stated that there is the possibility that current theory about binding energy is correct being that some of the energy is released as free electromagnetic waves in the binding of a new element. For this to be true, the energy nearer the particle, i.e. nucleons, which is greater near the particle must release mass which gathers the nucleons more tightly into the great sphere of the nucleus. Then this could alone account for the energy wave around the nucleus being able to bind the nucleus more strongly, however, that need not be the case and probably is not the case. Under this new theory, it could be a combination of the binding energy being held and added to the energy wave that binds the nucleus tighter or it could be the less amount of mass in the nucleus drawing the nucleus into the higher energy regions about the particles that binds the nucleus tighter. In the latter case, the mass could be lost to free energy while still binding the nucleus and in the former the mass is converted to energy held in the energy wave. This must be calculated from the pressure of the energy wave upon the individual particles in the nucleus and the pressure when the combined energy waves become a nucleus surrounded by a great sphere. The data has not yet been calculated from the spectral data attached. There is a lot of scientific doors that are opened by the data in the research article that need to be turned and calculated. The main concept is to keep in mind that an equilibrium must be maintained between the nucleon particle waves and the nucleon mass.

However, it must be taken under consideration that fission energy released need not be binding energy released. It could be the absorption of energy needed to balance the system of the nucleus upon bombardment such energy being absorbed from outside of the system and reemitted in the process. This would especially be true if fission atomic explosions result initially in an implosion of energy (other than the one artificially created) from the environment as energy is sucked out of the environment into the atom in order to accomplish the splitting of the nucleus made imperative by the invasion of the neutrons into the system. This concept will become clearer with the description later of the chemical cell or battery under this new theory.

The main point of this chapter is to show that under current theory, there is no real accounting for the unusual pattern for binding energy. However, under the new theory of a wave surrounding every particle as shown from the spectra of elements, it becomes clear why particles and nuclei are spherical and why the pattern for binding energy follows the pattern for sphere packing as if a spherical skin were surrounding the nucleus, because under this new theory that is exactly the case. Under the current theory, nuclei and particles are surrounded by a void that somehow makes them spherical which is another effect without a cause which is par for the course for the current theory of Quantum Mechanics. Let's put causality back into science.

11 Electron Shells

Quantum mechanics account of electron transitions: "And as Kierkegaard's stages are discontinuous, negotiable only by leaps of faith, so do Bohr's electrons leap discontinuously from orbit to orbit. Bohr insisted as one of the two 'principal assumptions' of his paper that the electron's whereabouts between orbits cannot be calculated or even visualized. Before and after are completely discontinuous."[82,p.76]

The entire Copenhagen interpretation of quantum mechanics is the interpretation that is most widely accepted being based upon the theories, ideas, and formulae of Niels Bohr, Werner Heisenberg, Max Born, Wolfgang Pauli, Pascual Jordan and Paul Dirac. This current theory of modern quantum mechanics holds to the idea that electrons make discontinuous jumps. And the Copenhagen interpretation "is still dominant within physics" according to the 2018 book by Adam Becker, "What is Real?" page 264. Even the alternatives to Copenhagen do not abandon the idea of discontinuous jumps. And every scientist still says conclusively without exception that the atom is mostly "empty space." Again, how is this depiction of electron shells scientific at all?

However, in this new visualization of the atom as is irrefutably shown from the actual data from NIST in the spectra of elements as shown in the research table at the end of this book, the electron shells are formed by the nuclear energy wave radiating from the nucleus of the atom that decreases in amplitude. It is contained in every atom according to the Balmer-Rydberg formula, not just in hydrogen as Bohr, Heisenberg, and Schrödinger found. This is empirical incontrovertible data of spectral light that we can actually see with our eyes.

So what does the data tell us? The electron shells are in the nodes of that coherent wave found around every particle and therefore every nucleus. Between each electron shell there is an antinode in the energy wave. Just as in the photoelectric effect a wave of high frequency can displace an electron, so the same is true in the atom. The energy in the antinodes of the standing wave from the nucleus pushes back the electrons. Since the energy in the wave is of varying frequency, the electrons are pushed back from the higher frequencies (note: I used frequency in my first editions when I pictured light as moving in a loop in the atom but now understand that the energy is non-moving because it is a partial quanta standing wave so "frequency" should be corrected throughout the book and replaced by "standing antinode of energy acting as a barrier") near the nucleus, but electrons are not pushed back from the lower energy antinodes further from the nucleus.

Due to electromagnetic attraction between the positively charged proton and the negatively charged electron, the electron tends toward ground state, but there are different ground states. Under the current model, ground state is defined as the closest to the nucleus. However, as will be shown from the empirical data of the atomic spectra, the ground state is the n-sphere shell that an electron can approach before the frequency of the energy wave barrier's energy effect is larger than the energy that the electron is carrying and therefore the electron cannot overcome the electrostatic attraction preventing it from further approach to the nucleus.

As concerns electron movement, the charged electrons would be changing position or moving in the nodes as they are constantly being affected by charges from surrounding atoms and by the energy in the environment. The electrons do not orbit in the sense of a planet because they do not move out of their node unless they have excess energy and the only movement appears from the spectrum to be transition movement. However, the node itself is spherical so that an electron can move around the node spherically without losing or gaining energy simply by the Coulomb force acting upon it. Where an electron is affected by a neighboring charge, it may be able to move inside a shell without gaining or losing energy if it does not meet with any resistance. This means an electron has the freedom to move freely around the sphere of the node it is in due to the energy that the electron carries. But if a charge acting upon the electron is strong enough to overcome the adjacent energy antinode of the proton wave, then the electron may transition, as will be further discussed. The electron radiates in this model of the atom under the laws of the classical electro-dynamical theory in that the energy radiated is equal to the adjacent antinode through which the electron must pass to transition. And it will be shown that electrons whether under classical theory or in the atom all radiate in the same manner. The universe acts the same in the Quantum and Macro realms. Only certain transitions are possible because the antinode barriers have different energies that must be overcome, but there are no discontinuous disappearing and reappearing electrons. The energy wave around the electron absorbs more energy from light and pushes through an antinode or the electron-wave emits the absorbed energy and falls back through that antinode. Therefore, the electron will not radiate away its electron-wave because that natural radiation that created the particle would destroy its mass. This is due to the law mentioned earlier in that there is a natural ratio between mass and surrounding standing light wave of a particle that is always maintained in stable particles. All light energy absorbed by an electron is excess energy that the electron seeks to eliminate as soon as possible to get back to equilibrium of particle to electron-wave.

As this new model of the atom is developed, it will be explained that the electron does not lose mass as the energy radiated is transferred to the energy wave of any subatomic particle like the electron and from the wave back to the electron to maintain an average constant mass for the electron. The radiation of energy by an electron is always from excess energy that the electron gains whether under classical electrodynamics in a vacuum tube, a cloud chamber, a particle accelerator, or in the atom. The laws of the universe hold in the macro and atomic worlds under this new theory while still keeping discreet energies in the spectra.

Bohr started from the basic assumption of a circular orbit where the centripetal acceleration is described by the distance between equal charges creating the orbits, so he assumed:

$$r = \frac{e^2 \cdot k}{m \cdot v^2}$$

where r is the distance between the electron and the proton in a hydrogen atom and k is Coulomb's constant and m is the mass of the electron and v is the acceleration.

However, under the theory here presented, Bohr neglected the presence of energy emitting or surrounding the nucleus that could keep the electrons at a distance. Instead Bohr postulated circling orbits of electrons with angular momentum that kept their exact quantized distance from the nucleus for no other reason than they simply had to which is no reason at all. Had Bohr postulated the presence of a force between the nucleus and the electrons he could have introduced a total pressure or force *p* exerted by the energy wave between an electron in ground state or the first shell and a proton in the hydrogen atom, and he would have gotten *r=p* in the formula above, because the radius to the electron would be the size of the energy quantum antinode or antinodes between the electron and proton and the energy carried in the antinodes of that standing wave. The strength of the force would be calculated: where **F** is the force under Coulomb's Law between the proton and electron in the hydrogen atom, where **W** gives the minimum threshold energy required to remove an electron from its ground state in the atom which is equivalent to its ionization energy when the kinetic energy of the electron is zero, then *p* is defined as *p=F*. However, this would give only the minimum force necessary between the proton and the electron in ground state. We have to explain why the nuclear (or particle) wave is non-radiating, either finite or infinite, and yet can form a wave without radiating. Then we have to try to put an energy level on that wave. No easy task as there are no laws existing for non-radiating light. However, since I first wrote that in the first edition, I found it was possible to prove these ideas from looking at the spectral evidence itself and that the equation for the proton wave had already been discovered in the Balmer-Rydberg formula. By an extension of the curve that formula follows, we see the inward curve of energy toward the nucleus and can understand how electrons stay in outer shells.

In the light wave surrounding a particle, the amount of energy is being held in a smaller volume of space for each antinode of energy than a radiating light wave. When the energy is emitted from the atom during electron shell transitions, it is radiated as normal light and the amount of energy emitted takes on the frequency of normal light. In the atom, in this new model, that same type of energy is concentrated into a smaller space due to the size of the atom and electron shells being so closely spaced. In normal radiating light we expect the frequency to be a specific magnitude according to the energy it carries and this frequency to continue as the light propagates through space. The wavelengths of light emitted by the electrons in atomic transitions is accurately known. Therefore, the stored energy of light in the atomic wave as a whole is a different magnitude than the frequency of emitted light. However, we know certain things about energy. We know massless particles move at the speed of light. Energy is massless even when under a full quantum. It is unmoving in a confined space because it is unable to break out of the confined space because it can't reach a full quantum for the wavelength necessary at atomic scales.

The reason this makes sense is that Planck's constant was used to explain why light is much more rarely emitted at higher frequencies in the ultraviolet. This would suggest that light energy at higher frequencies would more rarely emit light the higher the frequency which is scientifically and empirically seen. Therefore, if we take this phenomenon that light is emitted more and more rarely the higher the frequency to its expected conclusion, there must be a frequency of light so high that energy cannot be emitted. This is the explanation of the light wave surrounding a particle.

Max Planck had envisioned the surface of a perfectly radiating hot body (a blackbody that perfectly absorbed and perfectly radiated), that there existed tiny oscillators. He imagined tiny oscillators because he was trying to imagine some mechanism that could make an electromagnetic wave or radiation. If you oscillate a string, it makes a wave. At first, Max Planck thought the waves could be an infinite amount of energies of oscillation, but when this didn't work, then Planck imagined that the wave would not radiate or generate a wave with anything less than a full oscillation or quantum. He imagined that for each different frequency, the full quantum of energy for each was a constant. He then looked at the existing data of experiments of radiation at different temperatures and frequencies to mathematically calculate the amount of energy before the full oscillation of the tiny oscillator would make a complete revolution creating the wave. This begs the question, what is happening before the energy becomes h? Or before it becomes a full quantum? There must exist energy on the surface of the hot body that is not radiating because it has not made a full oscillation and is a partial quantum of energy that is non-radiating. Niels Bohr understood this concept and used it to come up with h-bar, the well-known quantum physics symbol representing ½ h radians. Why is this reduced constant used? As explained in the first chapter, Niels Bohr was trying to make his electrons not radiate so he used ½ h to describe ½ a frequency or ½ an electron orbit around the nucleus. He well understood that this could be a useful concept, but it was impractical in orbit mechanics of the electron, because the electron would have to assume to jump every ½ orbit at all times in order not to radiate. But Bohr understood what Planck was saying. There would exist energy that was a partial energy, which he described as ½ h, before it became a full quantum that could radiate. We will use this concept in describing the atom's nuclear wave because it is a postulate of Max Planck's theory of radiation. Bohr wasn't really describing a wave in the atom. We will be.

Again, the reason this explanation is necessary, and is given, is because the Davisson-Germer experiment showed that the atom itself was completely filled with something massless and undetectable, so that electrons at slow speeds would all be deflected from the atom, therefore, there was no void or empty space, in contrast to the Rutherford gold foil experiment which used high energy alpha particles. The only difference between the two experiments was the speed of the particles hurled at the atom. The kind of particle i.e. electron v. helium alpha particle should have made the reverse outcome. It should have been harder for the much larger alpha particle to pass through the atom than the much smaller electron. So there is only one conclusion, the speed of the particle projected at the atom when slow postulates a "something" in the void of the atom. "Something" that is massless. "Something" that is not detectable by its charge. "Something" that is permeable at high speeds, but also is able to deflect particles hurled at it at lower speeds. Therefore, there was not a void making up most of the atom, but this "something" in the permeable space was a massless "energy" that could repel electrons traveling at low velocity. What is known to repel electrons and is massless and without charge? Light energy in the photoelectric effect. Therefore, this model is necessary to explain the reason for this phenomenon. But light is equal to mass. $E=mc^2$. Therefore light radiating away from the nucleus would irradiate away the mass of the system, for example, how light radiating from stars makes the star eventually die. Therefore, we must postulate a non-radiating light. We have no other choice. All the physicists of the early twentieth century might have visualized this, but the experiments happened after the theory. If first, before the Bohr-Heisenberg-Schrödinger model of the atom was theorized, the Davisson-Germer experiment was done, this theory could have been visualized. However, the truth is that Rutherford and others pushed for a theory of the atom with scanty data saying that Planck's constant was of small enough size to fit in the atom and should therefore be used, so that, a too early model of the atom was hypothesized without first analyzing the Rutherford gold foil experiment against the Davisson-Germer experiment. Too long had the Rutherford assumption existed that empty space existed in the atom. This stifled and quashed the idea that anything else might be true. But as we have just reconciled the two experiments above, we can see there is no other conclusion than that energy exists in the atom that is able to deflect particles of sufficiently low speed. This gives causality to the distance maintained between the proton and the electron in the hydrogen atom, and by extension, to the nuclei and electrons of all other atoms.

So now it devolves upon us, to try to calculate how this system of light energy is mathematically described. Most of the experimental math has already been calculated related to the energies produced by the electron and its ionization and transition energies.

In order to explain what appears to be quantization states of the electron, we have to assume that the electron lies in the non-energy portions of light in the wave emanating from the nucleus. Where does a wave lose energy? At its nodes. Therefore, the electron particles have to be stuck in the nodes of the nuclear wave. So what energies exist between the nodes?

Niels Bohr did us the favor of explaining what light was emitted during electron transmission between shells using the Rydberg formula.

Bohr stated, where:
R is the Rydberg constant
n is the number of the shell with n_1 as the ground state shell
Transmission is given by:
$$\text{Energy} = R\left(\frac{1}{n_1^2} - \frac{1}{n_2^2}\right)$$

The fact that the electrons use just as much energy to transition away from the attractive Coulomb force of the nucleus as they take to transition toward the attractive force of the nucleus means that the Coulomb force is not dominating the atom. Rather the nuclear wave sets the energy for the transitions. The Coulomb force gives the direction for the tendency of the transitions, but does not describe the energy of the transition.

When we substitute energy for the wavelength of an antinode of a standing light wave, we get the light emitted or absorbed in the spectrum for hydrogen. Bohr was correct here. Schrödinger did not really address the electron transmissions in his equation. Rather Schrödinger was describing the changes in a wave over time when that wave was quantized or broken up into various energy levels by Planck's constant. Schrödinger's equation was describing a light wave that was quantized, that is, in picturing the electron as a light wave and picturing the quantum jumps as quantum movements of the light wave. This picture is too complicated when we see the electron as moving between the nodes of the nuclear wave that must fill the space in the atom, and when we see that an electron light wave is therefore not chopped up or quantized except that its energy level drops at a node and increases at an antinode which creates cycles which are described as frequency. We see that the electron is a matter-wave, but not in the sense of de

Broglie, but in the sense that it is a particle with a surrounding wave and that all particles have a surrounding wave. So Schrödinger's equation does not integrate Bohr's light emissions using the Rydberg constant. Although the Schrödinger equation was able to reproduce the Hydrogen emission lines which is why it works to a degree, but is not a true picture of the atom because Schrödinger was simply trying to picture the electron as smeared out to eliminate quantum jumps. This is a flaw, because the reason Bohr introduced quantization of shells was to show the light radiating from the spectrum which was the only reliable and thoroughly known experimental evidence of the structure of the atom in the early twentieth century. So a theory of the atom must incorporate the Rydberg constant and the light emitted and absorbed by the transition of electrons.

However, we come to the problem that the energy needed to transition gives light quanta much larger than the atom itself. In this new model of the atom, we must assume that the energy needed to transition is the energy necessary to overcome the antinode between shells. However, we have to account for the size of the atom and the difference between the wavelength of emitted light and the much smaller quantum size between each shell of this new model of the atom. We must assume that the atom has the experimental size that is known. We must assume that there is a wave in the atom. We must assume that the energies between the nodes of the wave are such that they can be overcome by the transition energies, therefore, we must account for all of this phenomena. So we have to consider what is known.

The size of the light emitted as radiant light from the transition in the formula above between the second shell and ground state is known to be 121.6 nm. However, the size of the entire hydrogen atom is about .1 nm. Therefore, in the hydrogen wave emanating from the proton, the energy of this light wave is contained in a much higher frequency (if we compare it to radiating light) or more accurately a smaller wavelength of unemitted light. So we have a space or distance between shell one and shell two that is a wavelength much smaller than would be expected considering the amount of energy is contains. We need to reconcile this idea.

Under Planck's theory, only a full quantum radiates, so necessarily, given the small wavelength space between shells and the small energy contained therein, it is described as a partial quantum and does not radiate.

The fact that it doesn't radiate freely through space does not mean it doesn't move. Under Special Relativity a massless particle must move at the speed of light. Under Planck's law, light can bleed from higher to lower frequencies before it radiates. See Isaac Asimov quote below.

Another reason for the warping or changing of the wavelength versus the energy contained in it may be caused by the wavelength constriction causing such a small wavelength with such a low energy content so that the light cannot radiate. The energy is measurable from the spectrum. The size of the hydrogen atom is known. Therefore, the quantum between shell 1 and shell 2 is a partial quantum and cannot radiate, so it is a non-radiating light wave around the particle. The only circumstance under which the nuclear wave can radiate is when there is excess energy to radiate. This is because the mass of the particle and its energy in the surrounding light wave of each particle is conserved.

(Where h is Planck's constant, f is frequency, c is the speed of light, and λ is the wavelength, the equation for wavelength is E=hf=hc/λ.). However, in the non-radiating light wave surrounding a proton in hydrogen, the equation E=hf=hc/λ is not true where the light is not radiating at speed c through free space. The c in the equation makes no sense in a partial radiating light wave emitting and returning to the particle. However, under Special Relativity, light being a massless particle does travel at c. However, due to the radiation being below the full quantum needed for radiation across space when the distance between each shell is viewed as one cycle of a wave, the wavelength versus energy cannot propagate as normal light. Instead, in the unemitted, or partially emitted, radiation of a particle, the wavelength has to be proportional to a partial h times c, so it would be other than the wavelength that would normally be seen in radiating light. I am going to examine the proposed theory of the wave emanating from the nucleus using the Bohr radius. It should be remembered that the Bohr radius is a "theoretical" radius and is dependent upon the Bohr Theory of the quantized angular momentum, so it is not the true experimental or observed radius of the distance between the proton and the electron in Hydrogen. In fact, the Weiss radius was calculated experimentally and was very different in size than the Bohr radius. However, as a starting place to try to derive a starting point for the energy in the wave, we will consider the Bohr radius.

Currently there are no wavelength laws for partial quantum energy. Therefore, we have to try to reconstruct what partial quanta unradiated wave energy might look like. How do non-radiating quanta behave under Planck's law?

In Max Planck's model of the Planck constant of radiation, where Einstein interpreted the quantum as a photon, the size of the quantum of energy was proportional to the size of the wavelength meaning that the larger the size of the quantum of energy, the smaller the size of the wavelength. According to Isaac Asimov, "Consider the consequences of this quantum theory. Violet light, with twice the frequency of red light, would have to radiate in quanta twice the size of those of red light. Nor could a quantum of violet light be radiated until enough energy had been accumulated to make up a full quantum, for less than a full quantum could not, by Planck's assumptions, be radiated. The probability, however, was that before the energy required to make up a full quantum of violet light was accumulated, some of it would have been bled off to form the half-sized quantum of red light. The higher the frequency of light, the less the probability that enough energy would accumulate to form a complete quantum without being bled off to form quanta of lesser energy content and lower frequency."[xxxii] Therefore, light at higher frequencies may be able to bleed out to lower frequencies. This shows the variability of the quantum of energy. Assuming the wavelength of the light between the proton and the electron in ground state is the Bohr radius of the hydrogen atom, the quantum must overcome the Coulomb force between the proton and electron. (In my later research of the spectra, I prove that the Bohr radius does not really exist as such, but that the energy of the wave forms an exponential curve toward the nucleus.)

The only empirical evidence is the energy given by the ionization energy in the atom and the transition energy in the atom when we view the atom as having partially radiating light that cannot propagate as normal light due to failing the Planck test for full quantum radiation.

The energy in the Coulomb force between a proton and an electron if contained in one quantum or photon of light would create light of a frequency and wavelength much smaller than the size of the Bohr radius in the atom. And at such a frequency, every electron would be ejected from the atom in radiating light and since many cycles would exist in the Bohr radius, we have to assume radiation, therefore we do not assume the quanta in the Bohr radius to be equal to the Coulomb force. And we assume the Bohr radius contains more than one partial quantum of light contained in each antinode increasing in energy and shortening in wavelength toward the nucleus. The geometry of the curved light wave of the proton particle creates a partial energy quanta series of antinodes between the proton and electron. The space between the current ground state of the electron in the atom to all existing outer shells is filled with 13.6 eV of short wavelength light which acts against the electrostatic force between the electron and proton at the first shell distance between the proton and the electron.

I originally searched for the energy of the antinodes between the first shell and the nucleus proton as below assigning the least amount of energy to overcome the Coulomb force between the electron and proton in Hydrogen, but I have since discovered through the examination of the spectra that much more energy exists in the space between the nucleus and first shell of each atom. I am leaving in this original work because it describes my thinking process as I worked to my final solution and was able to glean from the spectra all the data necessary to prove a model of the atom with conclusions I could not come up with entirely through logical thought alone. I needed the data from the spectra to tell me what the model looked like and to disregard much of what I had imagined and to accept what was irrefutable from the data itself and the equation of Balmer-Rydberg that had come before.

Originally I concluded, **where:**
h is Planck's constant
f is frequency
so that **hf** is the energy content of the quantum of light, or the photon, between the proton and the electron in a hydrogen atom, and
w is the energy required for the electron to break free from the atom where kinetic energy is zero;
and where:
k$_e$ is Coulomb's constant
F is the Coulomb force
q is the photon
q' is the electron
B is the distance between the proton and the electron in ground state, then:

$$F = k_e \frac{qq'}{B^2}$$

FB=hf Force times distance equals energy and the energy is described by hf which is the energy content of a quantum of light that would be needed to overcome the Coulomb force between the proton and electron.

However, this does not necessarily describe the partial quanta of energy inside the first shell. In fact, the nuclear wave emanating from the nucleus overcomes the Coulomb force by a completely different mechanism so that it is not the charge of the particles that rule the atom, but the energy in the wave so that the Coulomb force is a lesser effect.

Another way of looking at the content of the partial quanta energy between the nucleus and first shell is to look at it as equivalent to the ionization energy by examining it in relation to the photoelectric effect.

hf = w is the energy content of the quantum of light between the proton and electron;
λ = B

However, neither of these methods describe the resistance of electrons in the Davisson-Germer experiment by the resistance caused by the light wave in the vast area now considered empty space between the nucleus and the first electron shell. The above does not describe the deflection of electrons by the space between the proton and electron in the Bohr radius of the atom. Electrons can be deflected that are traveling at between 20eV and 200eV in this space under the observations of LEED electron diffraction in a crystal. Under Newton's law (F=ma) this gives the quantum in the radius between the nucleus and the electron ground state of between 20 eV and 200 eV where 200 eV would be equivalent to radiated light of a wavelength of 6.1992 nm in a crystal. Since hydrogen under current theory is a gas which contains air or space between the molecules, it is not possible to deflect electrons off hydrogen molecules. It isn't possible to get the energy in the Bohr radius of hydrogen by the method of deflection of an electron beam, because an experiment would have to be conducted on solid hydrogen at 14 K which can't be done.

To calculate the energy between the proton and electron in the space in the Bohr radius, it might be possible by calculating the energy necessary to deflect an LEED electron beam in an element with a low atomic number and mathematically correct for hydrogen. However, this hasn't been possible because of the spacing of hydrogen molecules. Hydrogen is a gas and therefore the molecules have space between them. However, a LEED electron beam may be able to hit some of the hydrogen atoms. The detector would have to be spherical as the beam would deflect in any direction. Hydrogen's wave is of a different strength because each element has a different strength permeability in the space between the proton and electron as each particle has its own wave and the waves superimpose as particles multiply in the nucleus. One can see how the nuclear wave gains energy when one considers that the Rydberg equation changes between hydrogen and helium by four times for helium with one electron. This means the energy in the nuclear wave is stronger in helium than in hydrogen. This effect is due to the four nucleons in helium each having their own nuclear wave and when these nucleons are in the nucleus of the atom, their waves become coherent and in phase. This will be discussed further in the chapter on The Formation of the Wave.

Under Newton's third law, the electron has to be held to the proton with an equal energy needed for ionization. In order for the electron to stay in ground state, the ionization energy has to be at the threshold energy (meaning kinetic energy of the electron is zero in the photoelectric effect equation). That means the force holding the electron in ground state is equal to the ionization energy. However, the energy contained in the quantum between the proton and the electron in ground state can be a higher energy than the ionization energy. The quantum of energy between the proton and the electron in the Bohr radius can be thought of as containing enough energy to form a barrier that cannot be crossed by the electron due to its high energy, but because it is not radiating, it does not eject the electron. The ionization energy of Hydrogen is -13.6 eV. This is the force needed to eject the electron through the outer shells of the atomic wave. This is not the force needed to get an electron into the nucleus.

Also, the quantum contained in the Bohr radius of the hydrogen atom must equal the binding energy upon the proton. As was discussed in the chapter on Binding Energy, the light wave surrounding the nucleus causes the nucleus to be bound together. Since binding energy is very powerful, we should also look at what energy a full quantum of light would hold if the wavelength was the Bohr radius. Current estimates of the Bohr radius are given as 0.0529177 nm while the size of the Hydrogen atom is estimated at 0.1 nm.

Therefore, **where:**
h is Planck's constant
f is frequency
E=hf=hc/λ

The energy content of a wavelength the size of the Bohr radius would be 23,430 eV. That is equivalent to 4.17678e-32 kg mass under e=mc2. However, this describes a full quantum and we began with the assumption that the quantum of light in the Bohr radius of hydrogen is a partial quantum. However, we can begin to see, even when looking at a fraction of this amount of energy, how much energy might be stored in a single atom in the Bohr radius so that this non-radiating quantum causes the pressure or barrier against the proton that makes the proton so stable and also makes fusion such a difficult process.

Next we have to guess how much of the energy in the Bohr radius makes a partial quantum if the Bohr radius were one frequency of an electromagnetic wave. Under Planck's law, any amount less than the total quantum would be not equal to a full quantum and therefore unable to radiate, but as we cannot have an arbitrary percentage of a full quantum and as Bohr already introduced ½ h in all the formulas and they appear to work pretty well so far in describing at least some atomic behavior, the best we can do for now is go with ½ h for the energy in the Bohr radius of hydrogen is 11,715 eV. If the Bohr radius was an observed length, we could assume this was the amount of energy in the first cycle of the non-radiating particle wave emanating from a proton. (It has to be non-radiating because if it radiated, it would have too much energy and eject electrons and it would lose energy by radiating.) But as the Bohr radius is a theoretical length dependent upon using the electron as orbiting around the nucleus of a planet with angular momentum described by ½ h, it is probably not reality.

Under current Quantum Mechanical theory, there is no reason why the ionization energy has a specified energy, because there is no force causing the electron to stay in its shell. In other words, the current Quantum Theory describes no force that holds the electron in a shell around the nucleus. Bohr just said it was there because it had to be there. Therefore, if there is no force holding the electron there, then how can there be an energy needed to ionize the electron? Neither Heisenberg's formula nor Shroedinger's formula explains why it takes a force to knock an electron out of an atom. Under this new theory, there is a light wave emanating from the proton (or nucleus) of every atom that exerts a force on the electron so that it takes another force to kick the electron off the atom. Einstein in 1905 mathematically explained the electromagnetic effect to show how a force exerted by a photon of light could knock an electron off a metal. Yet, in the atom itself no force was ever given under the current Quantum Mechanical model to explain why there exists an ionization energy necessary to ionize the electron out of the atom. Einstein showed that that a particular frequency of light could overcome the Coulomb force of the electron and proton so that the electron was emitted from the atom. Under this new theory, that same force exists in the atom keeping the proton and the electron separated. Therefore, the atom contains a force keeping the electrons in their shells so that it takes and equal and opposite force to emit an electron in the photoelectric effect. That is why this theory is more effective and useful in explaining the atom.

Unlike normal light that propagates at a single frequency, the light wave from a particle like the proton would propagate with decreasing energy before it returns to the particle. The original quantum would be "bleeding" out as Isaac Asimov described into the surrounding space creating smaller and smaller quanta as it bleeds from the original quanta. However, as we must assume that the proton never losses mass, the bleeding out of energy must always result in a partial quantum so that it does not radiate.

Hydrogen is the only element with a single light wave around the proton. All other nuclei are superimposed light waves of the nucleons. So in hydrogen the Rydberg formula is easily described and empirically seen. Each energy packet following the first cycle between the proton and ground state described above would be constrained by the Rydberg constant and show decreasing energy proportionate to the Hydrogen spectrum of light emissions seen in spectroscopy.

So, where all of the above symbols are true, and where:
d is the distance between the electron in ground state and the electron in the second shell of an anion of hydrogen where one electron is in each shell and,
R is the Rydberg constant
q'' is the electron in the second shell
hf_2 is the energy content of the frequency between ground state and the second shell
n is the number of the shell.
x is the distance between ground state and the second shell.

$$F = k_e \frac{qq''}{B^2 + d^2} - k_e \frac{q'q''}{d^2}$$

$$hf_2 = Fx$$

where the energy content of the quantum between hf_1, hf_2, hf_3 .... is proportional to **R**.

The light we see being emitted follows this pattern in hydrogen called the Lyman series because it shows the difference between shell 2 and shell 1:

$E_n = (13.6 \text{ eV})/n^2$

$E_2 = (13.6 \text{ eV})/4 = 3.4 \text{ eV}$

$E_1 = (13.6 \text{ eV})/1 = 13.6 \text{ eV}$

In dropping from the n = 2 state to the ground state the electron loses 10.2 eV worth of energy. This is the energy carried away by the photon.

Converting this to joules gives $E = 10.2 * 1.60 \times 10^{-19}$ J/eV $= 1.632 \times 10^{-18}$ J

So in this model, the energy in the wave between n=2 and n=1 is 1.63×10^{-18} J or 10.2 eV. Although that corresponds with the wavelength of 122 nm in radiated electromagnetic light, we are dealing with partial quanta of unradiated light in the atom. Therefore, the wavelength in the atom has to be shorter than the energy contained therein. This is the description of a partial quantum. Obviously, the small amount of energy contained in the wavelength described by the spacing of electron shells is a lot less than ½ h. However, that is only the energy that an electron uses to move from the node at n=2 to the node at n=1.

Note: In all of the above, I was simply straining my brain to come up with a way to calculate how the antinodes of a nuclear standing wave would look in the interior of the first shell. It wasn't until I wrote the next paragraph that I stopped and decided to try to find the riddle in the spectra itself. This proved the key to unlock the mystery. The data in the research article that was published in July 2019 speaks for itself, but I will continue to let the reader see how my logic progressed to get me to the answer.

Okay, we can empirically derive the energy between shells of the proton wave, because the spectrum indicates them. From ground state of the electron, the energy at the antinode between nodes is 10.2eV, 0.67V, 0.3eV, etc. So we can see that the energy levels drop per the Rydberg constant per each cycle of the nuclear wave. We have tried to describe the next transition to the nucleus as a single energy level because we do not see electrons transitioning in this region. However, the outer wave follows a well-defined curve as it decreases in energy outward so it must have a well-defined curve of energy increase inward. So if we continue to follow the pattern of the outer energy for each node inward, we get the proton nuclear wave. We will notice very high energies the closer we get to the nucleus. So the further from the proton or nucleus of Hydrogen, the energy is dropping off. We will have to address why the electron prefers to stay in ground state which is not the first antinode of the proton/nuclear wave. The inner energies of the wave are too strong, but not impossible to penetrate for the electron. However, in discussing the way the energy wave reacts to the energy of the electron below, we will see how the proton wave reacts to absorbing too much energy and energy at higher or lower than revealed by the spectrum and yet still results in the spectrum. The distance kept by the electron where ground state is actually an outer wave, is likely due to an equilibrium between the electron wave and the proton wave where the resistance of the two wave energies and the attraction of the Coulomb force cancels out. Ground state is therefore not defined as the inner most node of the proton wave in Hydrogen, but only the node that has the least resistance between the electron and proton wave for electrons carrying enough energy to pierce each antinode to reach into the atom as far as possible for the energy they are carrying.

Bohr's atom showed a single line or energy level as a single shell, but we know there are several natural Zeeman lines and Stark lines and natural splitting lines of hydrogen. So why are we describing only one series of the Hydrogen atom? Because unlike Bohr's hypothesis that all series of the Hydrogen atom appear in one atom, what we are seeing is basically different sized Hydrogen atoms. How is that? The temperature changes the size of the nuclear wave. The proportions of the wave would be the same, but the size of the Hydrogen atom would be different with temperature, therefore, it would change the transition energies of the electron. We don't see one size Hydrogen atom in the spectrum of light. We see billions of atoms in the spectrum, each atom is hotter or cooler than another. The spectrum is showing a system of atoms. We know that matter gains mass when the energy of the mass increases. The whole nuclear wave of the atom is storing energy when energy or temperature increases. So it comes as a natural conclusion when you put a wave in the atom and try to measure it that the series have to belong to different atoms of Hydrogen, or rather Hydrogen atoms in different energy states. When you choose causality in the atom, the pieces fall into place.

Okay, but there are not infinite series of electron transition series in Hydrogen. In fact, only certain series are allowed in Hydrogen. Therefore, the Hydrogen atom is only able to increase with energy or temperature to discrete sizes. I agree. We get another clue here. But one problem at a time. We'll have to come back to that. But the main problem with the current Quantum Mechanical theory is that both the Bohr atom and the Heisenberg/Schroedinger atom restricts the size of all Hydrogen atoms to one size. And they say the spectrum describes one Hydrogen atom with all the energy states described in one formula. Nature isn't like that. More energy causes more mass, so something about the size changes. Something about the energy level changes. There is a difference in mass between an energetic Hydrogen atom and a frozen Hydrogen atom. Therefore, maybe there is not a huge difference in the overall antinodes between shells of the Hydrogen atom, but there is a difference in the energy contained in the atom that is affecting its mass. This energy difference between a highly energetic atom and one of low energy is reflected in the series that the electrons take in transitioning through the atom. So a Humphrey's atom of Hydrogen is a low energy Hydrogen atom and a Lyman series atom of Hydrogen is a high energy Hydrogen atom. The ground states are different for each series, and the energy for making the very same transitions changes a little depending upon the energy of each atom. Again, everything is already proven by experiments of the past 100 years. It is just a matter of interpretation. Having the right model makes the experiments more logical, rational, empirical, and scientific rather than the mystical spin that has taken over physics since Quantum Mechanics was born.

The key points worth repeating are that the problem with the Max Born interpreted Heisenberg/ Schrödinger/Bohr model of the atom is that:
1. The spectrum proves that the electron takes the same amount of energy to transition away from the proton as towards the proton. What does this prove? This proves that the Coulomb force between the proton and the electron does not rule the atom. The Coulomb force is neglected between transitions as the energy to transition inward is the exact same energy as to transition outward.
2. We know that mass changes with energy. Therefore, there can be no one set size for transitions in the atom or one set size for the Hydrogen atom as currently described. In fact, there can be no single series that gives the spectral lines for Hydrogen. There

in fact have to be more series as the energy level of the Hydrogen atom changes with temperature or acceleration. The splitting of the Hydrogen lines shows that there are different sizes and energies of hydrogen atoms creating larger and smaller antinodes and therefore slightly different transition energies than strictly expected for an electron shell with a single energy. The Bohr atom and the Schrödinger wave equation and the Heisenberg matrix equation do not give splitting lines, but single energies for the Hydrogen transitions. (Current QM needs to make up more arbitrary rules to explain the splitting and keeps getting more complicated because they can't see that atomic shells are dynamic rather than absolutely one energy per shell.) And we see in the spectrum that there are several energies levels that the Hydrogen atom can take. There is no one set series of energy for electron transmissions except that they follow the Balmer-Rydberg formula closely. What does this prove? That there are different states of excitation for Hydrogen atoms with different energies in the spectrum. When the energy of the Hydrogen atom changes, the place where the Rydberg series appears changes, but the proportion of the transitions stays the same.
3. The high energy absorbed in the proton wave in a high energetic Hydrogen atom causes the electrons to transition using slightly more energy causing splitting. Therefore, there are no discrete shell energies that are there because they have to be there. There is a physical cause for the observed discrete energy lines that changes them dynamically according to energy.

Therefore, the Lyman, Balmer, Paschen, Brackett, Pfund, Humphries series are electron transitions in the state of different Hydrogen atoms that have different energies stored in their energy waves. Now the problem becomes why does the Hydrogen atom store only discrete amounts of energy? Why do the shells seem quantized or discreet at all? We need to look at the difference between the energy levels of each series to figure out why the Hydrogen atom only takes this much energy discretely. What this means of course is that the electron wave and proton wave each only absorb energy discretely. And seeing that will give us the bigger picture of a wave around entire atoms and molecules. We can immediately assume that the Hydrogen atom, like the electrons in its own proton wave is itself in the wave of other atoms. Actually, I've edited this book off and on, from time to time. As I write a little every few years, new patterns emerge. This reminds me of fractals. The pattern appears to be repeating. (In the end, I found these repeats in the spectra of elements themselves.) But let's finish describing the electron transitions in shells of a single proton.

Bohr calculated ionization energy for Hydrogen coming up with -13.6eV. But that is in the Lyman series state of Hydrogen as far as I can tell, because it appears to be calculated at n=1. The calculations have already been made by scientists using the Saha-Langmuir equation to compute ionization of hydrogen across temperature ranges.

Again, under current theory, there is no force holding the electrons in the atom away from the nucleus, but the ionization energies are measurable. Science has put the strong force in the nucleus to explain why the nucleus holds together, but there is no force to hold the electrons apart in shells under current Quantum Mechanics. The electrons are assumed in current science to be where they are, because they have to be where they are, as was explained in Chapter 1. Therefore, there should be no force necessary to remove the electron from the atom except to compensate for the Coulomb force. However, the ionization energy is not equal to the energy to overcome the Coulomb force. There is some other force holding the electron in the atom. There is no force under current theory holding the electron in its shell especially once the model of the Bohr atom with its angular momentum was replaced by non-orbiting electrons in the Heisenberg model that simply had a distribution probability pattern. The same is true of the Schrödinger electron that had a Max Born probability distribution pattern across a wave, nor even in the plain non-Born equation, there is no momentum or velocity or any other property exerting a force on the electron to stay in the atom in discrete shells. There are equations for angular momentum but they have no physical meaning in the interpretation of the Schrödinger model. This is the problem with the Heisenberg/Schrodinger theory both when taken as originally written and when taken under the Max Born rule of a square of the wavefunction.

Under the new theory hypothesized in this book and explained by the data in the research article at the end of the book, the size of the proton energy wave is in proportion to its mass. It follows that the magnitude and force of the energy of the wave of the proton is greater than that of the electron, because logically, it is the electron that is caught in the proton's energy wave. Also, this is logically true because it is easier to ionize an electron than to cause fission of the nucleus.

However, this whole supposition of circular orbits was based upon false assumptions. It was assumed by Rutherford (Bohr's mentor) that this would prevent the electrons from falling into the nucleus if they had planetary motion. In fact, the whole scientific community depicted the atom as a planetary system during the first twenty years of the twentieth century. Quantum mechanics actually formulates the atom without orbital motion but in an inconsistent manner, as it invokes it for magnetic moment when convenient. That orbits are in conflict with the observed evidence was made clear by modern quantum mechanics. Werner Heisenberg eliminated the electron orbit, but kept referring to it as if it existed when not seen. The inconsistency will be highlighted in another chapter. That orbits are unnecessary and contradictory to the evidence will be further clarified under this new theory.

Just as in the photoelectric effect, energy wave photons from electromagnetic waves can exercise force against electrons, so too the photons in the energy wave emanating from the nucleus exert a force upon electrons. The distance between the ground state n=1 electron shell and the next shell n=2 is defined by the shape of the energy wave and is the next node that the energy wave creates.

In the photoelectric effect, one finds experimentally that for a given light frequency f, the velocity of an emitted electron v is given by:
$$hf = W + \tfrac{1}{2} mv^2$$
where m is the mass of the electron, and the quantity W is called the work function.

In the atom, where the antinodes of the proton wave do not exceed the energy carried by the electron and no new light is absorbed by the electron causing it to eject the electron, the electron could approach the nucleus through electromagnetic attraction according to how much energy the electron holds, but would be denied approach to the higher frequency (antinode energy) waves near the nucleus due to a process similar to the photoelectric effect, i.e. the higher energy of the antinodes of the standing proton wave the closer the electron approaches the nucleus.

The true stroke of genius of Niels Bohr was to realize that emitted light from the atom in the spectrum could only be excess energy. Therefore he has electrons both absorbing and emitting photons to produce transition spectrum lines. As mass and energy are equivalent any emission of free energy results in an emission of mass, but the proton and electron have very long stable lifetimes. Therefore any atomic radiation emission is excess emission of light previously absorbed so that the energy wave holding the particle together does not disintegrate.

In this new model the electron doesn't have to "know" in advance what wavelength of light to emit before transitioning to a lower shell or node. Rather the light seen in the spectrum is from the nuclear wave. When the electron loses energy, the Coulomb force pulls it into an inner node if it retains enough energy to cross that antinode. It can lose any amount of energy up to its excess energy and that energy it loses is absorbed into the energy wave surrounding the nucleus and/or the particle. According to the energy loss, the electron will be pulled toward the nucleus into an inner node of the nuclear wave. The energy wave maintains a balance between the masses and energy waves of each particle under the law of nuclear equilibrium. Therefore to maintain the electron's mass to energy wave ratio, the energy wave will give back lost energy to the electron, if required, and thus emits the correct wavelength of light according to the lesser energy required for the inner shell. It must emit or absorb only the amount of energy to cross an antinode of specific energy. This creates the atomic emission spectra.

Where E_{en} is the energy of an electron in an orbit other than ground state, where E_{en1} is any inner orbit of E_{en}, where E is any energy and E_x is excessive energy loss than required for a lower orbital, then:

$$E_{en} - E = (E_{en} - E_{en1}) + E_x$$

The excessive energy loss is transferred back to the electron from the nuclear energy wave:

$$E_{en1} = E_e + E_x$$

and the nuclear energy wave emits:

$$\lambda = E_{en} - E_{en1}$$

Thus energy is released from an atom in only certain quanta due to a physical reason. Under the law of nuclear equilibrium a particle must retain its energy wave to mass ratio and cannot emit energy in excess of this ratio. Any emission of energy greater than the ratio set by the law of nuclear equilibrium is caught in the energy wave and given back to the particle or not allowed to leave the particle at all.

The outer shell n=2 and higher shells in neutral hydrogen are already in existence where there exists a proton, even shells still unknown exist whether we see them or not as the proton wave extends into space. The energy wave of the proton does not stop at ground state but continues outward from the nucleus weakening with distance. This means the size of the hydrogen atom in ground state is much larger than assumed by science. That is why there appears to be evidence for the Uncertainty Principle. But in fact, it is because the light energy wave of the particles extends beyond the particles themselves reaching further than one would expect. This is why science believes in Quantum Tunneling. However, the reason for tunneling is easy explained when we have the one assumption that a wave surrounds each particle.

Quantum mechanics also gives an unsatisfactory explanation of energy absorption by electrons. It says that energy must be absorbed in discrete amounts, also fabricated by Niels Bohr as there was no frequency in Bohr's atom to emit or absorb quanta besides an orbital frequency which did not match the transition energies (something Einstein didn't like at first and was surprised about). That energy must be absorbed in the exact frequency is also unnecessary under this new theory. Electrons may absorb any amount of energy as shown in Compton Scattering. They then pop outward to a higher n-shell most of the time unless the energy absorbed is so great that they ionize absorbing only part of the energy as shown again in Compton Scattering. In other words, the experiments to prove this new theory have already been done. The current QM does not give an adequate explanation, but this new theory with one single assumption makes everything clear.

The absorption spectrum is viewed across a continuous spectrum. Where E_{en} is the energy of an electron in any shell, where E_{en1} is any outer shell of E_{en}, where E is any energy and E_x is excessive energy gain than required for a higher orbital, then:

$$E_{en} + E = \lambda(E_{en1} - E_{en}) + \lambda(E_x)$$

Under current Quantum Mechanical theory: "When a photon with exactly the right wavelength encounters an atom of the cool gas, it is absorbed and its energy used to kick an electron into a higher orbit; if enough atoms of gas are present, all the photons of that wavelengths are absorbed, while photons with other wavelengths get through. The atmospheres of stars produce absorption spectra."[xxxiii] However, with a wave emanating from the nucleus of the atom and a wave around the electron particle, there is a place for storage of energy. Therefore, the excessive energy gain from an absorption of light is in fact re-emitted from the electron in its new outer orbital according to the excess frequency absorbed across a continuous spectrum:

$$E_e = E_{en1} - \lambda(E_x)$$

and the nuclear energy wave appears to only have absorbed:

$$\lambda = E_{en1} - E_{en}$$

This new model above predicts that the wavelength of excess energy will be random wavelengths. Millions of random wavelengths of excess energy in an element create a continuous spectrum. Only the specific wavelength of energy absorbed in the spectrum becomes conspicuous.

Therefore, the absorption spectrum appears to absorb discrete wavelengths, but in actuality the exact frequency of excessive absorption is re-emitted into the continuous spectrum background. Again, this is due to the energy wave maintaining a balance between the masses and energy waves of each particle. Therefore to maintain the electron's mass to energy wave ratio, the energy wave will release energy absorbed, if required, and thus appears to absorb the correct wavelength of light according to the energy required for the more excited state of the electron.

The explanation for the absorption spectrum appearing to absorb discrete wavelengths is that in actuality the exact frequency of excessive absorption is re-emitted into the continuous spectrum background at varying frequencies so is indistinguishable from the continuous background spectrum. Therefore, the absorption spectrum does not show indiscrete missing light because each atom is in its turn absorbing and replacing the excess light absorbed back to its origin. The excess radiation of energy is not always the same amount of energy that is excess energy, so that the excess energy of varying wavelengths does not disturb the continuous spectrum of light. Because atoms do not absorb light simultaneously, but each according to the distance from source, the return of energy by radiating the excess does not interrupt the continuous spectrum background.

Okay, there are going to be some Schrödinger quantum mechanical purists who say that under quantum theory, it is not that the electrons are making invisible jumps between shells as in Heisenberg's model, but that the electrons are vibrating at a certain frequency in their shells and that light of a certain special frequency in resonance with the energy needed to transition to another electron shell resonates at the exactly right frequency and causes the transition. This is the same theory whether under Heisenberg's model or Schrödinger's model. The exact right vibration or the exact right frequency of light has to be absorbed and emitted under the current model of the atom. This is not in accord with the facts, however, as can be seen from Compton Scattering.

The current quantum mechanical model of the atom is wrong and has been proven wrong over and over by Compton Scattering. As stated above, the current model of the atom states that "when a photon with exactly the right wavelength encounters an atom of the cool gas, it is absorbed." We know this isn't true. In Compton Scattering the wavelength directed at the electron is in the x-ray spectrum of light. It is light of a too energetic nature to be absorbed by the electron, yet the electron does absorb the light and is recoiled. But what is interesting is that the electron is only able to absorb a portion of the x-ray. This is in exact contradiction of having to absorb the exactly right wavelength model under current Quantum Mechanics. The x-ray after encountering the electron is deflected with slightly less frequency because a partial frequency has been absorbed by the electron. So it wasn't "exactly the right wavelength." And yet the electron was able to absorb some of the energy and transfer a partial amount of the x-ray into kinetic energy. Compton Scattering disproves the current quantum mechanical model.

In this new model of the atom, the energy wave surrounding the nucleus and the energy wave surrounding the particle can absorb any wavelength of light and compensate for its emission and absorption by renormalizing or conserving the energy of the particle and energy wave system as described above. However, the energy wave of the electron can only absorb a finite amount of energy, because the wave is finite, so the electron wave cannot absorb the entire x-ray beam in Compton Scattering. However, Compton Scattering shows the limit of the amount of energy that an electron can absorb. The difference in the frequency of the Compton beam before encountering the electron and after recoiling the electron is the amount of energy that the electron absorbed. This, however, upset the natural creation balance of the mass to energy wave surrounding the particle. Therefore, the excess energy as described above in the math is turned into kinetic energy. This new model explains why Compton Scattering is possible.

In this new model of the atom, the electron does not radiate as it moves whether in classic electrodynamics in the macro world or in the atom. Rather movement or acceleration occurs only when the electron is holding excess energy in its energy wave. This applies to electrons in a cathode ray tube, an accelerator, a cloud chamber, or the atom. In the atom transitions themselves, the radiation seen in the spectrum is discrete because it occurs as a balancing of the energy wave particle system in which the energy is re-distributed according to the energy level of the shell and the ratio of mass to energy wave of the particle. Therefore, there is no radiation during the transition and the electron moves without radiating during the movement. Radiation occurs during acceleration and transition but only as particles are trying to maintain nuclear equilibrium ratios and trying to expel excess light energy. Completely free subatomic particles move by acceleration due to holding excess energy. They radiate to release that energy and maintain their particle ratios and therefore slow down unless continually bombarded with new energy creating new excess energy which in turn appears to create constant radiation across a continuous spectrum. (The particles may also be forced to absorb energy by being pulled by magnetism which causes them to absorb energy while trying to radiate it.) Particles in nuclear waves in atoms move by transition in the energy wave of the nucleus, again, trying to maintain their proper mass to light wave ratio but because they stabilize in nodes of nuclear energy waves, they appear only to emit and absorb discrete wavelengths in the spectrum because they are not free to maintain any arbitrary distance from the nucleus' wave they are entrapped in. Rather they must remain in the other particles nodes and cannot sit in an antinode, therefore, the spectrum of emission and absorption is discrete.

In a manmade vacuum, the energy waves of particles that comprise the vacuum container extend beyond the mass of the atoms into the vacuum which explains zero point fluctuations in manmade vacuums. The extension of the particle energy wave into a vacuum means that the wave may release excess energy into the vacuum. Whether zero point fluctuations occur in the vacuum of space cannot be addressed without the presence of energy waves surrounding particles in space. But as explained, the current QM theory gets this completely wrong and creates a Vacuum Energy Catastrophe as explained in chapter 2 of this book.

In the current quantum atomic theory, a paradox exists in that the electron can only absorb certain wavelengths of energy below its ionization energy, however, at its ionization energy and above, it can absorb any amount of energy. In this new model of the atom, it can be seen that any energy is always allowed to be absorbed up to a maximum for electrons before it is radiated, but that is why particle accelerators can cause particles to absorb massive amounts of energy and not discrete amounts. Are we to believe that the particles change outside the atom? Under this new theory, all is explained. The emission of atomic energy is restricted by the law of nuclear equilibrium in downward shell transitions by the energy wave of the particle and is unrestricted in upward shell transitions as long as the energy input exceeds the radiation output.

In spectroscopy, under the new model of the atom, the reason for the proportionality of the spectral emissions become obvious. Viewing a spectral analysis, one can clearly see the change in energies have a pattern across the spectrum for a particular element as in the hydrogen spectrum. Clearly this relates to the change in energy of the light wave emanating from the proton as it propagates from the proton from higher energy to lower energy antinodes. Therefore, to calculate the energy of the light wave between each shell in hydrogen, the Rydberg formula should hold. (After writing this in my first edition, I did indeed find the Rydberg formula in the first twenty ions and neutral elements up to Carbon.) Bohr proposed this but could not find the equation in other than one electron atoms. In fact, all Quantum Mechanical equations cannot describe any atom or ion with more than one electron, but I found the Rydberg formula because I wasn't looking at the spectral lines emitted, but the Rydberg energy differences between lines because those are the antinode energies of the nuclear wave. If you don't have the right model, you don't know what to look for.

This is because for almost two centuries, the most significant feature of the atom that is easily detectable is the spectral light of atomic emission and absorption lines for each element that are distinctive and proportionately spaced by the Rydberg constant. Therefore, the nuclear light wave creates nodes and antinodes exactly with the same energy for each antinode of the light wave varying slightly with changes in the size of the nucleus. The spectral emission lines can then be explained by the introduction of this new model. Spectral lines are the most used and empirical evidence of the distance between shells in atoms, and is the light emitted in spectroscopy when electrons make transitions. Therefore to calculate the distance between the first shell of the atom and all the other shells the Rydberg formula must correlate to the forces and energies between each outer shell of the nuclear light wave.

A relatively recent 2013 experiment was said to photograph or reveal the s-orbital of the atom. What is interesting is that the s-orbital of the atom should be spherical. Instead, what this experiment reveals if re-interpreted by this new model of the atom is that the electron is surrounded by a light wave emanating from it. A picture of what is supposed to be a representation of the spherical s-orbital is shown in Physics World of May 23, 2013. It is easy to see that yet again another experiment is trying to conform to current Quantum Mechanical ideas while the results are clearly not a spherical s-orbital but a wave pattern surrounding the electron.[xxxiv] It is clear from the photoionization microscopy that there exists a wave emanating from the electron particle itself. The wave shows a distinct high energy crest close to the particle, a node, then a weaker antinode, a node then an even weaker antinode. This is the picture of what every particle looks like under this new model of the atom and what was first described in the 2006 edition of this book. The first antinode emanating from the proton, electron, or other particle carries the most energy and keeps other particles of the opposite charge at a distance from the particle. The strength of the wave drops off with distance. Clear nodes are seen in which other particles may be trapped and interlocked. This picture is said to interpret the standing wave of the first shell at ground state. However, the picture does not represent a standing wave in a sphere around a nucleus. The picture is clearly a standing wave emanating from the particle. The explanation given to the experiment states that this is an observation of Schrödinger's wave function. However, expected representations of a standing wave around the atom do not center on a dense particle and then spread out from the particle as is shown in this picture. In the Schrödinger wave function, the particle or density shown in the picture should not be seen. The density in the picture is not the proton. Only electrons were imaged as they were ionized. The scientists who did the experiment are trying to explain it in terms of the current Quantum Mechanical model, but the current model does not allow for the particle and the wave to show at the same time in the same experiment. There should not be the central density seen in the microscopy. Rather, this is a microscopy of an electron and the light wave that emanates from the electron itself. The experiments have been done. It is a matter of correctly interpreting them. Scientists try again and again to twist their experimental interpretation into the current Quantum Mechanics and the solutions contradict the current Quantum Mechanics, but no one says anything.

Under this new model, because the waves in the nuclear wave of Hydrogen continue on a curve proportion to the known Hydrogen transition series, I predict that there exists another series in the ultraviolet region of the spectrum above the Lyman series that is currently undetectable because our instruments are not sensitive enough to detect the difference in the high frequencies being emitted. Therefore, the current ground state is not the ground state. In fact, as mentioned earlier, the n=1 in the Lyman series may not be ground state in this new model of the atom even if no other series is found. The electron will find equilibrium in the state that is the lowest energy state in which the Coulomb force between the electron and proton is least, while the repulsive force of the electron wave is equalized against the repulsive force of the proton-nuclear wave. In fact, many shells in Hydrogen may be ground state if it is the energy level of least resistance.

This new model opens up understanding of the atom and many phenomena in nature as I learned while writing this book. Einstein was right. You start with causality.

12 Radioactive Radiation

Quantum mechanics account of radiation: "Let me illustrate these general features of quantum mechanics by means of a simple example: we shall consider a mass point kept inside a restricted region G by forces of finite strength. If the kinetic energy of the mass point is below a certain limit, then the mass point, according to classical mechanics, can never leave the region G. But according to quantum mechanics, the mass point, after a period not immediately predictable, is able to leave the region G, in an unpredictable direction, and escape into surrounding space. This case, according to Gamow, is a simplified model of radioactive disintegration."[62,p.333]

According to the above explanation by Einstein, under quantum mechanics, radioactive decay is unpredictable and, in reality, not possible under classical mechanics.

In this new theory, in atoms of some heavier elements, the external pressure becomes less spherical according to the number of nucleons. This can create a bulge in the great sphere of the nucleus which can cause helium-4 to be expelled due to uneven external pressure. Just as in the photoelectric effect, the nuclear waves of high frequency can eject particles. The energy from the nuclear energy wave causes a photoelectric effect pressure of high energy antinodes that creates a great sphere around the nucleus in what appeared to Bohr as a water drop model effect of the nucleus. (Bohr later dropped this hypothesis as there was no cause in Quantum Mechanics for a skin around the nucleus.) When there is an uneven bulge in the great sphere, the high energy of the standing waves encircling the nucleus can loosen nuclear particles. Also the higher frequency (antinode energy) of the waves inside the nucleus that are coherent outside of the great sphere expel the helium-4 particle when it transitions into a particular node in the great sphere of the nucleus. This is the cause of alpha decay in the known process of radioactivity. As particles in the nucleus shift in the nodes of each other's particle waves, this can expel a particle as the antinodes contain a great deal of energy near the particle itself as is shown in the research paper at the end of this book.

When neutrons disturb the spherical pressure shape that the energy wave tends to try to balance under the law of nuclear equilibrium, the neutron can create an uneven external pressure. The external pressure of the energy wave causes the nucleus to tend to retain symmetry so that the uneven mass of the neutron causes a part of the neutron that is bulging from the nuclear particle sphere to be expelled from the nucleus in the known beta decay process. Where there are two few neutrons created in nuclides, the protons exert too much electrostatic repulsion against the external pressure caused by the energy wave. This creates an asymmetrical warping of the nuclear energy wave by the protons which causes inverse beta decay to create uniform nuclear pressure. The proton captures a neutrino and then the external pressure releases a positron from the proton creating a neutron. The neutron would naturally have a greater mass deficit than normal at this point. The Standard Model does not correct for this imbalance, but instead uses complicated crossing symmetry to justify it. However, a balancing between mass and energy is necessary because the mass of the neutron and its energy wave must be constant if the newly formed neutron is separated from the nucleus. Stable atomic particles making up the known elements must be identical. An especially small free neutron does not exist. Neutrons do not come in various sizes as particles are identical. Although a proton would have a less great energy wave than a neutron, this predicts that if necessary the energy wave could absorb gamma rays or other radiation when necessary to compensate for lost energy or lost mass. Or if in the case that the energy wave of the particle is greater than necessary under the law of nuclear equilibrium, then the energy would be expelled. The types of observed decay can be explained from the tendency to maintain uniform pressure of the nuclear energy wave, depending upon the external pressure of the wave upon the nucleus and upon the configuration of nucleons.

These decay events appear to happen randomly, but produce for each element that decays a specific half-life. The half-life of radioactive decay under this new theory appears to show that, for certain elements that decay, the law of nuclear equilibrium is maintaining a uniform pressure upon the nucleus at an average fixed rate. Therefore, in the nucleus of atoms of the same element, there is a variability in positioning combinations that changes the shape and placement of nucleons. Therefore, once the nuclear energy wave is better understood, this process should be able to describe which atom will decay depending upon its state.

This is contrary to quantum mechanics which states: "This impossibility of predicting when the decay of a given nucleus will occur is an essential characteristic of the quantum world, and furthermore, in the words of the German physicist Werner Heisenberg (1901-1976) the 'cause for the emission at a given time cannot be found.'"[73p. 124]

Even with recent research on the shells present inside the nucleus itself for protons and neutrons, the foundation still rests on this principle of unpredictable decay in quantum mechanics.

But the fact that certain elements have a statistically reliable half-life implies that each element has a nucleus that is cycling in a particular manner. This implies that there is unique structure to each nucleus of each atom and each isotope. It also implies that there are nuclei within the nucleus. For example, since alpha particles are helium nuclei, then some heavier elements have bound helium nuclei in the great sphere of their nucleus. The nucleus of the atom then becomes as complex as the groupings of the electrons. The absorption of energy and transition of nucleons between energy waves of the nucleons according to their interaction with the great sphere of the nucleus is rhythmical. This is probably accounted for by the property of the particle wave of each nucleon being interconnected with the waves of the other nucleons in each element. A movement of a proton in the nucleus of one uranium atom probably affects the movement of another proton or neutron in the nucleus of its neighboring atom. In this way, the entire uranium ore describes a half-life because of the positioning in shells being interconnected between atoms in the ore producing configurations that eventually lead to the simultaneous ousting of a decay particle at pre-determined parts of the cycle. Each emission of decay changes the system and resets the cycle.

One can imagine the high probability that in heavy radioactive solids, the nuclei of neighboring atoms are linked in each other's nuclear energy waves so that there is a system of nuclear chemical bonds. This cannot mean that the nucleons of one atom are inextricably locked to the nucleons of another atom or else one decay emission would chain-react in a way as to emit all decay of the element at once with catastrophic results. Rather it is more likely that a nuclear decay reaction in one atom of an element reorganizes the nuclear energy wave of the particular atom changing energy levels which shifts the nucleons in nearby atoms thus creating a domino effect eventually causing another atom to decay depending upon its nuclear configuration. The decay products react in the same way but are affected according to their nucleon arrangement and binding energies until the decay effect reduces to an element that has a high enough binding energy i.e. the amount of energy created by the nucleon waves is more stable encircling the nucleus, so that it is sufficiently stable that when it is affected by reorganization of neighboring atoms, the nucleons merely change nodes in the nucleus without decay. Of course, the calculation of all the variables of the multi-body system would be extremely complex.

Perhaps it may be just as simple as measuring the equal and opposite reaction of the decay upon the atom and realizing that that energy is transferred from that atom to a neighboring atom which absorbs the energy and depending upon the nuclear configuration, the nucleons change configuration in their nodes with the emission and absorption and the energy is passed again to another neighboring atom until it hits an atom that has its nuclear configuration ready to decay. The nucleus of all atoms is under pressure as the binding energy of the nucleon waves does not escape but holds the nucleus together. The mass ejection involved in each decay causes a series of transitions in a chain of atoms depending upon the angle of ejection of the decay in the atom opposite to the decay emission. As the energy is passed through atoms by absorption and emission, the atoms through which the energy passes appear to utilize some of the energy. When the energy passes through fewer atoms before causing a decay, then the energy of the decay emission is greater and the half-life is shorter. If the energy passes through many more atoms before causing a decay in an atom, then the energy of emission of the decay is less and the half-life is longer.

There appears to be a cycle to radioactive decay as if a radioactive element is a clock with multiple gears or some other illustrative instrument that after so many changes in the gears causes a click or emission of a particle. This is the basis of atomic clocks. The crudeness of a clock as an illustration is only justified in the sense that it describes a pattern of energy exchange and movement that eventually leads to an effect that occurs precisely at a prescribed time i.e. a half-life. Obviously, the atoms are not so mechanical as the gears in a machine, but the idea can be visualized in this sense. Science only needs to fit together the puzzle of atoms in an element and then it can predict which atom will decay and in what sequence. The deduction of predictable emission is quite simple, that is, radioactive half-life is a very accurate and precise method of dating. Therefore, it is a reliable clock. A precision clock cannot occur from chaotic, random atomic decay with no definite pattern and process. Precision does not happen by chance. Consistent accuracy does not occur accidentally or haphazardly. Therefore, radioactive half-life is an exact process governed by methodical rules.

The hypothesis of an energy wave around each particle creating high energy transition nodes in the nucleus of every atom easily and simply explains radiation of radioactive elements.

13 Molecular Bonding

How does this pressured barrier view of the atom affect molecular bonds? Let's take a simple example of the bonding orbital. Sigma bonds are the strongest type of covalent bonds. The simplest sigma bond is that between the two "s" orbitals in molecular hydrogen H_2. Molecular hydrogen is composed of two hydrogen atoms sharing a molecular bonding shell. Each atom has one electron. When the two hydrogen atoms create molecular hydrogen, their energy wave antinodes overlap each other in such a way as to cause their electrons to share the same node. The electrons of each atom under valence bond theory share the same shell. In the case of each atom, each has an energy wave barrier with nodes that form the electron shells. So the node of the first atom's energy wave barrier would superimpose itself on the node of the second atom's energy wave barrier and the two electrons would share the same orbital node. One electron would be in the intersection of the nodes. Due to the model presented here, this would create greater antinodes where the combined orbitals intersect around this shared node due to the overlapping of the energy wave so that the molecular orbital of this type of valence bond would be very stable.

The likely explanation for why the two electrons in ground state of the helium atom have different ionization energies, is that the electrons each having their own energy wave become trapped or paired in a shell and more tightly bound together. After the higher energy required for the first electron to be removed from the bond, it only requires the same amount of energy to remove the second electron as it takes to remove an electron from a hydrogen atom. Therefore, in molecular hydrogen both electrons from both atoms are bound at the intersection in each other's energy wave. Again, theorizing that an energy wave of descending energy antinodes extends from every proton makes it clear that chemistry actually makes physical sense.

14 The Franck-Hertz experiment

In 1914, James Franck and Gustav Hertz performed an experiment which demonstrated the existence of excited states in mercury atoms. Electrons were accelerated by a voltage and the values of accelerating voltage where the current dropped gave a measure of the energy necessary to force an electron to an excited state.[2] This was interpreted as a proof of "quantization" i.e. multiplication by Planck's constant, however, under this theory it is explained by the separation of electron shells by the nuclear energy wave. In the theory here presented electron shells are discrete, but not "quantized". Or rather not quantized in the current sense of the word, quantization can only take place where there is a wave that creates it, as we know from Planck coming up with the concept when examining waves. Quantization according to Planck and Einstein was made up of energy packets. Quantization according to Bohr is made up of empty space between transitions. This is completely contradictory.

Much later, this experiment was interpreted as confirming the Bohr model of the atom. However, this experiment only showed that in mercury, energy was absorbed at a discrete amount, namely, at intervals of 4.9 volts. Of course, the emission of light would follow the discrete light shown in the spectrum using the formula for absorption and emission shown in Chapter 10. Radiated light is emitted and absorbed discretely, but the energy can either be absorbed or lost at any level so that after the electron finds equilibrium in the node, but the atom absorbs or emits only discrete wavelengths of light.

According to Wikipedia, "An additional advantage of neon for instructional laboratories is that the tube can be used at room temperature. However, the wavelength of the visible emission is much longer than predicted by the Bohr relation and the 18.7 V interval." It's amazing how an experiment that doesn't agree with the Bohr equation is said to prove it.

15 Suborbitals

It is assumed in Schrödinger's wave equation (and Dirac's equation) that the angular momentum quantum number l (lower case L) has various shapes for the electron suborbitals. The azimuthal quantum number shapes for electron orbitals are merely a mathematical construct from the Schrödinger wave function equation to differentiate different quantum states. The requirement that we be able to "distinguish" two electrons, and their wave-functions, from each other, thus requiring them to be labeled by distinct series of quantum numbers, was made mandatory by the Pauli Exclusion Principle. The quantum numbers were introduced through the study of spectral lines. Chemists identified the lines by their quality: sharp, principal, diffuse and fundamental. In 1904, Richard Abegg was one of the first to describe the model of eight electrons in valences of an element. This was later "grandfathered in" to designate the shape of the orbital. It was the chemists who recognized the properties of the atom and the electrons before the Quantum Theory was put forward.

The problem with the current Schrödinger wave mechanics of the atomic suborbitals is that the proposition that the electron is a standing wave advanced by de Broglie meant that the standing wave must follow a circular path with an integral number of nodes.[12] If the shape of the standing wave is not circular such as in the "p", "d" and "f" suborbitals, a standing wave does not form for any trajectory in these orbitals so that electrons in these orbitals cannot form standing waves. However, using these distinctions means the electrons do not form standing waves. Also, these suborbitals cross the "no-man's land" between n-spheres where there is zero probability of an electron being found. Therefore, the question of their validity is raised in the quantum model. This new model of the atom used in this book predicts that if it does not have a physical cause, it is not real. This new model is based upon the precepts of the original scientists whose view was that the world was first and foremost rational.

One must examine the introduction of suborbitals which seems to be lost in the historical development of quantum mechanics. Originally, the Bohr atom of 1913 showed one electron in each orbit on a plane. Physicists, Max Born and Alfred Landé working together during WWI (1914-1918) were studying crystals and assumed cubic and tetrahedral trajectories for electron orbits according to the planes of the crystal. This work interested Arnold Sommerfeld, later to be Werner Heisenberg's professor, and his associate Peter Debye, because they were working on a Keplerian model of the atom with elliptical electron orbits.[40] Sommerfeld later joined Bohr to create the Sommerfeld-Bohr model with elliptical electron orbits.

It was the chemist Irving Langmuir who in 1919 first suggested that several electrons be connected or clustered in the atom in order to explain the periodic table grouping of elements and chemical bonding.[41] Langmuir divided an electron shell into cells and arranged the electrons in these cells so that they were filled in ascending order. Langmuir says, "The first shell thus contains 2 cells, the second 8, the third 18, and the fourth 32…. Each of the cells in the first shell can contain only one electron, but each other cell can contain either one or two." Langmuir was not introducing shells but limiting the number of electrons per shell. He calls the shells "(nearly) spherical". His shells are three-dimensional and in a cubical atom based on Lewis' work. The Langmuir paper begins:

"The problem of the structure of atoms has been attacked mainly by physicists who have given little consideration to the chemical properties which must ultimately be explained by a theory of atomic structure. The vast story of knowledge of chemical properties and relationships, such as is summarized by the Periodic Table, should serve as a better foundation for a theory of atomic structure than the relatively meager experimental data along purely physical lines...."[41]

One must realize that the shape of the electron shells is not known even to this day much less the shape for suborbitals. It is true that in 2013, the first image of an atom's shell was said to be produced, but it is an image of electrons shot off of a hydrogen atom.[xxxv] It more probably represents the electron with its electron wave rather than the atomic shell. However, in either case, both the electron wave and the atomic shells are only seen to be spherical. There are no odd-shaped suborbitals. These weirdly shaped suborbitals are rather deduced theoretically. The atomic spectrum does not answer. Therefore, many shapes for electron trajectories were being suggested during the 1910s and 20s. Bohr's 1913 theory had introduced circular orbits.

Arnold Sommerfeld, working on elliptical orbits modeled after the solar system, showed mathematically that electrons moving as particles in orbits describing conic sections could only take on specific orbits. One must remember that all this work was done in developing the atom before 1923 when de Broglie introduced wave-particle duality and yet the atom kept all the data deduced from non-wavelike particles in classical planetary orbits in the development of quantum mechanics which still causes much confusion today.

An article describing Sommerfeld's contributions to the Bohr model says:
"The electrons can move only on some, allowed ellipses. He coined a second l [lower case L] number which was called the secondary quantum number or the azimuthal quantum number. The number defined the shape, the oblateness of an orbit. For n=1 the orbit can be only spherical (l=0), for n=2 there are two orbits of different shapes (l=0 - the elliptic one, l=1 - the spherical one). For any n there are n kinds of shapes of the orbits."[45] (Information in brackets added for clarity.)

The suborbitals finally adopted by quantum mechanics were rather given shapes passing through the nucleus, however, quantum mechanics was later to borrow and retain the limitation of elliptical planetary orbits [i.e. in current theory for suborbitals each of the different angular momentum states can take $2(2l+1)$ electrons.] and arbitrarily apply it to suborbitals that were non-elliptical.

Bohr took Langmuir's chemical interaction picture of electron grouping, and Sommerfeld's limited possible classical elliptical orbits for particle electrons orbiting the nucleus, and developed what he called "subgroups" that were elliptical within each shell. Bohr says:

"Thus for each group the electrons within subgroups will penetrate during their revolution into regions which are closer to the nucleus that [sic] the mean distances of the electrons belonging to groups of fewer quanta orbits."[43]

Here in 1921, Bohr did not introduce orbits crossing the nucleus that related to spherical harmonics, but rather Bohr was introducing Sommerfeld's conic section orbits. In fact, Bohr did not introduce a ground state orbit in his 1913 theory in which the orbit passed through the nucleus, but all angular momenta were multiplied by h-bar. The idea for a ground state orbit with zero angular momentum crossing the nucleus appears first to be introduced by R.B. Lindsay in 1927.[67]

Charles Bury, a chemist, showed how Langmuir's grouping of the electrons could be brought into closer accord to the chemical properties of the periodic table in 1921.[42] Keep in mind that this was before the Schrödinger Equation in 1925. Einstein seems to have been completely kept out of the loop on this and seems also not to have read the actual papers on the development of the grouping of electrons. He attributes the grouping exclusively to Bohr who was not a chemist but a physicist. It took Bohr two years to examine Langmuir's suggested electron grouping in 1919 and then to propose it be adopted into the atom, only to have it revamped by the chemist Stoner in 1924 (still before the Heisenberg and Schrödinger equations).[44] In reality the basis for Langmuir's octet theory of electron arrangement was actually proposed first in 1916 by Gilbert N. Lewis, but it was Langmuir who accused the physicists of taking no notice of the chemistry of the atom.[88] Langmuir had introduced an octet theory based on electron arrangement of 2, 8, 18, 32 in 1919. Once Bohr was able to see how the chemists had grouped the electrons to match the periodic table, he was able to predict in 1922 the chemical properties of an unknown element according to the grouping suggested to him by the chemists. This amazed Einstein who thought Bohr had come up with that prediction purely through quantum theory.[82,p.115] Einstein had said it was "a miracle" that Bohr had theorized an atom that explained chemistry when the truth was Bohr had used the work of chemists to explain the atom and the periodic table had been around for over 50 years. But more importantly, Bohr later introduced suborbitals from the Schrödinger equation that did not match the energy of the shells. Bohr had violated physics once again, but this time in order to introduce chemistry. Bohr did this against his conscience as we shall see.

That de Broglie is said to have saved the Bohr model of the atom by introducing standing wave orbits becomes invalid in a true physical sense when applied to subgroups or suborbitals or two electrons in a shell whose paths cross. Although electron grouping may be necessary to explain chemical properties of the elements, Bohr's introduction of variously shaped suborbitals meant they would necessarily be of differing energies from the energy of their shell. Differently shaped orbits require different energy even if their distance is the same from center—a fact that disturbed Niels Bohr tremendously though he introduced it. This introduction of orbital shape changes is unnecessary and defies the principle of the de Broglie standing wave.

The distribution of electrons in suborbitals was later refined by Edmund C. Stoner to more closely approximate the periodic table.[44] This is the arrangement of suborbitals as they appear in the periodic table today. The Sommerfeld planetary suborbitals, the chemist's arrangement of electrons not in accordance with the spectrum, and the calculation of quantized electron angular momentum by magnetic number according to the Zeeman effect were combined into modern quantum mechanics.

There is no doubt that groupings of electrons account for chemical properties of elements since it was the chemical properties of elements that was known first. The grouping of chemical properties was long known by the chemists since it was first devised in 1869 by the Russian chemist Dmitri Mendeleev. The chemist Langmuir convinced the physicists to adopt the obvious grouping of elements according to chemical properties. To account for this grouping, Langmuir grouped electrons into shells in the atom. Bohr adopted this approach and invented orbits of different angular momentum or shape for the same shell to make the grouping.

However, the introduction of suborbitals of various shapes is unnecessary to explain grouping in the new atomic model here presented in this book due to electron energy waves around the particles of the electrons and due to the nuclear energy wave both of which will account for chemical properties. These subgroups arising from chemistry do not need to be arbitrarily introduced into the atom as quantized suborbitals that pass through the nucleus and pass through the n-shells of the quantum atom in defiance of quantum theory as first postulated by Bohr. There need not be a paradox of electron orbitals intersecting the nucleus without the electron ever coming into contact with it.

In this new model of the atom with particles having an energy light wave, electron groupings can be accommodated as occurring in the primary electron shells whose spectral lines are empirically observed. All electrons inhabit a single type of spherical shell in every atom. They may be grouped by being paired in each other's energy waves so that they form an electron nucleus in the shell. Pairs of electrons in a shell form a tight bond within the nodes of the nuclear wave. The electrons may form a nucleus much as in Cooper pairs, but the electron nucleus is confined in the energy wave of the nucleus of the atom. When an electron is not paired, there is less ionization energy required. There may be multiple electrons in a shell, not just the paired electrons. The size of the node may constrain the amount of electrons able to inhabit each shell. The closer to the nucleus, the smaller the size of the node and the electrical repulsion of electrons may keep them apart as well as their own particle waves repelling each other. Therefore, Pauli's Exclusion Principle is explained because each electron attracts other electrons, but also repels them due to each particle having an energy wave surrounding it. No need for extra add-on rules, just a single hypothesis. All particles are surrounded by an energy wave. Thus we find the explanation for helium having two distinct energy levels. The higher energy level transitions are caused by the removal of the first electron from a paired bond. This is because two electrons bound inside their own electron nucleus or their own electron waves take more energy to transition. Once the electron pair are uncoupled during a higher energy transition, the single electron that remains needs less energy to transition. The weaker energy levels are transitions of single electrons. The higher energy levels are transitions of electrons from bound pairs in a shell.

The Zeeman effect which showed a splitting of a single spectral line in hydrogen to three lines by a magnetic field needed to be explained. It was assumed in the development of modern quantum mechanics that the splitting of the spectral lines was caused by the angular momentum of the electron suborbitals within each shell and by quantizing the angular momentum, one could get discrete, non-continuous lines. However, current quantum mechanics has no orbits. You see the contradiction and why Einstein and Schrödinger were fighting with Bohr. Einstein didn't like the abandonment of causality and Schrödinger didn't like the quantum jumps that were put back into his equation. No orbits, but suborbitals. No orbits, but magnetic moment. Crazy-making.

The Zeeman effect is not about suborbitals. It is about magnetic alteration of the waves emanating from the nuclear energy wave. A magnetic field will pull protons apart. Protons sit in each other's proton waves. When pulled apart, they experience an energy shift by being forced to pass through the antinode of another proton wave. This causes transition antinodes to be filled with more energy so making the electrons absorb and emit more energy to transition. It has nothing to do with suborbitals. It simply shows the effects of a magnetic field on the atom's nuclear energy wave. Hydrogen has three normal Zeeman lines because the nuclear energy wave only emanates from a single proton. However, in heavier elements, the effect takes on the anomalous Zeeman effect due to the additional pull of the magnetic field on the nucleons further pulling protons apart. In fact normal Zeeman doublets and triplets in heavier elements are proof that the nuclear energy wave cannot become completely coherent when there are many nucleons due to each nucleon having a separate energy wave.

Fine structure is given the spin attribute in quantum mechanics. Sommerfeld angular momentum elliptical orbits were attributed to the possible orbits of each separate spectral line of fine structure using Keplerian planetary orbits. Bohr used Sommerfeld's idea that there were only certain possible ellipses for each spectral line, but Sommerfeld was not talking about suborbitals in the modern sense. Sommerfeld meant that you could pick one of the possible conic section shapes to represent each shell. Bohr decided that each spectral line should include multiple suborbitals arbitrarily without direct evidence. These shapes were later radically changed by introducing spherical harmonics. However, there is no evidence whatsoever for suborbitals. The electron groupings were created by chemists due to chemical properties and Bohr just inserted them into the atom by copying the chemists methods, but the shapes given to suborbitals are imaginary. Suborbitals are just a mathematical construct to explain chemical periodic table grouping. The truth is there is no spectral evidence of quantum mechanics' suborbitals for any n-shell spectral line. Suborbitals such as p, d, and f would massively mess up the spectrum if taken literally. So science abandoned realism and went mystical.

Hydrogen fine structure is created by nuclear wave deformation due to the absorption of energy necessary to move protons through each other's antinodes. Valence electrons are those not paired. Grouping of electrons is due to physical phenomena of the energy wave emanating from the nucleus and the electron energy waves themselves. The confinement of electrons inside the nodes of the nuclear energy wave and the confinement of electrons inside each other's energy waves needs to be explored. Once the energy level of each energy wave of each particle becomes established, it will be known how many electrons may become bound in a single electron nucleus which is confined to a nuclear energy wave. The tables included in the published research article at the end of this book give the energy level of the electrons in the first 6 elements and their ions.

Bohr himself was not too pleased with the elliptical orbits that he had developed with Sommerfeld. Sommerfeld had introduced relativistic electron orbits. In Sommerfeld's model, electron shell n=1 could only have an s orbit which was circular; n=2 could have an s and p which was circular and elliptical; n=3 could have an s, p, and d which was circular, elliptical and another ellipse more oblate than p, and n=4 had the "f" orbital more oblate than "d". According to Heisenberg what bothered Bohr the most was not that these suborbitals were not in complete agreement with the number of fine structure in the spectrum which they weren't, but that as Heisenberg puts it:

"As to the first issue, I thought that I soon detected from Bohr's utterances that he believed less firmly than did Sommerfeld, say, in the applicability of classical mechanics to the motions of electrons within the atom. The fact that, on this assumption, the orbital frequencies of the electrons could not coincide with the frequencies of the radiation emitted by the atom, was felt even by Bohr himself to be an almost intolerable contradiction, which he tried merely to patch over in desperation with the idea of his correspondence principle."[74,p.40]

Bohr knew as did Heisenberg that such suborbitals as appeared in Sommerfeld's relativistic interpretation of electron orbits did not match the observed frequencies of the spectrum. Yet, they were not discarded but rather incorporated into quantum mechanics. The whole idea of inventing suborbitals was to account for spectral fine structure and to adopt the chemists grouping according to the periodic table of elements. Spectral fine structure is when one energy line of the spectrum is observed to be split into doublets or triplets. However, the frequency of emitted light and therefore the energy of the lines of fine structure are necessarily extremely close in value. One must remember that each spectral line represents a single shell of a single frequency and energy. Therefore, in quantum mechanics the suborbitals must be part of this single spectral frequency-energy line shell. They cannot have energies that vary too greatly from the frequency of the shell. But it doesn't take a physicist to understand that oblate elliptical orbits intrinsically are not going to have the same energy as circular orbits. Worse yet, the current model of spherical harmonics has the suborbitals creating very energetic shapes that cannot possibly be at the same frequency as the spectral line of the shell and therefore quantum mechanics' suborbitals could not produce line emissions from transition of electrons that were so close to each other as to be doublets and triplets. Bohr found this "an almost intolerable contradiction", but science accepts it today without the slightest objection.

Because of Bohr's fictitious suborbitals, the hydrogen electron is not placed in a "p" suborbital as it should be on the periodic table above boron so that oxygen is 4p, nitrogen is 3p, carbon is 2p, and hydrogen is 1p in grouping the non-metals. Instead, hydrogen is given an "s" suborbital arbitrarily when in fact there would be no problem in this new model of the atom, since there are no suborbitals of various shapes, but all shells are spherical, and electron grouping is caused by electron wave capture into nuclei and electron separation due to repelling electron-wave energy. Chemical properties are caused by configuration of electron grouping i.e. good conductors of heat and electricity are in a configuration that allows the electrons to absorb excess energy in many cases the electron waves touch each other making the passing of electrical energy fluid. Modern mainstream quantum mechanics leaves immense undiscovered structure in the atom that has gone unexplored due to the oversimplified and rigid quantum model of the atom with its arbitrary rules.

Schrödinger developed his wave equation from the arbitrary introduction of these various elements. An analysis made in 1969 showed that Schrödinger's equation is based on other equations, namely, the combination of the de Broglie matter-wave equation, Hamilton's principle on the curved motion of a system, and oscillations in a medium under classic Helmholtz and Maxwell laws.

According to one author on the subject (translation):

"Nevertheless, Russian physicists Ternov and Sokolov [9] in 1969 have found the logical inference of Schrödinger equations. It turned out that the Schrödinger equations are a system of three known equations. One of them is ratio for lengths of de Broglie's waves. Second is energy conservation law on orbit of Hamilton. Third one is general wave equation of oscillation of medium deduced for a sound by Helmholtz and for light in ether by Maxwell.

"For a mathematical formalism there is no difference what equations are joint in a system. But for physicists of a classic mechanics it is completely unacceptable to solve a joint set of equations, one of which is the law of motion of a rigid body and other is law of oscillation of medium. Therefore equations of Schrödinger and Dirac cannot be considered as physical. Any paradoxes, statistical solutions, 'vagueness' of an electron etc. occur from here."[57, 58]

The author is making the point that it is scientifically unacceptable to combine equations for the laws of motion of a rigid body and the laws of oscillation of waves in a medium such as sound in air or light in the aether because the electron is not known to be in a medium under modern quantum mechanics. (This begs the question of why the Schrödinger equation works at all which will be answered in this book. First it is important to understand it and deconstruct it.)

Another important fact about the Schrödinger equation is that the way it is interpreted in quantum mechanics is not at all what Schrödinger meant in formulating the equation. According to Heisenberg:

"For some time Schrödinger thought that the following picture of a discrete stationary state could be developed. One had a three-dimensional standing wave, which can be written as the product of a function in space and a periodical e^{iwt} of time, and the absolute square of this wave function meant the electric density. The frequency of this standing wave was to be identified with the term in the spectral law. This was the decisive new point in Schrödinger's idea. These terms did not necessarily mean energies; they just meant frequencies. And so Schrödinger arrived at a new 'classical' picture of the discrete stationary state, which at first he believed could actually be applied in atomic theory. But then it soon turned out that even that was not possible. There were very heated discussions in Copenhagen in the summer of 1926. Schrödinger thought that the wave picture of the atom—with continuous matter spread out around the nucleus, according to its wave function—could replace the older models of quantum theory. But the discussions with Bohr led to the conclusion that this picture could not even explain Planck's law. It was extremely important for the interpretation to say that the eigenvalues of the Schrödinger equation are not only frequencies—they are actual energies. In this way, of course, one came back to the idea of quantum jumps from one stationary state to the other, and Schrödinger was very dissatisfied with this result of our discussions... Born had made a first step by calculating from Schrödinger's theory the probability for collision processes; he had introduced the notion that the square of the wave function was not a charge density, as Schrödinger had believed; that it meant the probability of finding the electron at a given place."[74]

Schrödinger had created his equation with the express view of replacing quantum jumps. To Schrödinger the electrons in the atom were completely waves with the mass of the electron spread across the wave and around the nucleus according to Heisenberg above. Schrödinger then thought of the electron matter-waves as similar to Young's double slit experiment where light interferes with itself constructively and destructively. Therefore, in the atomic spectrum, light appeared as discrete lines where there was constructive interference when the electron passed between electron shells in the atom and where there was no light emission in the spectrum, this could be accounted for by destructive interference. Heisenberg explains Schrödinger's view of eliminating quantum jumps i.e. discontinuities:

"Now Schrödinger's interpretation—and this was its great novelty—simply denied the existence of these discontinuities. Thus when an atom passes from one stationary state to the next, it was no longer said to change its energy suddenly and to radiate the difference in the form of an Einsteinian light quanta. Radiation was the result of quite a different process, namely, of the simultaneous excitation of two stationary material vibrations whose interference gives rise to the emission of electromagnetic waves, e.g., light."[91,p.72]

Bohr explains Schrödinger's interpretation in this manner:

"In view of these results, Schrödinger has expressed the hope that the development of the wave theory will eventually remove the irrational element expressed by the quantum postulate and open the way for a complete description of atomic phenomena along the line of the classical theories. In support of this view, Schrödinger, in a recent paper, emphasizes the fact that the discontinuous exchange of energy between atoms required by the quantum postulate, from the point of view of the wave theory, is replaced by a simple resonance phenomenon. In particular, the idea of individual stationary states would be an illusion and its applicability only an illustration of the resonance mentioned."[94,p.75]

Firstly, one should question the Schrödinger equation's place in quantum mechanics since it was written with a completely different model of the atom in mind, a quite literal model of a single atom without quantum jumps.

Bohr explains that Schrödinger was following de Broglie's lead: "Already in his first considerations concerning the wave theory of material particles, de Broglie pointed out that the stationary states of an atom may be visualized as an interference effect of the phase wave associated with a bound electron."[94,p.73]

Now this was a resourceful point of view and as Heisenberg puts it, "so many physicists greeted precisely this part of Schrödinger's doctrine with a sense of liberation."[91,p.71] Unfortunately, Bohr and Heisenberg, not to mention Max Born, Pascual Jordan, and Paul Dirac who had all worked out matrix mechanics together, were not at all pleased with Schrödinger's interpretation and so Bohr had Schrödinger come to his house for a visit in 1926. Niels Bohr was in his forties and well-schooled in debate, even relishing it. He felt a good argument made things more clear, which is very true. Bohr had an interesting style of argument. Bohr was very soft spoken and mild in his arguments most of the time. With students and other colleagues, Bohr normally would simply question their opinion over and over again tiring them out until they came to his point of view, but he did this argumentation in a softly persuasive manner, always saying things like, of your point of view "but this is *very interesting...*" or "we practically agree..." or "there is no misunderstanding, but..." or "I say this not to criticize, but rather just to learn."[xxxvi] Bohr was a smooth talker.

But the situation was very different when he argued with Schrödinger. Bohr was only two years older than Schrödinger, but Schrödinger was no match for Bohr's keen intellect. Bohr saw and attacked the weakest point in Schrödinger's interpretation. Bohr saw that smooth transitions with changing frequencies that did not correlate to energy changes were against Planck's radiation law. Planck's radiation law was the solution to the problem shown by Wilhelm Wien in 1895. Wien "found that at a given temperature, the energy radiated at given frequencies, increased as the frequency was raised, reached a peak, and then began to decrease as the frequency was raised still further. If Wien raised the temperature, he found that more energy was radiated at every frequency, and that a peak was reached again. The new peak, however, was at a higher frequency than the first one. In fact, as he continued to raise the temperature, the frequency peak of radiation moved continuously in the direction of higher and higher frequencies."[55,II,p.128] Planck had proposed his constant to explain this discontinuity of radiation. Bohr jumped on this point against Schrödinger's wave equation showing that Schrödinger's interpretation did not account for this discontinuity. Bohr was merciless, relentless, and would give no ground. Heisenberg says of Bohr on this occasion: "And although Bohr was normally most considerate and friendly in his dealings with people, he now struck me as an almost remorseless fanatic, one who was not prepared to make the least concession or grant that he could ever be mistaken."[91,p.73] According to Heisenberg, where quantum mechanics was concerned, Bohr thought he could never make a mistake in his Copenhagen interpretation. This is astounding considering the many difficult discussions about the contradictions of quantum mechanics that Bohr admits being involved in both before and after this occasion. However, Bohr would not give an inch to Schrödinger. Heisenberg says that Bohr disagreed with Schrödinger's interpretation saying:

"You speak of the emission of light by the atom or more generally of the interaction between the atom and the surrounding radiation wave, and you think that all the problems are solved once we assume that there are material waves but no quantum jumps. But just take the case of thermodynamic equilibrium between the atom and the radiation wave—remember, for instance, the Einsteinian derivation of Planck's radiation law. This derivation demands that the energy of the atom should assume discrete values and change discontinuously from time to time; discrete values for the frequencies cannot help us here. You can't seriously be trying to cast doubt on the whole basis of quantum theory!"[91,p.75]

Schrödinger had based his interpretation upon that given by de Broglie who introduced wave-particle duality. Schrödinger was not as shrewd a debater as Bohr. Bohr himself had explained earlier to Heisenberg that upon introducing his model of the atom, he intentionally used the places in the radiation spectrum under Planck's law where the discontinuity of radiation existed. In other words, where the radiation law showed that there was no continuous rise of energy release, Bohr used those points as the points of radiation in the Bohr atom. Had Schrödinger wanted to turn the tables and use this against Bohr, he could have merely said that he was only using the points in the radiation law where the radiation was continuous and so choosing to do the opposite of Bohr. Neither argument would have been more correct. Bohr had explained to Heisenberg on their first meeting about Planck's radiation law showing that the elevation of radiation did not rise in a continuous fashion and it was this that led him to assume that the orbits of the electrons were in the positions that gave rise to discontinuous radiation rather than continuous. Bohr tells Heisenberg about the development of his model of the atom:

"The miracle of the stability of matter might have gone unnoticed even longer had experiments during the past few decades not thrown fresh light on the whole subject. Planck, as you know, discovered that the energy of an atomic system changes discontinuously; that when such a system emits energy, it passes through certain states with selected energy values. I myself later coined the term 'stationary states' for them." [91,p.40]

Bohr is speaking of Planck's radiation law having discontinuous energy states with selected energy values where the rise in radiation leveled out and these discontinuous energy states are the ones he selected from the others as his "stationary" orbits. Schrödinger's choice of the continuous states of energy in the radiation law would have been no less valid had Schrödinger chose to argue upon that line. Schrödinger's interpretation of his wave equation did not account for the discontinuous radiation in the Planck radiation law, but Bohr's quantum jumps did not account for the *continuous* radiation portion of Planck's law in the continuous background radiation of the absorption spectrum. Both views were limited. Schrödinger's interpretation did not allow for energy changes with frequency changes and Bohr's quantum interpretation did not allow for continuous frequency changes to emit during and across electron transitions nor for continuous radiation in a cloud chamber or vacuum tube. However, Schrödinger did not have the expertise to battle with Bohr. It stands to reason that if Einstein could not win an argument with Bohr, then Schrödinger didn't stand a chance. (Note: It is a Foundation Myth created after Einstein's death that Einstein lost those arguments. In actuality, Bohr had no idea what Einstein was getting at as Einstein was trying to argue about non-locality and entanglement. Bohr says he won. But Bohr never understood the argument. This can be seen from Bohr's answer to the EPR paper.)

Actually, both Bohr and Schrödinger were wrong about the interpretation of Quantum Mechanics because neither model incorporated both the continuous and discontinuous portions of the radiation law of Planck and Maxwell's equations. The new model presented herein accounts for all radiation laws both in the atom and as seen in the macro world.

In the end, Schrödinger's equation is not quantum mechanics of the real world and never was, but it was twisted arbitrarily in its interpretation to make it so. However, successful this arbitrary formulation of quantum mechanics has been, it is only successful as far as it has ascribed wave characteristics to the atom by the act of introducing Planck's constant as a limiting factor which is characteristic of wave energy and de Broglie's matter-wave equation which is also characteristic of waves. The Davisson-Germer experiment only proved that a wave existed in the atom. It did not prove that the de Broglie equation was correct as is shown from the data in the Davisson-Germer experiment which showed that the data collected did not meet the Laue condition nor the Bragg condition.[49] However it is certain that the de Broglie equation has a relation to the nuclear wave formation from the energy wave around each particle.

According to Davisson and Germer, the Bragg law was not obtained in the experiment:

"The single statement covering both reflection and diffraction is that for electrons of the speeds used in our experiments (bombarding potentials up to 600 volts) Bragg's law does not obtain; the wave-length of the beam of scattered electrons as calculated from the de Broglie formula is never the same (except in a special case to be mentioned later) as that of the corresponding beam of x-rays."[50]

Despite Davisson and Germer's admission in two scientific papers that their results did not correspond to the Bragg law, one finds that science textbooks, on the whole, say that the Bragg law was satisfied. Professor Nave says: "This peak indicated wave behavior for the electrons, and could be interpreted by the Bragg law to give values for the lattice spacing in the nickel crystal."[18]

By the definition of constructive interference, the separately reflected waves will remain in phase if the difference in the path length of each wave is equal to an integer multiple of the wavelength. Davisson and Germer found constructive interference where the spacing between the lattice was not an integer multiple of the wavelength. In fact, according to the Davisson and Germer data, the spacing between the lattice was continually increasing away from the expected Bragg peaks. The article says, "one notes a definite failure of the observed maxima to fall at the calculated positions." And further, "…electron diffraction beams do not coincide in position or in wave-length with their Laue beam analogues..."[50]

Yet, Physics Lab online reports: "With careful analysis, they showed that the electron beam was scattered by the surface atoms on the nickel at the exact angles predicted for the diffraction of x-rays according to Bragg's formula, with a wavelength given by the de Broglie equation, $\lambda = h / mv$."[59] This of course is not true.

Possibly the oddest part of the whole Davisson-Germer experiment suggesting that the de Broglie hypothesis was correct was the fact that de Broglie created his equation in 1923. This was before probability distribution, before probability clouds, before the Uncertainty Principle, and before the Schrödinger wave equation. De Broglie's final formula for the matter-wave was an exact replication of Einstein's formula "$p=E/c=h\nu/c=h/\lambda$" for radiation of light in 1916.[60] De Broglie started with the postulate that if $E=mc^2$ and if mc^2 is a wave, then $mc^2=hf$. But what is the math saying? E is substituted for hf (or Planck's constant and nu, the frequency of a wave of energy). However, science already knows that energy travels in electromagnetic waves whose energy is defined by hf. Therefore, substituting E for hf just means energy travels in waves and is interchangeable with mass which is saying nothing more than $E=mc^2$.[12] That the conclusions of Einstein and de Broglie were the same i.e. that h is inversely related to the wavelength is nothing more than saying that all energy is affected by Planck's constant in a way that the energy travels at c so that the wavelength determines the density of the energy. Neither equation is equating energy and mass except to say they are interchangeable, different forms of the same thing, but indeed different. Mass cannot travel at c or above. Light is incapable of having a wavefront speed greater than c due to Planck's constant. So what did de Broglie prove? That energy travels in a wave whose wavelength is inversely proportional to its energy. But Maxwell had already proved energy travels in waves and Einstein had already proved that energy and mass were equivalent but different forms of the same substance relativistically. So if one found an electromagnetic wave in the atom that emanated from the nucleus by bouncing electrons off of it, then it would appear to conform to the de Broglie wavelength as the de Broglie wavelength is based on Einstein's formula for light radiation. Therefore, one might think they had found de Broglie's matter-wave when all they had truly found was Einstein's quantum radiation of light. What one has is merely the electron beam interaction with a pilot wave.

However, de Broglie postulated that his matter-wave would take a circular path which in fact electromagnetic waves were not known to do. He postulated a fictitious wave—that did not carry energy—moving faster than light in a circle catching up with the electron and creating a phase wave or pilot wave that directed the electron in a single shell. Frankly, there is no empirical evidence for such a phenomenon. De Broglie postulated his matter-wave upon the principle of the Bohr atom that the electron was in a circular orbit in which the wave was strictly confined to the electron shell so that the wave lay strictly on the surface of the sphere of the electron shell. According to Davisson-Germer, the lattice is formed by parallel layers of atoms. What then is an electron lattice? Is it the pilot wave of the electron? What were Davisson and Germer measuring when constructive interference did not occur at equal intervals as the Bragg formula demanded but at gradually increasing intervals?

Furthermore, the Davisson-Germer data showed that the reflection pattern at a single scattering angle also show secondary maxima.[50] This data was never followed up on by Davisson and Germer as far as any search of the archives can tell.

The original Davisson-Germer article in which their Nobel Prize was based concluded that: "These turn out to be in acceptable agreement with the values of h/mv of the undulatory mechanics."[61] And what was that "acceptable agreement"? According to their later article, the agreement with h/mv was in their words, "The data obtained in our previous experiments yielded values of observed wave-length which, in a few cases, differed from the calculated values by more than fifteen percent." Although they were able to show in this later article that in two diffraction beams there was agreement with a wavelength of h/mv, the data still showed no agreement with the Bragg law.

The empirical evidence of the Davisson-Germer experiment proved that a wave existed in the atom that did not follow the Bragg law. This wave has the characteristics of Einstein's quantization of light radiation. The electron is considered to have zero rest mass so that the mass part of the de Broglie equation has no meaning and we end up with an equation for the electron that is exactly the equation for a wavelength of light radiation as stated by Einstein. Wavelength is Planck's constant divided by momentum. Therefore, if we empirically measure an electromagnetic wave in the atom it is the same formula as the matter-wave with zero rest mass! So the evidence for an electromagnetic wave in the atom is completely ignored in favor of the matter-wave theory.

So, the empirical evidence was ignored in favor of confirming a theory that physicists wanted badly to confirm. Waves travel at the speed of light, yet the electron does not. The evidence is that there is at least one electromagnetic wave in the atom. De Broglie said it is traveling in a circular path, in his words, "an electron turning in a circular orbit". However, if this were true, then how can two electrons occupy the same orbital without interfering with each other? How do the electron waves travel faster than the electron? How do they travel in a circle? How do they catch up to the electron? And how does this form a standing wave? What is scientific about this analysis at all when the Davisson-Germer experiment did not really prove de Broglie's formula. What ever happened to causality in physics? Why has mathematics replaced physicality when it need not do so? There are alternative ways of looking at the evidence in a scientific manner where one determines that where there is an effect, there is a cause.

It is no wonder that Einstein wrote in 1928 (although apparently speaking directly of the uncertainty principle and of the Copenhagen interpretation in general): "The Heisenberg-Bohr tranquilizing philosophy—or religion?—is so delicately contrived that, for the time being, it provides a gentle pillow for the true believer from which he cannot very easily be aroused. So let him lie there." (A. Einstein, letter to Schrödinger, May 1928)[46(p.281)]

What did Davisson-Germer find? Well an interesting aspect of the data points for the peaks where "the intensity of the reflected beam for angle of incidence 10 degrees—or rather, a certain function of this intensity—against the square root of the bombarding potential. What is plotted as ordinate is one less than the ratio of the current received by the collector standing in the direction of regular reflection to the mean of the currents received in two adjacent directions, one on each side of the beam....If the Bragg formula obtained, the maxima in this curve would occur at positions given by $V^{1/2} = n \times 3.06$, where n represents an integer."

The difference between the expected maxima according to the Bragg law plotted against the actual maxima peaks measured correlates with the ratios given by the Lyman series of hydrogen, namely, 4/3, 9/8, 16/15, 25/24, 36/35, 49/48, 64/63 for 2 through 8.

	Observed peaks	Lyman ratio	Observed peaks multiplied by Lyman	Peaks expected by Bragg Law using $n \times 3.06$
2	(unknown)	4/3	(calculated 4.59) 6.12	6.12
3	8.0	9/8	9.0	9.18
4	11.4	16/15	12.16	12.24
5	14.7	25/24	15.3	15.3
6	18.1	36/35	18.6	18.36
7	21.2	49/48	21.6	21.42
8	24.2	64/63	24.6	24.48

It appears that the observed peaks miss the expected Bragg law peak and are reduced against it by the ratio for wavelengths in the Lyman series of hydrogen. Unfortunately, due to lack of access to data other than available online, there are no other data to compare against this sole single series of points shown above that are from the two articles from 1927 shown as references 49 and 50 below. Whether this holds true for all electron diffraction and reflection data is unknown.

The data obtained in the 1927 articles by Davisson and Germer were no surprise to them as they had published earlier and say in their article, "These results, including the failure of the data to satisfy the Bragg formula, are in accord with those previously obtained in our experiments on electron diffraction."[49] And also, "A discrepancy of this sort was not unexpected. We had found in our first experiments that electron diffraction beams do not coincide in position or in wave-length with their Laue beam analogues, and it was anticipated that the properties of the crystal responsible for these discrepancies would manifest themselves, in the case of electron reflection, in a departure from the Bragg law." They reference their 1927 articles: Davisson and Germer, Nature, 119, 558 (1927); Phys. Rev., 30, 705 (1927).

The 1927 Physical Review article dispenses with data altogether concerning the misalignment between the peaks and the Bragg law except for graph 17 which does not allow calculation of the points.[61] The data for correspondence between de Broglie's formula and the wavelength is included in a table without Bragg law data saying that "it is preferable, however, to start at once with the idea that a stream of electrons of speed v is in some way equivalent to a beam of radiation of wave-length h/mv, and to show to what extent the observations can be accounted for on this hypothesis." Davisson and Germer then proceed to a table showing the wavelength to correspond to h/mv for various voltages of diffraction without giving any information as to displacement of the peaks away from the expected Bragg value. Finally on the twenty-sixth page of this article, they say: "While the dots representing the sets of electron beams fall along the plane grating lines, one cannot fail to note that they actually fall off these lines— and systematically; they are above or to the left of the lines as one cares to view them. At the time of writing our note to "Nature" we believed that these departures could be accounted for by imperfections in the geometry of the apparatus. At present, with more data at our disposal, we are less certain that this is true. ... We shall assume, that is, that the wave-lengths or voltages of the beams are correct but that the dots should be shifted to the right onto the lines. These shifts correspond to correction sin angle ranging from zero to about four degrees....There is a vague suggestion here that β [the variation factor] approaches unity as a limiting value. If this is actually the case, it means, of course, that at sufficiently high voltages (short wave-lengths) there is no difference between the occurrence of x-ray and electron diffraction beams."[61]

It is suggestive that the Lyman series as well as other series in the atom all approach unity at short wavelengths which is what Davisson and Germer expect of their data. The wave that Davisson and Germer found was not the de Broglie matter-wave. De Broglie based his wave on electromagnetic waves that follow a specific pattern and predictability. Davisson and Germer's electron beam hit a wave that followed the transitions of electrons in the atom. That means there is a wave in the atom that the electrons transition in and that Davisson and Germer's electron beam was transitioning in that wave.

Further, if the information contained in the following article is correct, then it would lend weight to the idea that scattering is being performed on the energy wave of particles and reflects the movement of the electrons in the spectrum as being their movement in the nuclear energy wave rather than the scattering actually being caused by encountering the electron.

"The Balmer formula for the spectrum of atomic hydrogen is shown to be analogous to that in Compton effect and is written in terms of the difference between the absorbed and emitted wavelengths."[86]

Although it is possible for scattering experiments to collide with electron energy waves, it is likely that when the scattering produces the same series that the electron transitions within, then it is most probably the nuclear energy wave that is being examined. In the case of electrons transitioning within the waves of other electrons, the electron waves would produce the same series as the series in which the electrons themselves transition. If the Balmer and Lyman et al. series occur both in the electron energy wave and in the nuclear energy wave, then the difference between the scattering effects would be the collision energies. Lower collision energies would produce diffraction or reflection by electron waves. Higher collision energies would pass through electron waves, but would still deflect off of the nuclear wave. Collisions that are higher yet would pass through both waves. Nothing can be said conclusively about which scattering experiments are hitting which particle waves without access to further data except it can be said that scattering experiments collide with waves that follow the electron series therefore the electrons in the atom are not transitioning in empty space but are moving through waves as they transition. These particle waves are not de Broglie waves and are not normal electromagnetic waves such as light. These waves have definite frequency (antinode energy) changes and drop in energy levels in a predictable, methodical series. This series is seen in scattering experiments. To speak even more plainly: In this experiment, since the electron beam collisions deviate from the Bragg Law in a definite pattern of decreasing energy, then the electrons are hitting a light wave that is causing a photoelectric-like effect and that light wave changes its energy in an ever-decreasing manner. In other words, Davisson-Germer found the nuclear light energy wave that is being introduced in this new theory where the light continually reduces in antinode energy as it extends from the nucleus. This is the reason the Bragg Law does not obtain. The Bragg Law measured light of a single frequency. The nuclear particle waves have decreasing energy antinodes. The electron beam is hitting curved light antinodes from the nucleus.

16 The Strong Force

Not only does this energy wave explain the barrier between the electron orbitals and the nucleus, but also this explains that the weak nuclear force is caused by this single barrier between the nucleus and the electron orbitals and it explains that the binding energy is simply caused by this same barrier. When the strong force is examined as a force created by the nuclear light energy wave antinodes emanating from the nucleus, instead of particle-particle interaction, then the theory of an energy wave completely unifies the forces of the atom.

Under this new theory of the atom, the Strong Force is created by the high energy antinodes of the particle waves near the nucleus. Because this makes the strong force as described under current Quantum Mechanical theory unnecessary as a separate entity or reaction, the current model of the strong force is removed from this theory and the observed force is defined by the pressure exerted by the light wave emanating from the nucleon particles.

Then there is the need to have causality for the particle waves not to radiate away. In writing this book, I have wavered back and forth. But I have concluded from an examination of Planck's and Einstein's original papers on the quantum that energy that is radiated is always radiated by a whole number n. Therefore, there exists less than a quantum energy that does not radiate. Einstein's paper on Specific Heat also makes this point. Therefore, if the wavelength is so short that the antinode does not contain enough energy to radiate, then the antinode of energy remains unradiated relative to the particle it surrounds. There is a balance of energy surrounding the particle. The energy stays around the particle because it is partial quantum energy and unable to radiate.

I am at all times in favor of causality. But I am also in favor of Ockham's razor, and this includes not adding forces that are unnecessary which I will address in the next chapter on the Standard Model.

This theory has no need for the strong force as a separate force as it exists naturally from the energy wave, because the partial quantum energy wave surrounding the particle gives the effect of what is considered by science to be the strong force. Space would be filled with this partial quantum energy from each proton so that space between nodes is filled with energy antinodes. As represented under an interpretation of Planck's constant, these energy filled antinodes of the standing wave of proton energy are explained by the inverse square law as the relation between the change in a surface of a sphere to the change in the volume of a sphere. This will be discussed further in the chapter on Electromagnetic Waves.

However, to briefly explain, if space is visualized as concentric spheres from a point of origin, then, in Maxwell's wave theory of light, the light radiates as it moves in concentric spheres losing energy according the inverse square law.

The same is true at the creation of a particle, partial quantum energy or the standing wave of light is created around the particle by leaking into the surrounding space, rather than radiating into the surrounding space. In the case of radiant light and in the case of light leaking out without radiation into the surrounding space, the light energy hits the surface of the next concentric sphere, then it fills the volume of the sphere. Thus the geometry of space slows energy or light at this area/volume change. Since this is true, we should be able to derive the inverse square law which is why the Balmer-Rydberg formula for the spectral light of Hydrogen is based on this law. That is not to say that light or energy has to fill the entire volume before proceeding to the next sphere, but light or energy must fill enough volume to push into the next sphere. With radiant Maxwellian light, the light fills the next sphere completely with a full quantum and is thus able to travel at light speed and radiate. With partial quantum energy light, the partial quantum wave energy that forms when the particle is formed also fills the surrounding space, but is unable to radiate, because it cannot fill the entire succeeding sphere with a full quantum of energy. For radiant light waves, the amount of energy or light that fills each consecutive sphere is equal to Planck's constant **h** times frequency i.e. it is a full quantum of light per sphere. Therefore, at light of extreme high frequency or short wavelength such as would exist at atomic sizes, the Plank constant **h** would create a very large energy packet because **h** is proportional to **f** which is frequency. Therefore, before the light wave emanating from the atomic particle could propagate through space, the energy content of **h** would have to be large. This would slow its propagation through space because of the geometry of space which is the entire precept for which Planck's concept of h was introduced. According to Isaac Asimov, he explains, "Planck proposed that the probability of radiation decreased as frequency increased." and "The greater number of high frequencies is more than balanced by the lesser probability of radiating at such high frequency. The amount of radiation begins to climb more slowly as frequency continues to rise, reaches a peak, and then begins to decline." In other words, the higher the frequency or shorter the wavelength, the lesser the chance of radiation. Asimov continues: "Consider the consequences of this quantum theory. Violet light, with twice the frequency of red light, would have to radiate in quanta twice the size of those of red light. Nor could a quantum of violet light be radiated until enough energy had been accumulated to make up a full quantum, for less than a full quantum could not, by Planck's assumptions, be radiated."[xxxvii] Therefore, it is possible for high frequency i.e. short wavelengths of atomic size to be unable to radiate, but to be localized in space. Under quantum theory of the Planck constant, Asimov continues regarding what would happen to such a high frequency or short wavelength wave, "The probability, however, was that before the energy required to make up a full quantum of violet light was accumulated, some of it would have been bled off to form the half-sized quantum of red light."

I am taking this assumption as true based on Planck's description of radiation. As I explain later, I attribute this to the geometry of space as described by spheres from each point in space. So then a light wave of a particle of a too energetic nature cannot radiate in its propagation before it filled the geometric volume of space and could punch into the next sphere in geometric space.

Some of the properties of the strong nuclear force and their explanation under this new theory are:

Property: At typical nucleon separation (1.3 fm) it is a very strong attractive force (104 N).

This theory predicts a similar force, however, it is caused differently. Rather than being a "very strong attractive force", the theory presented here shows that there is a very strong pressure exterior to the nucleus creating a great sphere.

Property: At much smaller separations between nucleons the force is very powerfully repulsive, which keeps the nucleons at a certain average separation.

Problem: There is no explanation in current theory why the strong nuclear force would suddenly become repulsive when it is the force that attracts nucleons together. Why wouldn't the nucleons be compacted together if the strong force were such an attractive force to the center of the nucleus? How could a single force reverse polarity and suddenly go from attraction to repulsion? Even the particle exchange theory of nucleon-nucleon interaction does not give a satisfying explanation for this since it suggests that the particles are held together at a distance being mediated by gluons. What causes gluons to work at specific distances? This is not answered.

Solution: Under the theory here presented the solution becomes obvious. Each nucleon creates its own energy wave, whether it is a proton or a neutron. That energy wave is strongest closest to the mass that creates it, so even though protons and neutrons may be at the center of the nucleus and combine their energy waves, the outward energy wave of each nucleon cannot be completely overcome. In other words, the nucleons are prevented from contact because their separate energy waves, emanating from each, prevent it.

Property: At short distances, the nuclear force is stronger than the Coulomb force; it can overcome the Coulomb repulsion of protons inside the nucleus. However, the Coulomb force between protons has a much larger range and becomes the only significant force between protons when their separation exceeds about 2.5 fm.

Problem: Why does the attractive strong nuclear force have such a short range of action if it is so strong? The particle exchange theory is complicated with many kinds of gluons.

Solution: Under the theory here presented, the energy wave is created by each nucleon. As each nucleon is separated from each other, their combined energy wave stops acting as one force and separates into its actual reality, that is, it separates into the two separate forces created by each nucleon and when it reaches a critical distance the separate energy waves of each nucleon push each other apart by their separate energy wave antinode effects acting upon each other.

As will be illuminated further in this book, the strong force when described as an overlapping of particle wave forces may be the only real force in the universe along with magnetism or the Coulomb force which is caused by the existence of charge.

Briefly, light wave energy when it becomes energetic beyond the point of being able to cross the Planck barrier (the Planck "barrier" will be clarified in chapter 18) fast enough, it becomes radiant light, a full quantum. But the energy in the universe has the property of forming spheres that become mass. As mass is formed the strong force or exterior wave energy forming the mass takes over and if too much energy around the mass has formed, it crushes the particle within seconds or milliseconds. If not enough mass forms so that the mass is surrounded by less than $e=mc^2$, the mass cannot be held together by the strong force from the wave energy due to the high energy antinodes of the energy light wave surrounding the particle then the particle disintegrates back into light energy. When exactly the correct amount of mass forms as in $e=mc^2$, the mass becomes stable as in the case of the proton with the strong force holding it firmly in just the correct ratio. The proton is the sole and only particle that is a long-life stable particle. (The electron will be described later as the other stable particle and how they differ). The proton is so stable that it is not seen to decay under natural conditions. An energy wave emanates from the particle as it does all particles constantly, but the energy antinodes of the wave keep the partial quantum light around the particle in a stable configuration. The law of nuclear equilibrium bars any further light energy or mass from being absorbed more than temporarily. This law keeps expelling excess energy causing entropy. Particles always attempt to lose excess energy which is heat, or if a full quantum, radiant light and send it to cooler particles that will accept it due to this law. This whole system is made possible by the by the energy wave causing a strong force antinode holding partial quantum light to the particles in an energy wave.

As we have seen, the data supports that if the space inside the nucleus along with the particles in the great sphere of the nucleus are added together in order to calculate density, then there exists a linear, proportionate, and consistent force emanating from the nucleus i.e. the strong force that is proportional to the density. There is no need to invent particle-particle interaction and gluons to account for nuclear binding.

If the nuclear wave field of the atom has extended antinodes they also loosely bind around atoms to form molecules, then there would be a very weakened wave field surrounding a molecule caused by a secondary great sphere that would try to become coherent around molecules, but would necessarily fail to bind with the binding energy of the antinode nearest the nucleus of the atom or directly around the particles from which it originates. Only the weakest part of the nuclear wave would create a secondary great shell. This could describe the concept of all the Van der Waals forces. The particle wave of the proton could also lead to a description of gravity as will be shown from the curve of energy that is created by this wave.

17 The Standard Model

In proposing this new model of the atom, it becomes very clear that it affects the Standard Model. That muons and mesons exist appears to be clear from experimental evidence. However, the role of certain unobserved particles becomes redundant in this new theory here being presented. It is therefore important that the original theory of Hideki Yukawa in 1935 be evaluated as the introduction to particle physics. In his Nobel prize lecture of 1949, Yukawa states:

"The binding energies of nuclei heavier than the alpha-particle do not increase as rapidly as if they were proportional to the square of the mass number A, i.e. the number of nucleons in each nucleus, but they are in fact approximately proportional to A."[25]

Here Yukawa is stating that after the element helium-4 the binding energies of heavier atoms do not increase by the square of the mass, but are instead approximately proportional to the mass. This is because the difference in binding energy per nucleon between the hydrogen-1 atom and hydrogen-2 (deuterium) is approximately ~1 MeV. The binding energy difference between hydrogen-2 and hydrogen-3 (tritium) is ~2 MeV. The binding energy difference between hydrogen-3 and helium-4 is ~ 4 MeV. This pattern does not continue after this but rather the binding energy appears to weaken after helium-4 and is only proportional to the mass of the atom of each heavier element.

Under the theory here being presented, this is no problem and is predicted by the density of the great sphere and the law of nuclear equilibrium as has been shown previously in charts relating the mass of the great sphere of the nucleus to binding energy. It is exactly what is predicted by sphere packing as previously shown in chapter 10.

However, being misguided because the Bohr model of the atom neglected to place a physical barrier between the nucleus and the electrons, Yukawa and others had to assume that some force or binding energy was missing in atoms heavier than helium-4. All that could be concluded was that the nuclear forces rapidly changed their physical attributes after helium-4. In his words:

"This indicates that nuclear forces are saturated for some reason."[25]

He describes a saturation meaning that the nuclear forces are somehow fully utilized and overwhelmed after helium-4 so that these forces cannot continue to gain strength which he assumes should continue to be the square of the mass. It was assumed that the binding energy should actually be the square of the mass even in heavier nuclei and that the binding force appeared to be missing the strength to hold heavier nuclei together. He states further:

"Heisenberg suggested that this could be accounted for, if we assumed a force between a neutron and a proton, for instance, due to the exchange of the electron or, more generally, due to the exchange of the electric charge, as in the case of the chemical bond between a hydrogen atom and a proton. Soon afterwards, Fermi developed a theory of beta-decay based on the hypothesis by Pauli, according to which a neutron, for instance, could decay into a proton, an electron, and a neutrino, which was supposed to be a very penetrating neutral particle with a very small mass. This gave rise, in turn, to the expectation that nuclear forces could be reduced to the exchange of a pair of an electron and a neutrino between two nucleons, just as electromagnetic forces were regarded as due to the exchange of photons between charged particles. It turned out, however, that the nuclear forces thus obtained was much too small, because the beta decay was a very slow process compared with the supposed rapid exchange of the electric charge responsible for the actual nuclear forces. The idea of the meson field was introduced in 1935 in order to make up this gaps."[25]

Therefore, it was assumed further that charge must pass between the nucleons themselves by other particles in order to hold the nucleus together. Since the binding energies didn't obey quantum field theory, he assumed there needed to be some correction. However, under this new model, the density of the great sphere indicates the pressure on the nucleus along with the law of nuclear equilibrium. Therefore, this rationalization by Yukawa was completely unnecessary under the new model being presented here. So Yukawa's particle exchange theory would have been unnecessary.

The fact that Yukawa predicted the meson and it was found is not surprising. There were by a 1964 count more than 80 subatomic particles discovered.[26] The 2005 government listing has 109 particles in the meson category alone.[28] In searching for Yukawa's predicted meson, physicists were bound to encounter something that approximated it. And actually, in 1937 what was actually found was a muon which is a lepton and really not what Yukawa had predicted at all. And even after the pi meson was found, Yukawa's theory proved to be inaccurate because it is now understood in the Standard Model that the strong force is mediated by the gluon. However, particle physics still held that the strong force is said to be mediated by particles as Yukawa had theorized even though his actual theory proved incorrect in the end.

What started the whole Standard Model mess is that Dirac tried to make the Schrödinger/Heisenberg equations relativistic. But those models were built on pilot matter waves of electrons not on a nuclear particle wave model. So the Dirac equation had electrons be massless which just isn't true. Yet, this was considered a great breakthrough. But totally unnecessary and incorrect because it was built on wrong foundations with no understanding of particles and their surrounding energy wave. Even the next step after the description of electrons and fermions in the Standard Model, it couldn't get mass into baryons or any other particle in the equations when describing the strong interactions under Yang-Mills, neither did the equation added to it for the electromagnetic and weak force have any particle able to have mass. This was even when the solutions all came out to infinities, and now you can cancel out whichever infinities you like to get to the right answer. But from its inception, the Dirac equation had to abandon particles with mass and invent negative energy. There is no such thing as negative energy. But in order to force the Dirac equation to be right, science was brainwashed to think that means negative charge. Negative charge is NOT negative energy. Opposite charges are not opposite energies. Nothing has less than zero energy. It would be less than zero Kelvin. Impossible. So scientists said, "Let's just change it to charge." That's cheating. Charge is not energy. The Standard Model as a whole is a very weak equation for explaining things although for numberophiles it is a beautifully symmetric equation, but a very uncompelling overly complicated way to get to any answer. It was only by adding a line to the equation that would break the beautiful symmetry of the equation which was introduced in 1962, some 34 years after Dirac started his beautiful symmetry equation, could the equation be made to show that mass has mass. For 34 years, science believed in an equation with massless mass. So clumsy!

Under this new model of the atom, the Standard Model is inherently flawed because it is based upon the wrong assumptions. The assumption that any particle is needed to mediate the strong force or the weak interaction is unnecessary. The energy wave field replaces it.

The craziness of the original Bohr model in not putting a barrier to account for the distance between the nucleus and the electrons has caused craziness to be piled on craziness. It is only because of the insight of Bohr to use the Balmer-Rydberg formula which does in fact appear in all atoms that made Quantum Mechanics appear so successful. And this Balmer-Rydberg formula is the basis for QED (Quantum Electrodynamics) leading to the Standard Model so that some solutions along the way to accurately predict atomic behavior.

But the craziness is not just my rude adjective. The first QED formula for the electromagnetic force showed that electrons have no mass. This didn't deter further work on the formula. Also, there was the problem of results with the answer of infinity. This is always a null result in physics up to that time anyway. So physicists invented a way to cancel out infinities that they called renormalization. That is a bit crazy to have to go through all that mathematical legerdemain to justify a theory where particles had no mass. Yes, I've seen the equation and yes it has beautiful symmetry, but a beautiful equation does not mean that it describes anything real, useful, or even helpful. From renormalization, we come to the Strong Force part of the Standard Model and again the abstruse mathematics can't make anything with mass although the physicists know themselves this is crazy. The particles have to have mass. Even Wolfgang Pauli interrupted the conference to brazenly point this out to Yang when he was explaining Yang-Mills theory. Pauli said, "What is the mass of the b field?" knowing perfectly well the answer. Yang equivocated and Pauli insisted. Yang sat down. Oppenheimer shushed Pauli and Yang got up and continued.[xxxviii] So physicists knew that the equations had no mass, yet in reality, the particles had to have mass, but giving the mathematics mass ruined their beautiful equations. Really!? Again, I am dumbfounded at what they call Science.

It was from this foundation of QED that the Standard Model was created in its entirety adding quarks, gluons, and various other particles as particle-particle interaction that was started by Yukawa particle exchange theory, Dirac mathematical beautiful symmetry, and then guage symmetry beautiful mathematical symmetry. Beautiful but complicated math symmetry does not make it reality. The universe tries to make so many particles until it gets the right stability for the proton $e=mc^2$ that of course there are hundreds of particles. And if you theorize a particular particle and smash protons, you are of course going to in all probability get something similar to that particle. But here we might mention that you can't get every kind of particle. The energy in this universe is particular in trying to make protons of stability. It will not make super symmetric particles to date. This is the attempt to create a particle that would include gravity into the Standard Model. But even if we could make a super symmetric particle that would make the Standard Model even more remote, unwieldy, and for all practical matters pretty useless. We don't use it to make electromagnetic predictions. We still use Maxwell's equations. We don't use it to predict electron (Fermion) systems. We still use the Schrödinger equation. The Standard Model is just one giant mathematical clunky infinities mess. Compare it to the model in this book where everything like the Strong Force, Weak Force, and a great possibility of gravitational force evolution is described by a single wave emanating from the proton in the proportion of $e=mc^2$. So much simpler! So elegant!

The book Quantum Story says this about the Standard Model: "Despite the model's obvious and unmitigated successes, its deep flaws have been painfully apparent since its inception in the late 1970s. The Standard Model has to accommodate a rather alarming number of 'fundamental' particles. Six quark flavours multiplied by three possible colour values gives 18 different types of quark. Add the leptons—the electron, muon, and tau, and their neutrinos, and we have 24 fermions. Then there are the anti-particles of all these, making 48, to which we need to add the field quanta: the photon, the W+, W=, and Z^0 particles and eight different types of gluon, making 60 in total. All of these particles have been 'observed' so to speak…making 61 particles in all. This hardly seems the stuff of a fundamental theory. These 61 particles are connected together in a framework that requires 20 parameters that cannot be derived from theory but must be obtained by measurement."[xxxix] The book goes on, but we need not. Enough is enough.

Under the new model shown here in this book, the new model is in exact agreement with the spectrum of the elements as shown in the paper that ends this book. The universe is full of energy that tends to enclose mass in a circular spherical fashion trying to become stable. Therefore, that energy creates many particles in the process that disappear in microseconds or it turns into light. We see this in accelerators. The only lasting particles are protons and electrons with the exception of neutrons when they are enclosed in the nucleus surrounded by the energy of the proton that protects them. So atoms are made of protons, neutrons, and electrons. And all the universe is made from these with the energy surrounding them to account for every single one of the phenomena known to Quantum Mechanics and the Standard Model today, but in a much simpler form. This is borne out in the paper at the end of this book.

18 The Formation of the Wave

A model of the formation of an energy wave that is only known at atomic levels is a difficult task. If this energy wave causes the interference pattern of Bose-Einstein condensate, then we know it must be formed by waves that emerge from the center of the atom NOT de Broglie matter waves which would wave perpendicular to the nucleus. The wave must not lose energy or else mass will not be conserved, therefore, I first thought the wave must be finite. But all that is truly needed is that the wave must not radiate. However, in current theory of electromagnetic and gravity waves we only see infinite radiating waves until eventual absorbtion. In sound waves and ocean waves, the reason that the wave stops is due to either friction or a barrier or loss of a medium to travel in.

However, this should not prevent an attempt to model this energy wave. After all, de Broglie said that a single electron traveling in a circle in one direction actually formed a standing wave in the allowed orbitals although, in reality, the only thing that we know of that causes standing waves is either a barrier at the ends of the wave or two separate waves directed at each other. In fact, when one reads de Broglie's argument for this we have this explanation from one of his original papers:

"Let us consider now the case of an electron describing a closed trajectory with uniform speed slightly less than c. At time $t = 0$, the object is at point O. The associated fictitious wave, launched from the point O and describing the entire trajectory with the speed $\frac{c}{\beta}$, catches up with the electron at time τ at a point O' such that $\overline{OO'} = \beta c \tau$."[27]

First, de Broglie says that an electron in an orbital launches "an associated fictitious wave" that "catches up with the electron". Only in this way does a standing wave form for the electron, de Broglie continues, "*only if* the fictitious wave passing O' catches up with the electron in phase with it".[27] Then he concludes, "In the case of an electron turning in a circular orbit of radius R with an angular velocity ω, one finds again for sufficiently small speeds the original formula of Bohr..."[27]

First point to consider: If the electron matter wave is only in phase and discrete when it is in a circular orbit, then what is contained in the suborbitals "p", "d", "f" and so on? An electron does not describe a circular orbit in suborbitals and therefore cannot be "in phase".

Secondly, de Broglie is not here actually picturing a standing wave. Waves propagating in the same direction that are in phase create constructive interference, such as in a laser, but constructive interference merely creates a "traveling wave" that is in phase *not* a standing wave.

Thirdly, when you arrange several electron waves onto a sphere, they no longer become sinusoids with n wavelengths around a circle. So what was the point if the atom isn't explained physically by quantum mechanics anyway?

So this will be an attempt to illustrate a dark energy unseen and unknown except by inference starting from the fact that the electron maintains a distance from the proton in a hydrogen atom. We have seen that a coherent wave emanating from the atom can be inferred from experiments such as the formation of BECs. And we have seen that it can be inferred from the fact that electron diffraction crystallography will still get a wave pattern while deflecting off the position of individual electrons therefore it is must actually be deflecting off of the nuclear energy wave as it would be impossible to only hit electrons each time and measuring the electron would collapse the wave function and there would be no wave result. See chapter 7 on the Davisson-Germer experiment as a refresher.

So this first attempt to model this energy wave of the type necessary will be illustrated here in a simplified model. As the wave is forming from its first cycle along the x-axis, it is creating an energy antinode as it moves. The energy cannot radiate away because frequency and wavelength are inextricably tied together and the wavelength is too short at atomic lengths to get a full quantum of energy. We know the equations of Planck and Einstein's photon show that only whole numbers can be used to radiate energy. Only full quantums. These are related to frequency, but just as truly, these are related to wavelength. There is therefore always a remainder of energy that is not a full quantum, especially with the amount of energy that it would take to radiate at atomic wavelengths. Free energy (not Gibbs free energy), but free energy as in a proton accelerator collision tries to reform in spherical shapes squeezing space to try for a balance and stability of $e=mc^2$. When this is achieved, a proton forms and the perfect amount of energy to keep the proton stable starts flowing away NOT radiating away, but slowing stretching out according to a decay curve so that the energy does not radiate away while it spreads through space. This continues always maintaining the ratio of energy wave to mass limit in accordance with the law of nuclear equilibrium. When does the energy stop spreading into space? That is where the spectra of elements comes in. For Hydrogen, we find the Balmer-Rydberg formula. It shows that the antinodes decrease in energy by an inverse square law infinitely. Every proton has the same spectrum because every proton is formed with the exact same energy around it to become stable. Our universe has energy that prefers the proton and becomes stable when $e=mc^2$ around the proton. The energy stays around the proton as it spreads in the same decay curve every time a proton is born. It strives to remain stable by maintaining this proportion of energy to mass of the proton so that we always get the Balmer-Rydberg lines in the spectra.

The energy is denser near the nucleus because it fills less volume as the wavelength is tiny comparable to a higher frequency of radiating light. But this is not full quantum light. It can't quite fill the short wavelength with enough energy to radiate so it is dark energy surrounding the proton. It is partial quanta energy. However, the system is dynamic especially in large nuclei. Large nuclei are caused by protons getting stuck in each other's inner antinodes that carry most of the energy. Neutrons get captured due to the principle of sphere packing to keep a more spherical shape to the nucleus. All particles having their own energy wave causes more pressure on some atoms, so that in radioactive isotopes of the same element some atoms are positioned or distributed differently on their mutual energy antinodes depending upon the distribution of nucleons. Therefore, the decay of individual radioactive elemental atoms varies. But the decay works like a clock as each nucleon sits in the mutual waves of other nucleons and each atom sits in the waves of other atoms, so that the shifting of one nucleon shifts all the other nucleons in the high energy waves near it. This eventually leads to systematic decay and very exact half-lifes as discussed in chapter 12.

By each proton trying to sustain an exact ratio of energy, the energy is thus conserved and the energy wave becomes a fixed closed system that is exactly the same for every proton. The charge of the energy wave is the charge of the particle. The wave is fixed about the mass that created it into a pattern of coherent waves. In the creation of mass, the emitted wave would be emitted equally in all directions from the source mass creating an infinite energy wave in accord with both e=mc2 and the Balmer-Rydberg equation once the electron shells are created by the spreading antinodes. The law of nuclear equilibrium governs how much of the mass of a particle is put into its energy wave. This law governs an exact ratio of mass to energy wave that a spherical free particle will experience. It also governs the balancing of the multi-nucleon energy waves. Thus, the atom itself is an asymmetrical decay curve antinode wave causing a pressurized environment with most of the pressure concentrated near the nucleus, because the force from the high energy/low wavelength antinodes is greater near the nucleus both from either an outward or inward direction as the wave antinode creates pressure.

Because the energy flows without radiating, because it is partial quanta energy compared to the wavelength, the total energy radiated is zero. Therefore, there is no measurable macro world energy from the energy wave of each particle. Only other particles are sensible to the energy emanating in each energy wave. But this has macro effects. This wave causes tunneling. This wave causes the Casimir effect. This wave causes Van Der Waal forces around atoms of elements. This wave causes covalent bonding. This wave causes entanglement of particles across the universe. This wave may even cause gravity and dark matter as it describes a secondary cumulative gravitational sphere around solar systems, galaxies, and galaxy clusters squeezing the clusters together to cause lensing. All the answers in one little wave around the proton. How can I prove it is there? I have. See the article at the end of the book. The wave antinode energies are in the spectra of elements for all to see.

If one pictures the strong force as an actual force rather than a particle-particle interaction mediated by gluons then it becomes obvious that the strong force is more powerful than gravity and the electromagnetic force because it is a great deal of energy surrounding a proton. What went wrong? Why didn't scientists catch this? Well, Bohr was antirealist (see Lee Smolin's books) and Einstein never studied the spectra. He was a pure ancient Greek theorist although a realist. He only studied how to put Maxwell's equations into Quantum Mechanics. He is the most respected genius of mine, but he should have studied the spectra. He relied too much on his own thought experiments and his own mathematics. But he got so much right and was so far ahead of his time, who can blame him? He was brilliant!

But Einstein missed the spectra. He missed the idea for this model. So where did we go wrong? First it must be restated that Planck's constant was introduced into the atom in an arbitrary fashion in order to presume some discrete placement between electron orbits without any physical cause before wave-particle duality. It was only successful because there is a wave inside the atom. Bohr introduced it without believing in photons at all! Bohr didn't believe in photons even in 1922 as quoted previously when Einstein won his Nobel Prize for them. So

Rutherford said the atom was empty space between the nucleus and the electrons. And Bohr said the electrons stay there for no reason at all.

We start with a wave surrounding the particle. As with any energy wave the energy must fall off with the distance squared. This is because the energy is released from a spherical object at a set rate and as it moves away from the spherical object, it describes larger and larger spheres, therefore, the same energy must occupy a larger volume. We see this in the Balmer-Rydberg lines.

In this new theory, each electron in the atom need not oscillate or orbit the nucleus. The movement of electrons whether by oscillation or movement within the node of the nuclear wave is only dependent upon the energy of the electron and the contact with surrounding charges. Therefore, the movement of the electron will appear random while inside the atom unless its energy or relative position with respect to other particles is known. The movement of electrons causes them to radiate and lose energy, however, that energy is recaptured in the electron wave. However, when the electron absorbs energy in excess of that which is required to keep it in its electron shell or to keep its mass to energy wave ratio constant, it will release the energy as either heat by electron oscillation or electromagnetic radiation.

When a positive energy wave overlaps with a negative energy wave as in a proton-electron pairing in an atom, the wave becomes neutral where the waves overlap although the particle retains the main portion of its charge just as a particle retains the main portion of its mass. When a free electron's negatively charged energy wave comes in contact with a free protons positively charged energy wave, the two particles slowly begin to move toward each other. Their speed toward each other accelerates as they approach each other, because the charge of the energy wave is proportional to the square of the distance between the particles. If the velocity of the relative approach is strong enough, the electron may become trapped inside the nucleus. However, this electron-proton nucleus is not stable because there is too much mass to energy wave. Instead of a proton-electron nucleus strengthening the combined energy waves, it weakens the outer wave so that the two particles fragment. This is because of the law of nuclear equilibrium. The electron-proton nucleus has no particle separation because the particles retain their respective charges although the wave they create is neutral. This electron-proton nucleus is therefore particle-like. Because of the law of nuclear equilibrium, the electron-proton particle is now a particle larger than a proton so the law of nuclear equilibrium apportions the energy wave to mass ratio so that the energy wave is given less energy and the proton-electron particle is given more mass in the form of a neutrino. This completes the sphere and mass of the newly formed electron-proton particle which we call a neutron. (The neutron considered as a combination particle does not necessarily have to be the case, but should be considered as a possibility which in all probability is simply a matter of semantics since for all intents and purposes if the energy wave becomes completely unified by all three of these particles and the mass becomes completely bound, then the neutron becomes its own particle.) In this case, the neutrino mass binds to the proton-electron particle due to energy wave pressure on the newly formed particle. This transfer of energy wave to mass weakens the energy wave and fragmentation decay occurs due to the law of nuclear equilibrium. Particles with a larger mass than a proton have a weaker energy wave. When the neutron decays the neutrino is sent out of the nucleus only to be reabsorbed by the proton's energy wave.

In neutron decay, as the electron is thrust outward at a high velocity, if it is inside another proton energy wave when it is cast out then it can only gather speed racing toward the proton according to the initial distance from the proton. In this way, the electron will eventually only be able to accelerate to reach ground state around a proton without being able to reach the nucleus. This is because of the concentrated energy of the wave nearest a particle. The concentration of energy near the particle is independent of the charge of the wave, although the energy carries the charge of the wave, it is so dense near the particle that it acts as a barrier through use of the photoelectric effect even as the wave itself is pulling oppositely charged particles together.

When a proton wave overlaps an electron wave, the wave no longer attracts other electrons. Other electrons with sufficient energy can become fixed inside the protons wave. The wave of this second electron is combined with the wave of the proton and the initial electron and the overall wave becomes negatively charged. This is normally caused by an imbalance in the system.

The ground state electron shell is of a different radius for every atom of every different element since the asymmetrical combining (because the nucleons occupy distinct positions in space in the nodes of their respective nucleon waves) of energy waves of various nucleons causes orbitals to be spaced differently in each atom of each element. The combining of nucleon waves causes different energy levels of electron shells for each element. The number and arrangement of nucleons into a spherical nucleus creates further seemingly chaotic spacing of electron shells. This accounts for the fact that x-ray diffraction shows that every crystalline compound has its own characteristic x-ray pattern. Science knows there is a pattern to every element's spectrum, however, they haven't been able to find them all or account for them since the Schrödinger wave equation shows Bohr's data that the electron shells are regularly spaced after ground state with angular momentum put as integer multiples of h-bar. The Schrödinger wavefunction equation, later enlarged upon by Dirac to incorporate spin and special relativity having a Max Born probabilistic interpretation both still keep this regular spacing of orbitals, but the spectrum does not show regular series spacing in heavier atoms.

"The description and understanding of the simplest atom, the atom of hydrogen, was achieved in 1926 by Schrödinger. He was frustrated, however, when the next more complex atom, helium, failed to yield his equation."[73]

Rather one uses only the Ritz combination principle to define transition series in atoms that have more than one electron. Although Dirac's equation incorporated what little that could be deduced from quantum mechanics in the 1920s, it had no physical representation. "Einstein granted that the equation was 'the most logically perfect presentation' of quantum mechanics yet found, but not that it got us any closer to the 'secret of the Old One.' It neither described the real world phenomena that Einstein wanted to understand nor proposed new concepts that would make the real world accessible to understanding."[46p.286-7]

In this new model, the nuclear energy wave has waves that are regularly spaced no matter the number of nucleons. This is because the nucleon waves become coherent near the nucleus when they form a great sphere about the nucleus. In hydrogen-like atoms with one electron, the

Rydberg formula holds because the only interference with the nuclear wave comes from a single electron wave. When additional electron waves are combined with the nuclear wave, there may be various patterns of destructive and constructive interference that appear and there is definitely various energy levels to be taken into consideration when electron bonds pair and unpair. Therefore, the Rydberg formula no longer holds, but the Rydberg-Ritz combination principle holds due to each electron wave being identical to each other. However, it is no secret to science that in the helium ion, although the Rydberg formula holds one must take into account the mass of the nucleus. Why this should be so under quantum mechanics is only weakly and unconvincingly explained. However, if one considers that the electrons are moving in the nuclear wave, then it becomes mandatory that the more nucleons, the stronger the wave and the mass of the nucleus very much controls the energy of the electron transitions as the strength of the waves changes. As Bohr explains:

"During the following months the discussion about the origin of the spectral lines ascribed to helium ions took a dramatic turn….An answer to the last point was, however, easily found, since it was evident that the mass *m* in the expression for the Rydberg constant had to be taken not as the mass of a free electron but as the so-called reduced mass $mM(m+M)^{-1}$, where M is the mass of the nucleus. Indeed, taking this correction into account, the predicted relationship between the spectra of hydrogen and ionized helium was in complete agreement with all the measurements."[90,p.42]

Since gravity is not a factor in quantum mechanics, the argument that the mass of the nucleus has any bearing on the electron transitions becomes tenuous indeed especially due to the circumstance that the mass of the nucleus has no bearing on the charge where neutrons are involved. This turns out to be simply more unexplained crypticness yet added to the substantial heap of quantum mysticism. Under this new model, one understands that the mass of the nucleus is an essential factor in describing the nuclear energy wave of each atom.

However, although it seems certain that the size of the proton wave and the electron wave is the same, meaning, the radius of the finite sphere each creates is the same length and the charge is the same magnitude, it appears that the proton's energy wave dominates and the electrons transition in the nuclear energy wave rather than the converse. It may be that there is the same magnitude of charge but possibly energy differences when comparing the electron wave to the proton wave. The small volume and high density of the photons near the nucleus of each wave keeps particles separate even with unlike charges.

The fact that light is known to be emitted at a single frequency from a single source in all directions and yet falls off as the distance squared would suggest that all other forms of energy that fall off obeying the distance squared law would be a single frequency. Only the intensity of light falls off as the distance squared, not its energy or frequency. Where energy or frequency falls off as the distance squared then the wave itself is changing frequency with distance because frequency defines energy. Light redshifts in a gravitational field. The strong force is greater than the force of gravity and creates a greater force on light. A redshift is a shift in the frequency of a photon toward lower energy. So in the atom, the energy in the energy wave of a particle falls off at each node or shell and therefore is not of one single frequency. However, the beginning frequency near the particle i.e. the highest frequency of the particle is higher for more massive particles. Therefore, the energy of the proton energy wave starts from the particle with high energy rays, compared to the electron energy wave, for every proton and the variance in transition energies from gamma ray emissions to x-ray emissions is due to the frequency of the proton energy wave falling off with distance so that the transitions become easier with distance. The energy wave of the electron has a frequency that is very weak perhaps extremely weak. Therefore, the transitions are made with almost no energy and the spectrum of transitions appears continuous in free electrons. The placing of an electron in the energy wave of a proton enhances the energy needed to transition in the electron energy wave. Electrons transitioning in the waves of other free electrons appear to emit a continuous spectrum which is technically untrue. It is just that there are so many available transition energies with such a weak field that the spectrum appears continuous. In protons in a particle accelerator on the other hand, the protons transition only in gamma ray fields and therefore the transitions can only be made by x-ray and gamma ray energies.

The higher frequency of the proton near the particle does not knock the electron from the atom due to the Coulomb force, but it does keep the electron at a distance using the principle of the photoelectric effect.

We can imagine isolating one light wave emanating from a particle creating the energy wave about the particle in the following fashion. (Remembering always that the energy wave about the atomic nucleus is the same as the energy wave about a particle such as a proton.) We can imagine one light wave emanating from the surface of the particle. It leaves the particle at 100% frequency as proscribed by the Law of Nuclear Equilibrium. With each wave it loses frequency due to the Strong Force of the particle acting upon the light. If the loss were 1% of the initial frequency per new wave cycle, then at 50% of the initial frequency, the light wave would turn in upon itself at the horizon of the energy wave. The light would then continue to lose 1% frequency per cycle until it reached the particle again to be reemitted as a new light wave. This would create a loop in which the inner cycles would have a much stronger frequency leaving the particle and much weaker frequency reentering the particle. On the other hand, the outer cycles of the light wave would have a much lesser frequency. That is to say, a light wave leaving the particle would have at its first cycle a 100% strength frequency of light leaving the particle and a 1% of this initial strength frequency returning to the particle. The main thrust of the force of the photoelectric effect would keep most particles from entering into the region near the particle or nucleus. After several cycles however the light wave would have 75% initial frequency strength being directed outward from the particle or nucleus and would have 25% strength frequency for its returning cycle. This loop of returning light cycles would weaken the outgoing light by decreasing frequencies. The overall effect would be that the particle's inner frequency would be high enough to use the photoelectric effect to pummel particles away from other particles or nuclei and that the outer frequency of the energy wave would allow particles such as the electron to stay at a distance from the nucleus without falling into it.

However, it is entirely possible that the light might actually blueshift on its return approach to the particle or nucleus as it is not known how light behaves inside a black hole. To repeat, the atom would have a "black hole" effect without absorbing other particles that approached since the photoelectric effect of the light radiating from the particle or nucleus would bar approach. However, if the light did indeed begin a blueshift as it neared the nucleus, this would account for the "great sphere" effect of the atomic nucleus. It seems reasonable that although curved light has been observed to redshift that in a closed system around a particle the light wave would have to obey the law of conservation of energy, therefore, the light would need to stop redshifting as it turned back toward the particle and then blueshift in order for the an equal amount of energy to return to the particle. This is because under Planck's radiation law, the frequency describes the energy of the system. The returning incoming light, if it increased frequency as it neared the nucleus, would create a photoelectric effect inward, trapping particles that came near enough to other particles overcoming the repulsion of like charges under Coulomb's law. This would create a two-particle nucleus. Any further particles, traveling fast enough to break through the photoelectric effect of the outgoing energy wave could come near enough to the inward blueshifted light that they would form an even larger nucleus. The inward light effect would cause a "water drop" nucleus or "light bubble" around the nucleus. This would cause the balancing effect upon the great sphere of the nucleus.

For the hydrogen atom, where:

f is the frequency of the light emanating from the proton,

f_R is the frequency of the light returning from the horizon of the energy wave back to the proton,

p is the force of the photoelectric effect on the electron by light prior to causing the effect i.e. pre-photoelectric effect pressure,

T_f is the threshold frequency for the photoelectric effect to occur to the electron due to the light emanating from the proton,

F is the magnitude of electrostatic force between the proton and the electron,

n is the number of the shell of the electron which is a node in the energy wave,

then:

Where $T_f = F + p(f_R)$, $n = 1$

19 Electromagnetic Waves

The energy of a photon is described by **h * n-cycles per second** where h is Planck's constant. Even though light would be considered as being made up of photons, one shouldn't necessarily picture the photon as a particle. For example, if one feels the pulse of one's wrist, one feels definite intermittent energy surges, but this does not mean that the blood has become particles. So too waves of light travel in energy surges and these individual surges react as would particle-like photons. But this does not mean the flow of the wave is discontinuous.

Photons of energy burst from the source mass in a particular size and volume, however, they soon meet with a larger space in which to fill. That energy is required to fill consecutive spheres is evident from the inverse square law which shows that at each equal radius from the point of origin there is equal energy. Photons must reorganize to fill the larger space. The larger space acts like a barrier making the photon slow, then resurge into the next space barrier while continuing at the same relative overall speed as in the case of electromagnetic waves traveling in vacuo. **This is due to the fact that the surface area of a sphere increases more slowly with increasing radius than does the volume of a sphere.** That is to say that the Galilean Square-Cube law creates Planck's quantum as a physical geometry of space. This causes a threshold effect i.e. a volume barrier. (This effect was seen and used in the atomic bomb.[82]) Pushing more energy into the volume of a sphere when the surface of the sphere does not increase proportionately with the radius causes a bubble skin effect that must burst before the energy can enter the next spherical volume of space. In electromagnetic waves traveling in vacuo, every photon carries exactly the same amount of energy, only the volume and density of the photon changes and there are more photons per second in higher energy waves. The volume barrier acts as a resistance to energy. The faster the energy hits the barrier, the more is the resistance of the volume barrier. The more slowly the energy approaches the barrier, the less is the resistance created by the barrier. It is easier for waves to travel through the volume barrier at slower velocities and lower energies therefore the probability is higher that there will be more lower energy waves. Photons tend to take the path of least resistance. The volume barrier is the natural "skin" created by energy traveling through space which is spherical.

This can be described as follows: Energy is released at certain velocities greater than the speed of light filling a particular volume of space in the shape of a sphere described by that velocity of release. The energy adjusts between the volume of the sphere and the area of the surface of the sphere. As the velocity of release increases, the length of the radius between the source energy and the surface area of the sphere decreases. At certain velocities, the adjustment to the smaller surface area of the sphere between the volume of the sphere becomes increasingly more difficult to exceed. The smaller surface area of the sphere becomes a barrier in which the released energy is forced to slow itself in order to overcome the barrier or more precisely, the barrier of the lesser area of a sphere acts to slow the energy. However, as the velocity again increases, a velocity is again reached in which it is possible to overcome the difference between the volume of the sphere and the surface area of the sphere and the higher velocity energy is able to break the barrier. This creates a fluctuation in velocity between surface area barriers resembling discontinuous energy levels for the waves traversing space in concentric spheres i.e. instead of a system emitting energy in a linear fashion, energy is emitted as if it meets levels of resistance as it increases because in fact it does due to the difference between the volume of a sphere and the area of the surface of a sphere in space. Energy is therefore not allowed to increase to infinity due to its emission in concentric spheres whose volume is greater than its surface area.

Also, because the lower energy waves such as infrared compared to ultraviolet are actually moving more slowly, they spread out more before they hit the volume barrier. And there are less of them per second. However, the ultraviolet waves are actually traveling faster and cannot spread out very quickly before hitting the volume barrier. In a given time, the number of times that a wave hits the volume barrier and has to reorganize slows the wave down. All electromagnetic waves travel faster than the speed of light but it is the ratio of stuttering between reorganization at each volume change barrier that creates the same speed known as "c" for all electromagnetic waves. *The volume barrier is described by Planck's constant.*

Electromagnetic waves must continue infinitely in vacuo because they have no mass to form a loop back to the source (without mass there is no light loop), so they have no finite wave but weaken over time with distance because the photon is constantly reshaping itself into ever larger volumes of space. Light is transverse but each cycle increases in volume without losing velocity therefore without losing energy. The photon of one cycle enlarges efficiently, spreading in ever increasing spheres which are flattened by the volume barrier into convex oblate spheroids perpendicular to the source creating a cone from the source of the energy wave. The angle of the cone depends on the velocity of the prime photon. For higher energy waves, there is little time to spread out very far before the velocity of the wave again thrusts the prime photon forward into the next cycle. The energy of the wave dictates the size of the cone and the wavelength due to the pressure created by the space volume barrier i.e. Plank's barrier. In electromagnetic waves traveling in vacuo, every photon carries exactly the same amount of energy, only the volume and density of the photon changes and there are more photons per cycle. Energy can only be lost to mass and the mass (i.e. particle) energy wave. It is merely the number of photons per cycle per second that create the amount of energy. Using any length of time to get the energy of an electromagnetic wave we have **cycles/time * h = E** and velocity of the wave = c, the speed of light. The only variable is the cycle which describes the energy. Cycles are shorter i.e. smaller at higher energies and described by smaller wavelengths. Smaller is more energetic only if it is denser. Energy is denser at smaller wavelengths. Therefore, energy has volume. So what is Planck's constant? The volume barrier. It is described by the spherical shape of space. It keeps everything in the universe from traveling at speeds greater than "c" and is the explanation for special relativity.

Therefore, it is inaccurate to say that the light from distant stars would have reached us by now had the universe existed forever. Actually, we are seeing through the Hubble telescope light from galaxies that has in fact reached us, but is so weak due to the distance of the light that it would never appear any brighter no matter how much time passed. This is because the prime photon is spread over such a large sector of space by the time it reaches earth. After traveling the same distance from a single source, the prime photon of a more energetic wave is going to be smaller in volume than the photon of a lesser energetic wave, because it hits the space volume barrier, Planck's constant in a shorter period of time. Light is not "tired light" because energy is not lost hitting Planck's constant, but it is slowed to the speed of light. The space volume barrier is not a solid barrier, it is not tangible. It is created by the spherical shape of space causing a difference between increasing volume and spherical area with radius. It is caused by the fact that the energy must reorganize, the same way that water must reorganize when it comes out of the narrows and into a broader space except that the energy is from a spherical source in light. The energy from the energy wave created by the mass of subatomic particles is formed from a point at the source and travels along the x-axis in a straight line. If we imagine a point at the center of a sphere and lines intersecting that point that intersect every point on the surface of the sphere, even if there were infinite points being intersected on the surface of the sphere, a greater sphere encompassing the first sphere would have a larger set of infinite points on its surface and the lines intersecting the smaller sphere would not intersect every point on the larger sphere. (Although the points do not need to be infinite if there are actually finite cells in phase space, however, this shows that there can be different infinite sets if necessary.) These lines from the center of the smaller sphere represent the source x-axes of the waves and do not cover all points in space but the distance between lines widens the further from the source mass they are. Therefore, the energy which is constant in wavelength and frequency constantly meets larger volumes of space to fill between the ever-widening x-axes. Eventually nodal lines will form around each prime photon electromagnetic wave cone. Each wave emitted from the source sphere, hits the same amount of empty volume to fill at the same time. The ratio of times meeting the volume barrier according to the velocity of the wave keeps the speed of the wave constant. A more energetic gamma wave has more energy and travels faster than a radio wave but the volume barrier described by Planck's constant is hit more often creating a constant speed of c for all light.

The lessening of density means that the signal weakens through the near vacuum of space no matter if it is interfered with by mass or not. The energy remains constant but the volume changes. The speed remains constant due to inertia. No matter the speed of anything in the universe, the Planck barrier slows it to **c**. Energy dilates like mass at the Planck barrier as velocity increases the resistance of the Planck barrier. That creates the wave. This creates the oscillation of the wave. It also creates a particle-like photon stutter effect as the wave breaks through the Planck barrier at each cycle.

Energy could be modeled or envisioned as a set of energy particles or a liquid or a gas or in various other ways. But however one envisions energy, when the various parts of the energy or volume of energy are thrust forward with equal speed and momentum in a straight line and then hit Planck's constant in the first cycle, the energy becomes spherical and then flattened. As the energy approaches Planck's constant it slows. The smallest velocity occurs at Planck's constant however the overall speed averages to c according to how many times the wave hits Planck's constant in a particular time interval. After hitting Planck's constant it resumes its original speed due to not having lost any energy in vacuo. If a portion of a photon is absorbed, the rest of the photon is still traveling at light speed and will continue to do so. Therefore portions of a spread-out photon that arrive at earth from a star can be captured in a dish of a telescope. Since a photon has a specific amount of energy which is exactly h or Planck's constant and because it has a specific size and volume and density, then it is a particle. The energy is allocated to Planck's constant energy by the fact that faster moving energy is cut-off into smaller chunks or packets while slower moving energy is allowed to be carried in larger volume packets.

However, the small packets and the large packets carry the same amount of energy in different velocities, volumes and densities according to how Planck's barrier has divided up the energy. This is probably best illustrated with water waves. Water appears to blend into itself and merge and become homogenous and create fluid waves, but in fact water is tiny particles called atoms that maintain a separate particle existence and the molecules of water remain H2O despite its apparent fluidity. Therefore water is made up of distinct particles that only appear to flow into each other, but are distinct. Light is similar but light photons can actually flow into each other while still remaining distinct unlike water molecules which do not superimpose. In 1911, Planck showed that energy is emitted in discrete amounts but absorbed in any amount.[48] This leads to the conclusion that part of a photon may be absorbed.

If the wave interacts with another wave, gravitational field, magnetic or electric field, then portions of the prime photon are affected because the size of the prime photon is affected. If any portion of the prime photon exceeding *c* is caused to take a path less oblate than its natural volume, that portion of the prime photon hits the Planck barrier and the wave is distorted.

Diffraction occurs through matter interacting with the volume of the photon. Photons going through a single slit will be forced inward where photons collide with the mass at the sides of the slit. The energy wave of the atoms either results in an elastic or inelastic collision with the photon. Either way this causes the photon to spread through either loss of energy or reaction to force. This happens at both sides of a single slit. Double slit diffraction occurs because of the same compression of the photon when it is traveling in a coherent wave front of convex oblate spheroids. When compressed the photon will bounce back outward and enlarge inversely to the compression. Photons are not solid although they have volume. A portion of a photon may be captured or reflected. The reason science has had to deny this is because it is assumed that only a quantum (photon) of energy can be released from the atom at a discrete wavelength. This is because Einstein had shown that all light travels at the same velocity and h is inversely proportional to wavelength, so in order for light to have more energy between different forms of light, then the whole photon must be emitted or absorbed. However, if we view light as actually traveling at different velocities between the Planck barrier, then even a portion of a photon of an x-ray has more energy than a portion of a photon of a microwave. That light does not have to be absorbed or emitted in whole quanta in the atom has been discussed. Atoms only appear to absorb and emit whole quanta, because any amount of energy may be released or absorbed by the atomic or electron energy wave and only the excess energy will be emitted. It has already been proven through experiment that individual photons may appear to travel and tunnel faster than light.[63,64,65,66] The reason these experiments do not invalidate special relativity is that the resistance of the Planck barrier keeps the wave front at **c**. The reason for the faster than light effects that have been observed can be explained if one views wave energy as traveling faster than light between the Planck barrier.

However, photons travel in vacuo at a certain volume inversely proportionate to distance traveled. However, the angle of enlargement is very small and can only be seen through astronomical distances not necessarily in lasers on earth. Although the explanation of pulsars seems highly plausible, an alternate explanation would be that the high energy photons emitted have large gaps in the nodal lines after traveling through space. The photon having a volume still permits constructive and destructive interference.

Unlike electromagnetic waves, the nuclear energy wave is constantly connected to a source mass in a loop formation and is therefore finite. Electromagnetic waves are all created from the energy wave surrounding nuclei and particles and are all forms of the same excess energy of the wave. Stable subatomic particles do not emit excess energy, but keep a set ratio between their waves and their mass under the law of nuclear equilibrium. However, when decay happens, subatomic particles have less mass and expel electromagnetic waves. When electrons become unexcited, they release the energy from their energy wave as electromagnetic waves. In the atomic nuclear wave, the amount of excess energy is released from the atomic energy wave as electromagnetic waves of any energy amount that is in accordance with the law of nuclear equilibrium to keep the subatomic particles at the proper ratio of mass to energy. According to the amount of energy released, the electromagnetic wave forms. The length of the energy along the x-axis is dependent upon the amount of energy released. The velocity of the energy released describes the energy of the wave. The waves are released in pulses of excess energy. The wavelength depends on the velocity of the energy, because Planck's barrier is hit more often at higher velocities and prevents all waves from exceeding **c**. The volume of the wave is defined by its velocity which is always greater than **c** between **h** barriers. Otherwise, there is no wave. Otherwise, there is no particle.

What shows the energy of the wave is its wavelength alone. Intensity as measured by amplitude only shows how many of the same waves there are that are in phase or it shows how spread out the photon has become over time. Therefore, since frequency determines energy by the photoelectric effect according to Einstein, then wavelength alone determines energy then where **λ = Hz,** where **E= λ/sec*h**, then **λ/sec = E** and **h = E(λ),** therefore photons are all the same energy and have the energy of Planck's constant. The intensity shows either the volume of the photon or the constructive interference of photons or both, and the wavelength shows the speed of the photon between the Planck barrier and therefore describes the energy. The photoelectric effect proved that every wave of the same wavelength has the same energy, because every wave of the same frequency has the same wavelength.

Where v is the velocity of the wave, **v>c**, for every energy wave: **v/(h/t)=c** therefore **hv/c=E/c** and **E=hc/λ** because h is the volume barrier that reduces the overall speed but not the energy of the wave. The wavelength is inversely proportional to the energy because of the Planck barrier which describes the change in the volume of the sphere occurring more rapidly than the change in the surface of the sphere. The fact that a light wave changes velocity, which means it changes energy as it travels, does not make it a particle. It is still a wave. It is continuous. It has frequency. It simply has changing energy. So you have "h" Planck's constant as volume energy i.e. intensity energy and then you have velocity energy which is frequency energy or the force energy. A wave carries two kinds of energy. Planck's constant is the volume energy or the volume of work per area that can be done and that is constant for every single wavelength. The intensity energy gives the age of the wave i.e. the Doppler effect and the density of the wave. However, the velocity energy, which is described by the frequency, gives you the head-on energy of the wave or more precisely the wave's ability to do work with a certain forcefulness. Planck's constant multiplied by the frequency gives the velocity energy or work ability of the wave. The inverse square law governs the volume energy, therefore, Planck's constant over time gives the intensity. When the volume energy is less dense, the signal is weak and less amount of work can be done, however, the forcefulness of the work remains constant due to its velocity energy being unchanged.

Any energy change in a body that is described by the radius of the body in such a way that at equal radii in any direction, the energy change is the same implies what Newton thought i.e. that a force emanates from the center and loses energy in concentric spheres. Although Einstein eliminated the force of concentric spheres from gravity, he was unable to curve spacetime yet again for electromagnetism which is still assumed to be a force emanating in concentric spheres from its source. How to explain Einstein's variation will be later attempted. Niels Bohr explained the atom as obeying the inverse square law due to both Rydberg and Ritz working out the connection between the spectral series of hydrogen. The Lyman series, the inner series, of hydrogen is the inverse of the number 1. The Balmer, second series, is the inverse of the number 4. The Paschen series is the inverse of the number 9 i.e. there is an inverse square law that Bohr describes in this way:

"In the simplest case of the hydrogen spectrum, the terms are with great accuracy given by $T_n=R/n^2$, where n is an integer and R the Rydberg constant."—T referring to the transition energies of electrons from one shell to another.[90,p.37] Bohr attributes the Rydberg-Ritz combination principle entirely as the basis for his model of Rutherford's atom. He says, "A starting point was offered by the empirical regularities exhibited by the optical spectra of the elements, which, as first recognized by Rydberg, could be expressed by the combination principle, according to which the frequency of any spectral line was represented with extreme accuracy as the difference between two members of a set of terms characteristic of the element." [90,p.84] Bohr says Lord Rayleigh realized that if the spectral lines represented the emission during an electron orbit, the sequence should be quadratic, but that surprisingly it was linear which suggested to Bohr that it was the energy of the transitions. [90,p.37] This was Bohr's great achievement: he recognized from the work of others that the energy in the spectrum was not being emitted by the orbit, but was the difference between orbits. The fact that he interpreted other's work makes the achievement no less extraordinary as indeed Einstein, Newton, and even Aristotle did the same. It is seeing "the big picture" that counts in making the giant leaps in science. The interesting thing about this is that Heisenberg quotes Einstein as making the discovery that the transitions emit the light in the spectrum. Heisenberg quotes Einstein as saying: "As you know, I suggested that, when an atom drops suddenly from one stationary energy value to the next, it emits the energy difference as an energy packet, a so-called light quantum."[91,p.67] One wonders if this predates Einstein's 1917 paper.

Light travels at the same wavelength when viewed as head-on energy (not wavelength in the convention sense of cycle wavelength). The energy measured is head-on energy going directly from the source to the measuring instrument in a straight line. The width of the energy viewed in this way never changes. The velocity of the energy between wavelengths never changes unless absorbed. The width that is not measured is the parallel energy, or the increasing volume of the photon. The photon is shaped like so many contact lenses ever widening from the source. The source does not appear larger although the photon enlarges because it acts like a lens focusing the light back to the source. Because a photon is a particle that slows to a stop at the Planck barrier then slowly gains its initial speed, then slows down again toward the next Planck barrier, it appears as waves of energy when viewed head-on. The photon never loses it speed, therefore, it appears to be a point particle. But it is not a point particle as it loses its intensity over time. Therefore it is a spreading particle. It is shaped by the volume barrier which presents in spheres of differing wavelengths to the photon depending upon the speed of the photon. The velocity of the photon describes when the sphere volume, that is the Planck barrier, will be hit. The velocity of the photon describes both its volume and its density. The Planck barrier describes the shape of every photon which is a convex oblate spheroid or fat lens.

Bohr introduced Planck's constant into the atom without a wave. Yet, it is obvious that Planck's barrier cannot be overcome in nature by mass under special relativity, but only a wave of energy has sufficient velocity to overcome the Planck barrier. That is why nothing can travel faster than c under special relativity, because it hits Planck's volume spherical space barrier. Planck's constant, itself, is only a description of moving energy. Planck's constant is approximately a third of a billionth of an erg per second. It describes traveling energy. If Planck's constant can be used to describe the separation of the electrons from the nucleus, then there is traveling energy between the electrons and the nucleus. Therefore, since Planck's constant is successful in restricting atomic electron shells to discrete distances from the nucleus then there is a wave emanating from the nucleus toward the electrons.

Although the measurable energy drops to zero at the Planck barrier, the wave is merely flat, but still traveling at the speed of light. A completely flat wave would be absolutely dark energy and have no measurable energy, but could conceivably move at the speed of light. The only impossible wave is one that does not contain enough energy to make the volume change where the volume becomes greater than the surface of the sphere at the Planck barrier. That happens at ½ h where a partial quantum is trapped below the Planck barrier. It is no longer a traveling wave.

At some point traveling across the expanse of space the volume of a gamma ray will be the volume of a microwave. However, the volume and density tell us nothing about the measurable energy. The measurable energy is related to the frequency. The volume and density are related to the intensity. The intensity is related to the strength of the signal. The intensity is related to the distance the signal has traveled. The intensity is related to constructive interference. However, the frequency is related to the energy because at whatever intensity, whatever volume, and whatever density, the energy is traveling at the same speed between Planck barriers. Therefore a far-traveled gamma ray with a photon of the same volume as a fresh microwave still will have its initial velocity between Planck barriers under the law of inertia. Therefore, the energy it carries will be measured as a gamma ray even though the intensity is measured as weak.

However, unlikely any view is, it should be considered in order to rule it out if necessary. Whether an electromagnetic wave is time-varying electric and magnetic fields should be evaluated. As has been described, Planck's barrier creates the wave and what is called an electromagnetic wave is merely free excess energy released and absorbed by matter. Since all particles have an energy wave surrounding them and that wave is identical in that it is energy, then the charge of the energy wave becomes saturated by the charge of the particle and takes on its charge. However, emission of excess energy from the wave does not cause the particle and its wave to lose charge. Therefore, what is termed electromagnetic waves is merely a form of pure energy that is free and excess energy not needed under the law of nuclear equilibrium between a mass and its energy wave. Therefore it does not need to be electromagnetic in nature. It forms waves according to its velocity between Planck barriers but does not have charge. It remains at a constant speed due to inertia. Therefore it need not be electromagnetic. It need only be energy in a wave form. The only seeming empirical evidence that the electromagnetic wave may be truly an electromagnetic wave is the fact that a gamma ray can split its negative and positive electric waves to form in pair production the mass of a positron and an electron which can in turn become gamma rays. This process should probably not be called annihilation because at the atomic level the transition from mass to energy appears to be a natural ongoing process since each particle is constantly changing mass to energy around itself in its light energy wave. This would make it appear that a gamma ray has both a positive and negative electric field which would mean the energy wave of the atom is also an electromagnetic field. A possible model of the creation of mass from light could be that due to the velocity of a highly energetic gamma ray between the Planck barrier being so great that it closes the gap too tightly and the wavelength nears zero so that the Planck barrier squeezes the energy into mass thus slowing it down below light speed in the process. There is however another strong argument in favor of the particle energy wave being electromagnetic and that is the Zeeman effect and the Stark effect. However, this may have to do with the saturation of the energy wave with the charge of the particle rather than charge being intrinsic to the energy wave. It is a difficult concept, but the energy wave must be the same substance, i.e. energy, for all particles, the difference must be only related to the charge.

Einstein's rendition of Planck's light quanta hadn't destroyed Maxwell's wave, but rather clarified it. It can be compared to the blips on a heart monitor of someone in a hospital room. The blips sound like distinct points or particles, but just because the blood is pumped through the heart in bursts of energy doesn't make the blood less of a fluid. One sees quantized blips on the heart monitor yet there is a continuous flow. One sees photon particles of light yet there is a continuous light wave.

Particles are particles and waves are waves. Energy is a wave. Particles have particles and waves, but they are separate things. There is no duality—there is no particle measured sometimes and a wave measured another. There is only wholeness of a particle with its wave. Each separate and working together.

For electromagnetic waves, the electron is the oscillator. If an electron oscillated once, it would send a wave the same way that a rope tied to a pole when oscillated sends a wave down the rope. When the wave hits the end of the rope, there is a pulse of energy equal to the oscillation. This does not change the wave into a particle even though the photon pulse ran down the wave. It behaves like a particle in that it can do work, but the photon is an energy pulse sent by each oscillation of the electron down the wave. Einstein's $1/2mv^2=hv-w$ means that the wave-pulse, the photon, does "work" in the mechanical sense. It doesn't mean the photon is a particle.xl It means the hump under the cycle that moves down the rope contains the energy that pulses down the rope. A tidal wave hitting the beach has energy without being a particle. We know this. A wave can do work without being a particle. Energy waves have no structure. Things can't be built with waves or energy. Particles have structure. Things can be built with particles. Energy traveling in pulses at the speed of light has no structure.

Models are important in science to understand the underlying reality. If it is useful to describe the photon as a particle, then it is a good model. But the photon does not need to be a particle to do work. Energy is a wave and mass is a particle. Einstein used the photon as a particle in a heuristic approach. This was and still is a useful shortcut, but it doesn't make structure out of energy. As Einstein proposed, it helps to more easily define the light wave if "the energy of a light ray spreading out from a point source is not continuously distributed over an increasing space but consists of a finite number of energy quanta which are localized at points in space,..." xli This is nothing more than Max Planck's formula did to the electromagnetic wave. From the time that Max Planck calculated the ratio of the pulse under each cycle, this created discrete energy pulses in the wave. A photon is the energy in one electron cycle of oscillation and it carries the energy of the electron oscillation, the same way that oscillating a rope carries energy under the wave. This electromagnetic radiation retains the energy and frequency of the oscillation of the electron that created it. This may be measured as a particle, but it is not a particle. A full quanta electromagnetic wave does not have structure. A wave can only have the structure of a wave. Energy is a wave that can do work like a particle. A particle and a wave are distinct and have distinct properties. Waves ripple and create interference. Particles do not. It is only the wave portion of the particle that appears to create wave behavior. The mass portion of the particle behaves like matter and may have structure built on the structure of the particle.

. At the point where the electric field changes to the magnetic field after one cycle, the energy of the wave drops to zero. This creates the pulse and the photon particle-like behavior in full quanta radiation. In a standing partial quanta wave, the wave has nodes and antinodes.

20 Electromagnetism

Magnetism is described quite differently under this new theory of an energy wave surrounding the nucleus of each atom. Using a bar magnet, one feels the energy wave of the atoms since the electrons and protons polarize, their individual waves separate as the electrons are pulled away from the nucleus in a magnet. The electrons do not separate fully from the atomic nucleus, but are merely pulled further apart than normal causing a distinction to appear between the two waves because their energy waves no longer overlap as much. Therefore at one end of a bar magnet, the electron energy waves are further separated from the nucleus than normal with the electron waves extended toward that pole. On the other end of the bar magnet, the positive nuclear energy waves are polarized away from the electron waves and the positive nuclear energy wave extends toward that pole. In the middle of a magnet the fields are neutral. In a spherical magnet the fields are neutral toward the magnetic equator and pull apart from each other at the poles. If the magnet is cut in two, the separation at the poles of the magnet still remains, it therefore pulls apart the neutral atoms that used to make up the center of the magnet. The charge at the poles is dependent upon the separation of the negative electron fields from the positive nuclear wave.

This effect is described to some degree by science as being the alignment of electrons in the same relative directions toward one pole, but the cause is not described. The cause is the energy waves of the particles being pulled apart from each other creating partially separated waves. The energy waves of protons and neutrons in an atom are only neutral where they overlap. Naturally with the electron wave always being separated to some extent from superimposing exactly on the nuclear wave, every atom is potentially a magnet. However, in neutral elements the electron waves and nuclear waves of each atom are constantly overlapping at their extremities with other atomic waves. Therefore most elements are naturally kept neutral. Because there is a partial overlapping in neutral atoms, every macro world object has some magnetism although most objects are essentially neutral where they have equal numbers of protons and neutrons in various alignments of atoms and molecules. However, aligning the negative electron waves in opposing orientation to the nuclear proton waves causes an exposure of part of the waves at opposite poles. This exposure of the waves increases with successive layering of oriented atoms, with an increase in distance between the electron waves and the nuclear waves, the farther they are from the axis of the magnet. There isn't really any interference at the center of a magnet, as science now assumes, but rather the atoms are neutral at the center and radius of the electrons from the nucleus becomes larger as the distance from the axis of the magnet increases. That is why the magnetic field merely curves past the neutral axis of a magnet rather than creating zigzagging interference lines around the axis.

Therefore, when one pushes two like poles together in a bar magnet, one actually feels the particle energy waves of either the electrons or the protons. Pushing together two unlike poles pulls the polarized proton waves toward the polarized electron waves. However, science and humans have imagined through the centuries that unlike poles attract completely which certainly seems to be the case. However, this is untrue because of the energy wave. When unlike ends of magnets hold each other in a bond, they do not touch completely. Rather one has basically a chemical reaction similar to an ionic bond. The electron energy waves that are distanced from the proton energy waves when the waves reach a strength and distance close enough to each other, their particles speed toward each other as they invade each others waves as if forming a new atom from the two unlike poles of a bar magnet. The minimum distance between the magnets is the atomic radius. This is not visible to human sight or human microscopic technology because we cannot see the atom. The electron waves do not penetrate the nuclear proton waves completely. There is always a gap because of the energy wave density near the particle using the photoelectric effect. With the energy wave in place about the atom, the orientation of all the electrons in the atom in polar opposition to the nucleus would not change the permeability of the atom to other atoms.

So Niels Bohr, Rutherford, Einstein, Sommerfeld, Heisenberg, Schrödinger, Coulomb, and all the great physicists thought that unlike particles attract completely without any gap which started the whole unlikely saga of the quantum atom. Without knowing of the photoelectric effect of particles from their energy waves, it was assumed an electron would fall into the nucleus of the atom. The complete attractiveness of south and north poles of a magnet seems extremely logical and apparent from what is seen in the macro world. But even in the current model of the quantum atom, one must assume that the poles never invade the Bohr radius. However, had it been known that the atomic radius was never violated by unlike magnetic poles due to the nuclear light waves, then the Bohr radius would never have been postulated in the first place as a discrete law-defying radius created despite empty space in the atom. If humans had realized there has to be a gap between unlike charges unless those charges attract each other at enough velocity to penetrate each other's energy waves, then there would be no quantum mechanics as it exists today. But because of the apparent macro-world complete attraction of opposite charges, it became necessary to invent an imaginary gap in the atom between electrons and protons using Planck's constant. This was a lucky guess by Lorentz, Haas, and Bohr. It was lucky because there is a wave in the atom which obeys Planck's constant, so the experiments seemed to confirm their theory that Planck's constant should be used in the atom due to it being the proper "length". It was unlucky because the current model is very restrictive, unlike nature, which quantum mechanical formula keeps science from progressing because it has a formula that appears to work. It is restrictive despite probability distribution because the formula says there is zero probability that the electron would be located at the nucleus, yet the electron frequently collides with the nucleus in scattering experiments and natural high-energy collisions. Therefore the current theory is rigid in that it creates rules for the atom which must constantly be broken i.e. the electron must be in a standing wave, but it travels in suborbitals that are not absolutely spherical and in which two electron standing waves may inhabit but do not have interference. The rules are so limiting in order to describe one phenomenon of the atom, such as the fact that electrons do not fall into the nucleus, and then because the rules are so restrictive, they constantly need to be broken to account for other phenomena. Nature does not behave so.

We cannot tell from the current model of the atom why elements such as iron would be more susceptible to magnetic effects. This is because the grouping of the electrons in the atom was done from an arbitrary arranging in accordance with the periodic table. The arrangement has many merits, but does not describe in detail any exact chemical properties of any elements except perhaps the tendency of certain elements to be less stable and the tendency for certain elements to ionize. But the current quantum model of the atom does not explain tendency toward magnetism. Perhaps in these elements, most electrons are actually physically located on one hemisphere of the atom naturally. Under this new model it is no longer necessary to have electrons equally spaced about the atom to account for solidity as the nuclear energy wave accounts for solidity.

Science explains a magnet as the alignment of electrons and protons, however, this violates quantum mechanics as it positions electrons in the allowed orbitals in order to distribute them on the same hemisphere of the atom. There are 26 electrons in neutral iron. With two full p-orbitals with six electrons each, there is no way to push the electrons to one side of the orbital without violating the exclusion principle unless the electrons are all in highly excited states where only two electrons at one time in any p-orbital or d-orbital or perhaps even f-orbital with the standing wave of the electron cut-off to one hemisphere of the atom only. Even with trying to explain that an unpaired electron in each atom of iron is in the same orientation as in other atoms of the metal does not account for the standing wave aspect of the electron nor the probability cloud, but gives the electron a position violating uncertainty. And then what would keep the iron solid since there would be a definite failure of electrostatic repulsion in one part of the probability cloud of the atom? This possibility is not allowed for in quantum mechanics which only allows a relatively even distribution of electrons around the atom. Solutions to the Schroedinger equation (psi squared) and Heisenberg matrix mechanics describe either a wave surrounding the atom or points in a probability cloud. A standing wave does not have a position in one hemisphere of an atom. And it hardly seems likely that magnetism can be accounted for with electrons surrounding the entire atom.

Science explains magnetism by reverting to the Bohr model of the atom. "The net magnetic moment of an atom is the vector sum of its orbital and spin magnetic moments."[75] "From the classical expression for magnetic moment, mu = IA, an expression for the magnetic moment from an electron in a circular orbit around a nucleus can be deduced."[2] This is convenient but has nothing to do with quantum mechanics. As Werner Heisenberg explains matrix mechanics which is the first modern formulation of quantum mechanics, "After my return to Göttingen, I showed the paper to Born, who found it interesting but somewhat disconcerting, inasmuch as the concept of electron pathways was totally eliminated."[74,p.46] Heisenberg had totally eliminated electron orbits from the atom as being non-empirical. Instead he stuck to anharmonic oscillation for transitions of electrons without orbital pathways for electrons.

Although in view of his principle of Complementarity, Bohr ascribes orbital pictures to "general asymptotic correspondence" with the macro world, Bohr describes matrix mechanics thus:

"Shortly afterwards, an advance of fundamental significance was achieved by Heisenberg, who in 1925 introduced a most ingenious formalism, in which all use of orbital pictures…was avoided…. In fact, by representing the mechanical quantities by hermitian matrices with elements referring to all possible transition processes between stationary states, it proved possible without arbitrariness to deduce the energies of these states and the probabilities of the associated transition processes."[90,p.55]

Obviously, Bohr is saying the only anharmonic oscillation is described by the transitions and matrix mechanics avoided "orbital pictures" which he calls "ingenious".

There is no such thing as an electron orbit in quantum mechanics, so magnetism cannot be explained as a magnetic moment arising from the orbit of an electron around the nucleus without contradicting quantum mechanics. One might question whether the Schrödinger wave mechanics has an electron in a periodic orbit. The answer is no. According to Heisenberg, "The concept of an electron pathway was lacking in Schrödinger, just as in the Göttingen quantum mechanics..."[74,p50] Also, the explanation above of using the "spin magnetic moment" described as the magnetism arising from intrinsic electron rotation is completely debunked from current understanding. As one physics textbook states, "This picture of the electron as spinning is, however, wholly discredited today. We cannot even view an electron as a localized object, much less a spinning one."[76] The truth is quantum mechanics gives no satisfactory explanation of magnetism at the atomic level.

The placement of electrons in suborbitals by physicists happened quickly over a few short years evidently with little thought to magnetic effects. How electrons could be aligned in polarization to the protons cannot be explained by solutions to quantum mechanical formula, but nature doesn't follow quantum mechanical rules.

The magnetic moment of the electron arises in this new model of the atom not from the intrinsic spin of the electron. When Pauli heard about the idea of the electron having intrinsic spin, he criticized it severely, noting that the electron's surface would have to be moving faster than the speed of light in order for it to rotate quickly enough to produce the magnetic moment associated with it.[92] Pauli told Kronig, "it is indeed very clever but of course has nothing to do with reality".[93] This new model of the atom shows an energy wave about each particle that carries the charge of the particle and is in constant motion creating magnetic moment.

21 Electromagnetic Fields

It is a given under this theory that energy, without mass to create a loop, moves under inertia at a constant speed and cannot be stopped or slowed less than c unless absorbed.

The electromagnetic field is different than an electromagnetic wave although they both appear to drop off in intensity. The electromagnetic wave in vacuo is not actually losing energy. The energy carried by the wave is rather becoming less dense and increasing in volume as it travels in the electromagnetic wave which is infinite. A field is caused by mass. In a field, the energy drops off proportionally to the distance squared. The energy of the wave field surrounding a particle drops off in an inverse square law, because all energy waves emanating from a sphere fill larger and larger spheres whose surfaces are proportional to the square of the radius. Therefore, the energy wave field between two spherical particles is proportional to the inverse of the distance squared as well.

In a bar magnet, the electrons align and emit a loop of energy. As this loop of energy hits the Planck barrier it reorganizes and then proceeds into a larger volume. This is clearly seen in the field around the magnet. Since the field around a bar magnet is finite it does not lose energy. Every line of force extends back to the mass. It is created from the mass of the protons and electrons as a unit and is therefore created from mass and is therefore finite. Magnetic fields do not cause the body to lose mass. Radiation in the form of infinite electromagnetic waves causes the body to lose mass. Electromagnetic waves are on the contrary massless and therefore never loop back, but are infinite.

In the case of fields then, it is not the velocity or frequency that determines the energy, it is only the density or intensity that determines the energy. This has to be the case because the velocity of the field loops back on itself. In the case of particle energy waves, the density near the particle determines the energy of the field near the particle with each progressive spherical volume increase with radial distance increase the energy drops off inversely to the distance squared. Therefore, the electromagnetic field and the particle energy wave field follow this formula for their energy, but the infinite electromagnetic wave never loses its energy except by matter absorption. Due to the loop configuration of the particle light wave, the energy of each photon or cycle is different per cycle from origin to the point where the wave returns to the source particle. This could mean that at some point the wavelength of the finite wave becomes nearly energyless and extends far beyond the atom without creating any interference or measurable energy.

22 Electricity, Classic Electrodynamics, and the Atom

Using this new theory of the atom, there is no difference between classical electrodynamics and the atom. Classical electrodynamics is the measure of the perceived energy of the electron through spectroscopy. The spectrum of each element appears to have quantized light emissions. This perceived energy is merely excess energy greater than the measure of energy needed in the electron's energy wave to mass ratio that is released as electromagnetic waves. The electron or any other particle will not absorb excess energy unless forced to do so. And the energy will not be released into any system unless that system has less excess energy or that system is forced to absorb the energy. The reason the second law of thermodynamics holds is because of the law of nuclear equilibrium. A particle must retain its mass to energy wave ratio. Therefore in a colder balanced system i.e. a system where the particles are at nuclear equilibrium with the correct mass to energy ratio, no heat will be released into a system with excess energy i.e. a hotter system. All excess energy retained by particles whether electrons or nucleons will tend to a state of balance of nuclear equilibrium. The particles will tend to give off excess energy into a lesser energy system.

Under the law of nuclear equilibrium, the system of the energy wave around each particle and the mass of the particle tend to maintain equilibrium. Therefore, the system of the nuclear waves of atoms tend to maintain equilibrium. This means that the energy waves in atoms are exposed constantly to cosmic rays and radiation on earth, however, sometimes the atom reflects this excess free radiation and sometimes the atom absorbs this radiation if the system needs the energy or if the atom is forced to absorb the energy by the magnitude of the energy not allowing it to be reflected or in other cases the magnitude of the energy is low enough for it to pass through the particle waves without disturbing the system. The atom will absorb radiation from the environment (outside of the system) when it is put into a system that requires such absorption to maintain equilibrium. Therefore in crystallography at times the atoms are forced to absorb the energy and at times the energy is low enough that the atoms can reflect the energy or allow its passage through the energy waves.

An example of forced absorption to provide equilibrium is the case of heating an element to incandescence in order to observe its spectrum. The element is forced to absorb and emit the heat energy, which is a form of electromagnetic energy, in order to equalize the system. By absorbing the heat, the atom can move the electrons to higher energy levels and utilize the heat, thereby equalizing the system. When the heat is reduced, the system is equalized by emission of the excess energy.

Another example of a system tending to maintain nuclear equilibrium is that in a chemical cell. Atoms of metals and salt solutions that form a battery induce imbalances in the system that require the absorption of radiation from the environment (outside of the system) by the atoms in the battery in order to induce ionization energies to bring the entire battery system to equilibrium which happens when the battery dies. The absorbed radiation from the environment used by the atom to produce ionization to equalize the battery system in the form of excess ionization energy can be drained off by the battery system as voltage. This implies that in a complete void where there is no matter and no electromagnetic radiation, a battery will not function. As long as there is even UV light outside the battery circuit that it can absorb to transition the electrons, a battery will produce energy, otherwise, it cannot.

The electron when bound in an atom is bound by the energy wave to its electron shell and cannot move through the energy wave outward without gaining energy. It cannot move to another atom without gaining energy in excess of ionization energy. This is true even if the atom is an ion in a battery. This is because although there is no longer an electrostatic attraction to the nucleus for an electron in an anion, the electron still must break through the energy wave to exit the atom.

Electricity and magnetism were first described by Michael Faraday in his experiments. It was James Clerk Maxwell who then united four theories of magnetism and electricity to give a picture of a field produced by the shifts from the electric to the magnetic wave that Faraday observed empirically. This led Maxwell to determine that the field shifting from electric to magnetic waves traveled at the speed of light. From this it was deduced that light was an electromagnet wave. But what has not been said and should be said is that Maxwell was calculating what electricity was in view of the fact that magnetism could cause electricity and electricity could cause magnetism and that you can't have one without the other. Therefore, the point that the textbooks omit to state is the obvious. Electricity is a form of electromagnetism. Electricity travels at the speed of light (or it would in a vacuum). However, most of the time, electricity is carried in a conductor or through a solid and this slows down the speed of electricity to slightly less than the speed of light, but when electricity jumps across from a cathode to an anode in a vacuum, it does so at the speed of light. This does not mean the electrons travel at the speed of light, but the electricity does.

Electrical current stems from absorbed electromagnetic radiation that transitions electrons in the system. Once the electrons transition, however, the electrical current does not behave as does normal light. The electrical current is a type of unbalanced light in which a portion of the light has been utilized by the system. Light can split into oppositely charged particles as in electron-positron pairs. It may be that light can split into opposite charges without creating particles. However, it may happen in an electrical circuit, what we see is an imbalanced form of light being unable to leave the system and searching to neutralize itself. Just as the energy in the energy wave field of a particle is light that loops back to the particle and carries a single charge. Electrical current is singly charged light. Just as light may separate its charge into two oppositely charged particles that carry an energy wave field of a single charge, so too light can separate its charges while still remaining a form of light. Electrical current is light of a single charge created by transitions of electrons that utilize a single charge and discharge its opposite.

All electrical current generated by electrons is excess energy emitted as electromagnetic waves. Electromagnetic waves are neutral so they do not provide the magnetism of the current. Voltage is electromagnetic energy. The waves instead of dispersing outward into space in all directions in natural free spheres from the source stay in the system that utilizes them. The fact that light can travel without dispersing is seen in a laser beam. What can be done in a laser can be done in a wire. In other words, even if the mechanism is different than laser technology, the idea is that light can stay within a system. A system for generating electricity maintains a constant atomic imbalance that forces constant excess absorption from the environment outside of the electrical circuit into the system. However, due to the constant imbalance in the system, the excess energy in the form of electromagnetic waves is never released from the system back out into the environment but is carried in a current along the system usually along metal wires, such as copper wires, that carry the systems imbalance. Ionization energy is being absorbed from the outside environment along the entire wire all at once which is why a magnetic field is created all along an electrical wire that has a current i.e. electromagnetic wave running through it. However, the electrical current is actual electromagnetic waves that travel through the system at the speed of light being slowed by not being in vacuo. The electromagnetic waves do not travel in their normal spherical release in all directions outside of the wire because they are trapped inside an unbalanced system so that they stay in the imbalanced system. Light only radiates in quanta, but the energy from magnetically pulling electrons away from the nucleus, creates too much energy in the nuclear wave field that needs to be released in the quickest fashion possible to maintain nuclear equilibrium. In a solid, the quickest fashion is through physical contact of the nuclear wave fields.

The fact that the energy of an electrical current must come from the environment is due to the law of nuclear equilibrium. If moving electrons actually released their own inherent minimum energy, their mass would disintegrate. This is not allowed. When a system needs energy due to imbalance, the electrons absorb the energy from environmental free radiation instead. Fortunately, the universe is filled with light. Where ionization is occurring along a wire, creating electrical energy, the electron energy wave fields are being separated from the nucleon energy wave fields as the electron absorbs energy and this creates the magnetic field along the wire because the charges are being pulled apart as in a magnet. In a cathode ray tube, the electrons can speed from cathode to anode, but this is not true in a copper wire. There are two system imbalances that are created in the electrical circuit. There is the primary imbalance created by human intervention, then this itself creates ionization along the entire wire which sets up a secondary system of imbalance because the ionized free electrons are drawn to all the cations that are created in the system as well as to the imbalance created by humans at the ends of the wires. Therefore, the electrons ionize due to the primary imbalance by gaining free electromagnetic radiation from the environment, then continually are drawn back into the atom to nearby copper cations instead of being able to pass through the system as would happen in a cathode ray tube. As soon as the electron is attracted to the copper cation caused by the secondary imbalance it releases its gained energy, however, once neutral, it again attempts to balance the primary system by gaining more energy, ionizing away from the atom once more in an attempt to fix the primary imbalance. There is a constant release of excess ionization energy into the system along the wire while the copper atoms remain in the system of the primary imbalance. This means there are constantly cations and free ionized electrons in a copper wire in a closed electrical circuit. These cations and free electrons are positively and negatively charged energy wave fields that create what are essentially small magnets when they are pulled apart so that they align and cause magnetism along the entire wire according to the right hand rule.

When the circuit is broken and the loop is no longer closed, the primary imbalance in the system no longer exists and the absorption of energy from the environment of free electromagnetic wave radiation into the system stops. The free ionized electrons in the wire all return to ground state in the copper atoms releasing energy at the transitions and this causes an electrical surge as the ionization energy is emitted all at once when the circuit breaks.

The same is true for a cathode and anode connected to a vacuum cathode ray tube. Electrons receiving voltage which is excess electromagnetic energy, break free from atoms by absorbing this electromagnetic radiation. The electron is free to travel at high speed through the vacuum tube. This is because in a vacuum, the electron does not encounter as many other atoms in which it may expel its excess energy. Not every atom will accept excess energy from the electron. Under the law of nuclear equilibrium, a system can be forced to absorb excess energy such as in a system that is either forced by high velocity to absorb excess energy, or is unbalanced by charge differences, or is deficient in energy, it automatically seeks to absorb the excess energy either from free radiation or from nearby systems with excess energy. Therefore, any accelerating electron is carrying excess absorbed energy and while it is accelerating, it carries the energy. Electrons tend to be constant emitters of excess energy and therefore to continue acceleration they emit and gain energy continuously from the environmental free radiation. When the electron hits mass, it may release that energy to the mass. When it does so, it loses acceleration, but must regain it because it is in an unbalanced system. Free electrons seek positive ions to balance the system. Electrons are drawn to ground state. Therefore, the electron will absorb energy during its passage toward positive ions. If it is emitting excess energy and still accelerating, then it is also absorbing energy so that it keeps its mass to energy field ratio. There is gamma radiation even in a vacuum from which it can absorb energy if necessary to balance the system. All electromagnetic waves emitted by the electrons traveling through the tube are excess ionization energy that has been absorbed and in every case the electron keeps its mass and its energy wave field ratio under the law of nuclear equilibrium. In other words, the electron behaves no differently in the atom than in the macro world of electrodynamics.

In the system of the human body, when we run, i.e. accelerate, we burn energy from within our system and have to restore that energy by eating. However, an atom, a particle, and an electron are under the law of nuclear equilibrium. They must feed constantly from the environment to keep accelerating especially if they are emitting while they are accelerating. Particles do not have an internal store of energy such as does the human body once they get down to their nuclear equilibrium limit. If electrons emitted their intrinsic energy while passing through a cathode ray tube, they would disintegrate. Science knows this but does not explain it thoroughly. Here is the explanation. The energy is constantly being drawn from the free electromagnetic waves in the environment while the electron is emitting. The electron absorbs electromagnetic energy and reemits it in order to keep moving along the cathode ray tube. The rate of absorption and emission depends upon the imbalance in the system between cathode and anode caused by the voltage which creates a system in which the electron returns to equilibrium in the system going from the part of the system which has too many electrons to the part of the system that has too few. The voltage creates the magnitude of imbalance and the electrons compensate by moving across the system to balance it. In order to do so, the electron may not use internal energy that would reduce its mass and destroy it. It therefore must absorb from free energy in the environment. By free energy, one is referring to the infinite electromagnetic waves that have been released by other atoms in the universe as excess energy.

To be more precise: an electron absorbs energy and reemits in the direction of absorption. In a system where the electron must travel from point A to point B due to the imbalance of the Coulomb force in the system, then the electron absorbs random radiation. The electron is then able to accelerate, but not in exactly the straightest path from A to B since it absorbs from any direction. Rather it is the combination of the imbalance in the Coulomb force creating the necessity to travel from A to B that causes a force that brings the group of electrons to eventually pass from A to B. However, the path is not a single straight thin beam of electrons which would be the case if the electrons knowing they must pass from A to B only absorbed energy from the direction of A in order to get to B as humans do when they send up a rocket. Humans emit the energy in the rocket in exactly the correct opposite direction to aim at the correct goal. Electrons emit and absorb from all angles in the cathode ray tube, but the Coulomb force of the imbalance of the system keeps them in the correct path.

In 1831, Michael Faraday demonstrated how a copper disc was able to provide a constant flow of electricity when rotated between the poles of a horseshoe magnet, although he used an electromagnetic in other experiments.[70] This can be explained under this new model of the atom. When a copper disc is put into a magnetic field, there are immediate unseen effects whether the disc is revolving or not due to an unbalance in the system i.e. a difference is created between an atomic system in a neutral field and an atomic system in a magnetic field. The electrons of the metal are affected by the magnetic field and therefore attempt equilibrium by absorbing energy to shift position away from the negative pole of the magnetic field. This absorption is not observed by humans. However, when the copper disc is made to revolve, at a revolution of 180 degrees, the electrons emit the absorbed radiation that is no longer needed and absorb more radiation to position themselves according to the new magnetic state. At another 180 degrees, the electrons again emit this radiation and absorb radiation. This continues as long as the copper is made to revolve. This is actually an oversimplification. At every degree and fraction of a degree that a copper disc is turned in a magnetic field, all the electrons in the copper are forced to move and realign in the field. No electron may move without absorbing and emitting energy since there are no vast energy stores in the electron. What science does not seem to realize is that when an electron must move, it must transition. To transition, it must absorb energy. The Coulomb force may appear just to be able to move an electron without the electron doing any work so that it might appear as though the electron neither has to gain nor to lose energy, but this is not true. The electron is sitting in a nuclear energy field in an atom of copper so that in order to move the electron needs energy to pass through the antinodes. Using a magnet to rearrange the electron in the atoms of copper causes them to absorb energy and transition and then to reemit that energy as the magnetic field varies.

In a wire in which the atoms are forced to ionize by the imbalance of the system thus creating electrical current, the electron energy wave fields are separated from the nuclear energy wave fields and are not aligned in their separation in any particular direction. Thus the magnetic field created around the wire is cylindrical due to the energy wave field of each particle carrying its own charge and the separation of the fields causes magnetism. The fields are not separated as in a bar magnet, but rather in any orientation according to the orientation of the atoms in the wire. Therefore, the magnetic field is oriented around the wire in all directions. The strength of the magnetic field depends upon the ionization energies that have pulled the electron energy fields apart from the nucleon energy fields.

The energy is absorbed from the background radiation that exists throughout the universe in the form of infinite electromagnetic waves. The energy emitted is in the form of an electrical current which is itself originally excess energy from another atom in the universe. At no time does the electron or the atom lose its intrinsic mass or energy as this is conserved in their energy fields. Electrical current is absorption of universal free energy and re-emission in a specific system instead of outward in all directions due to the law of nuclear equilibrium tending to keep a system in balance. The reason certain materials such as copper are more susceptible to a magnetic field and are more likely to absorb and emit excess energy is unexplained in quantum mechanics. There is much hidden atomic substructure yet to be discovered.

Let us consider the point of electricity and magnetism further since it is actually a difficult concept. In one physics textbook it is explained:

"In his attempt to produce an electric current from a magnetic field, Faraday used the apparatus shown in Fig. 21-1. A coil of wire, X, was connected to a battery. The current that flowed through X produced a magnetic field that was intensified by the iron core. Faraday hoped that by using a strong enough battery, a steady current in X would produce a great enough magnetic field to produce a current in a second coil Y. This second circuit, Y, contained a galvanometer to detect any current but contained no battery. He met no success with steady currents. But the long-sought effect was finally observed when Faraday saw the galvanometer in circuit Y deflect strongly at the moment he closed the switch in circuit X. And the galvanometer deflected strongly in the opposite direction when he opened the switch. A steady current in X had produced *no* current in Y. Only when the current in X was starting or stopping was a current produced in Y. Faraday concluded that although a steady magnetic field produces no current, a *changing* magnetic field can produce an electric current! Such a current is called an induced current. When the magnetic field through coil Y changes, a current flows as if there were a source of emf in the circuit. We therefore say that an **induced emf is produced by a changing magnetic field**."[76,p.590] [Bold and italics in original.]

This is a lovely recounting of the experiment, but it gives no explanation. Science understands electron transitions, yet fails to use the obvious explanation in the textbooks. Here what Faraday was seeing is that the electrical current, caused by the imbalance in the battery, caused the atoms in the wire connected to the battery to have electron ionization which causes a charge imbalance i.e. a magnetic field. A magnetic field is an atomic charge imbalance. You have two kinds of atoms: neutral and charge imbalanced i.e. ions. Neutral atoms cause a neutral magnetic field i.e. undetectable. Charge imbalance i.e. separation of the energy wave fields carrying the charge of the electrons from the energy wave fields carrying the charge of the protons is a magnetic field i.e. a charge imbalanced field. When there is electron-proton energy field separation, there always is a charge realignment i.e. always two poles. So the electrical current in the battery wire ionizes the electrons away from the protons in the wire, i.e. the electrons transition, and thus causes a charge imbalance and then the electrons and protons align according to the system which is a cylinder magnet which means the electrons and protons separate their energy fields at a distance forming small magnets that form a circle, end-to-end, around the circumference of the wire which alignment naturally continues in all the atoms to the core.

However, the cylinder magnet has a definite direction, so one imagines that the field is rather analogous to a bar magnet fixed at the center of the wire and spinning like a bar magnet pinwheel.

In an electrical magnet, there is not a stable, unchanging position of all the electrons as in an iron bar magnet. Rather there is a constant routine of: gain energy, ionize, go to ground state of nearby cation, lose energy, repeat. Therefore, the magnet of an electrical current caused by a wire attached to a battery is in constant electron motion. This creates the right hand rule. Now back to Faraday's second wire designated Y above carrying no electrical current at a distance from the wire X connected to the battery, but wire Y is in wire X's magnetic field. The second wire Y realigns its electrons (by transition) in accordance with the magnetic field the moment its electrons feel the effects of the magnetic field which is the charge imbalanced field that the wire has been placed in. This realignment of electrons by transition, i.e. absorbing and emitting electromagnetic energy, need only happen and remain in effect while the field is active i.e. while the first wire X is creating a magnetic, charge-imbalanced field. The electrons realign in wire Y by ionization or at the very least by transitions. Transitions take energy. Where does the energy come from? The universal excess background radiation. The electrons in wire Y absorb energy from the environment and realign for the magnetic imbalanced charge field produced by wire X. Some transitions are the gaining of energy and movement of electrons away from the nucleus while this sets up the need for other electrons to lose energy and move to ground state depending upon their positions in the atom. So you get an electrical surge in the galvanometer due to the necessity of electron transitions for the atoms in wire Y trying to realign. Once they realign, there are no further electron transitions that need to take place as the system has become balanced in wire Y as all electrons are in position in balance with the charge field of wire X. So, no more electrical current in wire Y. No more transitions. No more absorbing and emitting electromagnetic free radiation. But turn off the current in wire X and you have shut down the magnetic charge field imbalance of wire X and wire Y is now in a neutral field again. Wire Y's electrons must rearrange for the neutral field so back the higher energy electrons go to ground state and the electrons that transitioned to lower states transition back to higher states so that the atom returns to its previous configuration for a neutral charge field. Whether the electrons in wire Y transition to higher or lower energy levels depends upon which side of the nucleus they are in relation to the magnetic field of wire X. These transitions are done through absorption and emission of free electromagnetic radiation and once again you get a burst of current. However, once the atoms in wire Y are positioned in balance for a neutral magnetic field, they no longer emit electrical current.

Since the galvanometer measures energy surges, then the electrons cannot only transition outward at the first change in magnetic field environment around wire Y and then transition back inward with the reversion to original status quo neutral magnetic field. Therefore, energy is both absorbed and emitted during any change in the positioning of electrons in a magnetic field and that is registered by the galvanometer. This means there are both inward and outward electron transitions to reposition according to the environment of the exterior magnetic field.

The reason that electrical current looks different from light in the atmosphere is that electrical current is not emitted in quanta into space. Electrical current is what light looks like when it is traveling from atom to atom through the energy wave fields of the particles. We don't get lightning in space because there are no atoms close enough to allow the passage of light from wave field to wave field. Electrical current is light being passed through the molecules by direct contact instead of into free space. When light is in quanta in free space, it takes the fastest route not being bounded by the necessity to touch atoms to transfer energy because the light quanta each have enough energy to fill a frequency and self-propel. But electrical current is light that is transmitted by atomic contact, so when it runs out of atoms, it stops propagating.

How is this different from the current theory? It is interpretation of the theory. Richard Feynman said, "Millions of electrons moving through a wire represent an electric current."[xlii] In fact, no one can describe any more precisely than this what electricity is. That is because there is no barrier in the atom between the electrons and the nucleus where energy can be stored. Therefore, even though the Maxwell equations clearly teach that electrical current is the very same thing as light, modern Science does not say that it is. Instead, they don't understand their own mathematical equations. The problem is light coming from the sun does not look like an electrical current in the air which behaves like lightning.

Current Quantum Electrodynamics does not interpret or teach that electricity is actually the same exact thing as a light ray coming from the sun, even though the mathematics of Maxwell's equations teach us that they are exactly the same thing. That is because they look and behave differently, but they are, in fact, exactly the same thing and they just travel differently. Light and electricity are taught to be electromagnetic waves, but no one explains that they are exactly the same thing. The energy self-transports in quanta in light and the energy transports by atomic contact in electrical current. The theory is there in the mathematics, but no one gives the interpretation. I believe the electrons in a cathode ray tube are what confuses scientists because they react differently than electrons in an atom. But electrons in an atom when viewed as the movement of electrons through the nuclear waves of the nucleon particles, then we can see why electrons in a cathode ray tube are not transitioning through the nuclear waves of atoms and behave differently.

So why can we see electrical current if it does not radiate, but travels by atomic contact of the energy wave fields? The reason is that some of the electrical current can release in full quanta as light. The fact that we see lightning means that some of the current is being released as light in the electromagnetic spectrum. The fact that a light bulb works means that some of the current is released as light by emission of full quanta. But the current itself is light traveling by atomic contact.

So what would be empirical evidence that magnetism pulls apart an electron from its orbital shell and therefore causes the nucleus to have a greater distance between the electrons and the nucleus so that the distance necessarily has to fill with energy to keep the balance of energy to new electron excitation state? We would expect to see in the spectra of an element an effect showing that a magnet pulls the electrons into higher excitation states. The transitions of electrons caused by a magnetic field should show more electron absorption and emission simply due to the magnetic field causing the higher excitation of electrons in the atom.

This view of electricity as the electron being pulled away into a higher energy state by the magnet so that the atom has to absorb energy to account for the higher state of excitation represented by electron transitions is shown in the light spectrum and has been known since the 1800s.

Early experiments in spectroscopy show how the spectrum is changed by the introduction of the magnetic field. Everyone focuses on the Zeeman Effect of the splitting of the energy lines of the electron shell, but the observations of the intensity of the lines goes unexplained. Let's examine how this view of electricity explains the spectrum.

First, let's examine what should be expected if this concept of electricity is true. When a magnet pulls on atoms of a conducting metal, electricity is generated if the magnet moves. Electricity is therefore created as a result of pulling electrons into higher shells by magnetism. When the electron is pulled into higher shells by magnetism, the atom must absorb light energy to conserve or account for the change in the energy state of the atom. The atom is in an excited state because it is being forced by the magnet to transition outward and therefore forced to absorb energy. As the magnet moves, the electrons drop back toward ground state and the energy is released into the conductor not as light, but as electrical current by transfer of the energy between touching nuclear waves around surrounding atoms. The passage of this current passes through the nuclear wave field of the atoms at the speed of light because the photon is massless. So electricity is created by the dropping of the electron toward ground state by the motion of the magnet.

Therefore, the prediction is that a magnet should excite atoms and create a quantifiable absorption spectrum. The lines in the absorption spectrum where the electrons are being released from the higher excitation state caused by the magnetic field should become more intense as the magnet is creating higher excitation the same way that raise in temperature creates higher excitation of electrons. However, with a magnetic field, the temperature won't make a difference, because the magnet is causing the excitation of electrons.

In William Frederick Schulz 1909 paper on "The Effect of a Magnetic Field Upon the Absorption Spectra of Certain Rare Earths," we see this effect actually happening. Schulz recounts the experiments of Jean Becquerel, one of the pioneers of spectroscopy.[xliii] He says, "J. Becquerel studied the effect of a magnetic field upon the absorption banks of the transmission [emission] spectrum of these crystals, both at ordinary and at very low temperatures. It was found that some of the bands were not at all affected, others simply becoming broader and changing in intensity in the magnetic field, whereas in yet others the effect was similar to the Zeeman effect in vapors. The magnitude of these effects was found to be proportional to the strength of the field and independent of temperature." P. 384

The lines got intense where the transitions occur more frequently. A magnet makes the electrons absorb energy despite temperature. This pulling out of electrons causes electrical current where the interactions take place in a conductor. Where the interactions take place in free space, the atoms radiate light instead of electricity.

The poles of the magnet shift the proton further away from the electron. Or relatively speaking, the electrons and the nucleus pull apart. This effect of changing the magnetic field with regard to the polarization of the field shows this effect in the spectrum as the transition lines appear toward the opposite magnetic pole. This article states that the Zeeman lines appear in opposite directions to the magnetic field. This means electrons and protons reverse position. Fortunately, in this new model of the atom, the electrons can have a position relative to the nucleus without the need to collapse an electron cloud or collapse a quantum state. The nuclear wave protects the atom at all times so that the electron can maintain a position in the atom with respect to the magnetic field.

So spectroscopy shows that the magnetic field is creating an excited state for the electrons, and a movement of the magnetic field drops the excited state causing a release of energy through the system. In a solid, the molecular bonds mean that the nuclear wave fields of the atoms are physically in contact. This allows the current to flow through the system via the wave fields of the atoms rather than radiating as light. This is electricity.

This model of electricity predicts why the magnetic field has to move to create electricity. While the magnetic field is stationary, the electrons are held in place in higher excitation. There is an equilibrium and the electrons cannot lower their excitation state because they are held in higher shells by the magnet. As soon as the magnet moves, some of the electrons are freed from the magnetic field and may drop to a lower excitation state emitting energy through the nuclear wave fields of bonded atoms in the solid conductor. To release the electrical current of the excited electrons, the magnetic field has to move away from the electron held in higher energy shells.

Therefore, in solids where the nuclear wave fields of the atoms are not all in contact with each other, "resistance" will be experienced as the light or electrical current will have to re-route to find a path to flow by contact with the energy wave fields of other atoms. And in superconductors where the crystal lattice pattern of the atoms are manipulated by temperature so that the atomic energy wave fields of the atoms are all in contact with each other, there is no resistance and the electrical current flows freely through the material because every atomic nuclear wave touches every other. This free flowing electrical current pushes the full magnetic field to the exterior of the material because of the free flow of the electrical current. As long as the circuit is closed, the excess energy stored in the atoms moves freely trying to come to equilibrium by each atom attempting to release the excess energy as electrical current through atomic contact of the wave fields. This effect of freely flowing electrical current is because the electrical current is actually light and it behaves as light does. It is merely light traveling through atomic wave fields versus light traveling by radiation.

This is an uncomplicated explanation and one that stems from what we already know. Until science gets the big picture, until they unite classical mechanics with the quantum wave of the atom, there can be no real progress, only more and more abstract mathematical equations. While science does not understand the atom, they are like children playing with matches who cannot foresee such things as the implosion of Bose-Einstein condensate. That had minimal consequences. Something else might not.

This new model of the atom is so simple to picture. It is less easy to mathematically calculate as we must calculate a non-radiating finite wave system, but as with all new models, this can be a springboard for scientific advancement.

Kinetic Energy

Kinetic energy is the measure of energy that forces the electrons toward the nucleus. By transitioning into the inner higher energy antinodes of the atom toward the nucleus, the electrons absorb energy. The electrons must absorb more energy to transition toward the nucleus than to transition outward from ground state. However, to return to ground state which is now in the outer portion of the nuclear-wave with respect to the electron, the electrons have to release energy as they transition outward. The act of releasing energy and the transition outward pushes with an opposite and equal force to the force applied. This is Newton's opposite and equal reaction.

Electrical Energy

James Clerk Maxwell showed that light and electricity were the same thing.

Another form of electron-wave energy absorption and emission is electrical energy. In electrical energy, a magnet pulls the electron to an outer node, forcing it across several antinodes. To maintain equilibrium, the electron absorbs radiation energy in its electron particle-wave equivalent to the energy of all the intervening antinodes that it passes through in order to gain energy for the transitions caused by the attraction of the magnet if the magnet is positive. If the charge of the magnet is negative, the electron is forced to the opposite side of its node and forced through the antinodes of the proton wave that are in opposition to the charge. As long as the magnet is stationary, the electron holds its place in an outer node of the proton wave having absorbed energy equivalent to the transitions of the antinodes. The electron has the tendency to return to ground state of its electron-wave, but the magnet is holding it in a high energy node.

When the magnet moves away, the electron immediately falls to ground state. If the proton wave is not in contact with the proton wave of another atom, the energy is released as radiated light. Where the proton-wave of the electron is in contact with the proton-wave of another atom, the energy is released as non-radiated light transmission through contact transmission into the waves of surrounding particles. The energy absorbed by contact excites the electron to higher energy nodes so that the surrounding atoms are now in a high energy state and tend to the quickest path to ground state, and thereafter, release the light by contact transmission through to the next surrounding atoms.

When the magnet moves back, the magnet again raises the electrons to higher energy states and the electron wave absorbs energy for the transitions to higher nodes of the proton-wave. By sustaining the movement of a magnet, the electron is able to be raised to a higher excitation node in the atom which causes it to absorb energy, and then when the magnet is moved away, the energy stored in the electron wave is allowed to release the energy that has just been absorbed from surrounding radiation. This is already described by Maxwell's equations.

Electricity is light transmitted by transmission of energy through proton-wave contact. Radiation of light occurs when excess energy is released without contact of proton waves. The higher the separation of the proton waves from each other, the more likely the energy will be released as light radiation as in a gas. Solids are better conductors of electricity because their proton-waves are in contact. The better contact between the proton-waves of the solid, the better the conductivity. Molecules whose proton-waves are so arranged that the proton-wave of every atom is in contact with every other atom will perform as superconductors. In a superconductor, the transmission of light energy through contact is unimpeded because every atom proton wave is in contact with every other proton wave so that there is no resistance. Resistance is the phenomenon where the path of the light energy moving through contact with proton waves is rerouted continually due to areas of the structure where the proton waves are not in contact and the light must reroute using the shortest possible path per Fermat's Principle though the proton waves that are in contact.

Both the phenomena of electricity and radiation happen simultaneously in most materials. The electron wave tends to ground state releasing energy both as contact light transmission and radiation light transmission. One we call electricity, the other we call light. Light radiation released inside a solid is recaptured by surrounding atoms and reabsorbed. However, on the surface of solids where the atoms meet a gas, the radiation of light may progress without immediate reabsorption. Because the motion of a magnet pulls electrons to different outer nodes depending on the magnetic field of the electricity, then electricity contains light energy being transferred by contact at all allowable frequencies of all allowable energy levels for that particular element all at once.

Because the electron is able to absorb any amount of energy per the equations above for light, electricity can be produced from absorption of any radiation present in the environment from radio waves to gamma rays. Even in a vacuum, radio waves, gamma rays and cosmic rays are always present, therefore, electron-waves can always absorb energy to transition and electricity can be produced in a vacuum.

23 Doppler Shift

When the electromagnetic wave nears a gravitational field its prime photons increase in volume and lessen in density mimicking tidal effects. Although the prime photon retains a constant velocity, this causes a red shift. The gravitational field pulls the prime photons toward it causing parallax. The Galilean transformation and addition of velocities would apply before the Planck barrier was hit. In electromagnetic waves traveling toward earth from atoms from a distant star, if the star was speeding toward earth, the velocity of the energy released from the atom would add with the velocity of the star, therefore, the energy of the prime photon would hit the Planck barrier sooner and the light would appear to us on earth blue-shifted. In a body traveling away from the earth, the Galilean transformation would appear to slow down the energy of the prime photon before it hit the Planck barrier and the light would be red-shifted. Because there are blue-shifts there cannot be "tired light" per se. Rather it is a combination of both the lessening density and larger volume of prime photons that makes the red-shift appear along with the fact that a larger red-shift would appear if the body emitting the light were traveling away from the earth. Therefore, the red-shift is a combination of factors. The blue-shift on the other hand cannot be explained unless a body were moving toward the earth. If the red-shift were tired light, then the blue-shift would have to be more energetic light which is illogical that light would become more energetic after traveling from a distant galaxy.

If this theory of light is correct i.e. if light naturally red shifts after great distances by increasing in volume, then this natural red shift must be subtracted from the Doppler red shift to give the true velocity of galaxies traveling away from us. In fact, the universe would not be expanding as rapidly as science supposes although it would still be expanding.

Planck's constant describes this red shift. As previously discussed, Asimov described Planck's calculation as explaining why light radiated more easily in the red, saying, "The probability, however, was that before the energy required to make up a full quantum of violet light was accumulated, some of it would have been bled off to form the half-sized quantum of red light. The higher the frequency of light, the less the probability that enough energy would accumulate to form a complete quantum without being bled off to form quanta of lesser energy content and lower frequency."[xliv] The Doppler Effect should be corrected for the quantization of light radiation.

23 Special Relativity

It is true that particles when they are accelerated close to light speed and push against the Planck barrier and enlarge in mass in an attempt to increase volume into the next spherical space, thereby changing their energy wave to mass ratio and extending their life. However, this does not prove that humans or living cells would age less quickly when made to hit the Planck barrier. Clocks slow down because atoms are pushed against the Planck barrier, but the changing of a clock does not stop time. Aging itself must slow down. If a cell being smashed into the Planck barrier enlarges, do its processes slow down? Aging is the measure of time, not whether a clock runs slow. For instance, a cesium clock traveling at a higher speed than another cesium clock will transfer mass from its energy wave to the particles. There is a physical change in the atom that results in the change of time that the atomic clock keeps. This does not mean that time itself is slowing, but merely the clock.

Mass is slow energy. Energy is fast mass. Energy accelerated against the Planck constant may produce mass in the same way that gamma rays may create pair production. In this way, there may not be conservation of baryons in the universe. There may be only conservation of energy and mass. Energy may actually be creating baryons in the universe which slows the rate of expansion.

As for prolonging the life of subatomic particles, a particle is not life, so increasing the time before it decays does not show that aging slows. If cell division slows, that is, if bacteria live longer near light speed, then time slows. This may be the case, but it has not yet been proven.

Time is a necessary fourth coordinate to describe a position in space, but is it really another dimension? When one takes a length and multiplies it by itself, one gets a new dimension going from one dimension to two i.e. going from a line segment to a square. When that length is multiplied by itself again, one goes from the square to a cube. Multiplying that same length by itself again gets a shape of a fourth dimension, but that is not what time is doing in special relativity. Time is not the first length again multiplied by itself, so time is not a fourth dimension in this literal sense. Rather time is its own number and it needs to be considered in order to find position in space so it is a fourth coordinate, but strictly speaking, not a fourth dimension.

24 Feynman Diagrams

Because Richard Feynman used a photomultiplier, he observed that light is a particle rather than the non-empirical philosophical route at which that conclusion was arrived at in this new theory. Feynman explains: "If you put a whole lot of photomultipliers around and let some very dim light shine in various directions, the light goes into one multiplier or another and makes a click of full intensity. It is all or nothing: if one photomultiplier goes off at a given moment, none of the others goes off at the same moment (except in the rare instance that two photons happened to leave the light source at the same time). There is no splitting of light into "half particles" that go different places. I want to emphasize that light comes in this form—particles." And then later Feynman continues: "For many years after Newton, partial reflection by two surfaces was happily explained by a theory of waves, but when experiments were made with very weak light hitting photomultipliers, the wave theory collapsed: as the light got dimmer and dimmer, the photomultipliers kept making full-sized clicks—there were just fewer of them. Light behaved as particles."[71]

The assumption that light is a particle because one hears clicks in a photomultiplier is analogous to saying that blood has turned to marbles because one hears distinct blips on the heart monitor.

The problem with Feynman's empirical observation, is that it does not take into account the shape and velocity of the particle. That a photon is constantly changing velocity which gives it its wave-like quality is not considered. The photomultiplier only hears the single click of each photon, but partial reflection in a mirror sees the change in velocity of a single photon. Because the velocity of a photon between hitting the Planck barrier and moving back up to inertial speed and reducing back down to the Planck barrier gives an overall velocity of light as "c", there does not occur to science that a particle can move like a wave. Feynman's photon particle does not need to be a point particle, neither does it need to be a spherical particle, neither does it need to have all parts of the photon moving at the same speed in order to make the photomultiplier click once for each photon at full intensity. As long as the photon is energy separated by multiple stops, then it displays as particles. But the particle itself is a movement of various velocities in the direction of travel. The photomultiplier hears the multiple stops of light. And no matter how small a photomultiplier is, it will not sense the spread into a convex oblate spheroid of a freshly emitted photon because the photon spreads so slowly. It is only through the lightyears of space that the redshift of the photon is observed. (This redshift is different than the Doppler effect.) So by all instrumental appearances, the photon appears to have only one width and one overall speed that is the speed of light that is punctuated by distinct spaces between photons where the energy falls to zero at Planck's constant and therefore the photomultiplier hears stuttering called clicks.

In Feynman's introduction to QED he starts with a paradox. The fact that apparently light is made of only distinct particles called photons that apparently have no wave-like qualities according to the empirical evidence of the photomultiplier and the fact that light appears to partially reflect from glass in waves of reflection that depend upon the thickness of the glass. He concludes: "Try as we might to invent a reasonable theory that can explain how a photon "makes up its mind" whether to go through glass or bounce back, it is impossible to predict which way a given photon will go." And later, "The situation today is, we haven't got a good model to explain partial reflection by two surfaces; we just calculate the probability that a particular photomultiplier will be hit by a photon reflected from a sheet of glass."[71]

A collision of two particles in a Feynman diagram where the particles do not actually collide but are repelled by each other and then one of the particles pulls back and runs parallel to the other particle is simply two particles joining into a nucleus. Any collisions of particles that occur in Feynman diagrams where repulsion takes place with only virtual interaction is due to actual interaction with the energy waves of the particles. Bohr postulating that there was no physical barrier between the nucleus and the electrons has led to virtual interactions and virtual particles. There is no need to invent "virtual photons" when each particle is emanating real photons in its energy wave.

From the beginnings of science when a phenomenon is not understood, it has been called "a force". Since the early twentieth century, when a phenomenon is not understood, it is called "a probability". A probability puts the possibilities everywhere and then tells you where they are the highest. This is an avoidance of an actual physical explanation just as quantum mechanics is an avoidance of explaining the actual phenomena of the atom and resorting to probability. When no model can be conceived to explain physical phenomena, probability models become useful. That is all that is needed for a good theory. It must be useful. And there is no doubt that quantum mechanics and quantum electrodynamics are useful in explaining physical phenomena. So they are not bad theories. However, the science of future ages will undoubtedly call the twentieth century "the rebirth of the supernatural and the elevation of the mystical to the scientific level" because probability denies direct causality. In response to Heisenberg in 1926, "Wilhelm Wien, [a Nobel laureate] who held the chair of experimental physics at the University of Munich, answered rather sharply that one must really put an end to quantum jumps and the whole atomic mysticism..."[82,p.129] When there is no model, when there is no available answer, probability theory provides an indirect answer. But the truth is that there is an answer. There are causes for every effect. Otherwise there is no such thing as science which is founded upon the belief that we live in a rational world. Probability can never be disproved because it is the most vague manner to answer any question. In everyone's life, there comes a time when that person tries to cover-up their ignorance and, at that time, their answer is always "maybe – probably ". This is because probably and probability are the weakest arguments. Probability can never be proven completely wrong and it can always be proven partially correct. Probability theory cannot be disproved, but one day it will be replaced. That is a certainty. Probability theory is the last refuge of physicists who lack imagination. That is not to say QED is not ingenious and highly clever. So is quantum mechanics. But to say that it reflects the true nature of the universe is questionable.

QED describes electrical current in this manner: "There are only two states of polarization available to electrons, so in an atom with three protons in the nucleus exchanging photons with three electrons—a condition called a lithium atom—the third electron is farther away from the nucleus than the other two (which have used up the nearest available space), and exchanges fewer photons. This causes the electron to easily break away from its own nucleus under the influence of photons from other atoms. A large number of such atoms close together easily lose their individual third electrons to form a sea of electrons swimming around from atom to atom. This sea of electrons reacts to any small electrical force (photons), generating a current of electrons—I am describing lithium metal conducting electricity. Hydrogen and helium atoms do not lose their electrons to other atoms. They are 'insulators'."[71p.113]

Side point: This explanation is not in most physics textbooks used in grades K-12 which is annoying as most schoolbooks teach pre-1911 physics when there is no need because an elementary grasp of physics as is known currently can be gained by children. It is just as easy to teach a child that a photon is a particle as it is to teach that a photon is a wave, an atom is not a miniature solar system, etc.

At any rate, this explanation is weak and unsatisfactory. Here it is explained that ionization takes place by photon exchange of electrons. But QED uses "virtual photons" for particle interactions. "Such an exchanged photon that never really appears in the initial or final conditions of the experiment is sometimes called a 'virtual photon.'"[71p.95] Therefore under QED electric current is caused by virtual energy. The electrical current is described as "a sea of electrons" and "a current of electrons", yet an electrical current moves nearly at the speed of light through a wire and the electrons do not. Scientists have known since the first experiments with cathode ray tubes that traveling electrons produce electromagnetic waves and the ionization between different metals produces an electric current in a cell, but no one explains where the energy comes from and this example is no different. The question is not answered by the fact that there is a "sea of electrons swimming around the atom" as Feynman says. Feynman assumes that where there is a current of electrons, there is electrical current, but QED doesn't answer where the energy comes from. How can an electron radiate energy without losing mass? That is what science fails again and again to answer and QED does not answer it but sidesteps the issue yet once again. Here the "electrical force" is the photons, but there is no explanation as to how virtual photons create electrical force. It doesn't explain how a current of electrons that are not supposed to lose mass by expelling energy actually expel energy in the form of electrical energy. This is a completely dissatisfactory explanation. Electricity can hardly be described by science at the level of the electron and most attempts by teachers and professors to describe it fail.

What is surprising is the lack of humility of most physicists. As early as 1925, Erwin Schrödinger's paper is said by Paul Dirac to contain "much of physics and all of chemistry."[72] When Dirac himself published his 1928-30 paper unifying the spin quantum number, quantum mechanics and special relativity, "Dirac boasted that his theory described 'most of physics and all of chemistry.'"[46, p.286] And Richard Feynman says more than once of QED, "the theory describes *all* the phenomena of the physical world except the gravitation effect, …and radioactive phenomena, which involved nuclei shifting in their energy levels."[italics are his][71] He goes on to describe the "laws, out of which we can make the whole world (aside from the nuclei, and gravitation, as always!)."[71p.86]

Can even one physicist describe exactly what is brittleness at the atomic level? What makes a certain configuration of electrons cause a material to crack and snap and cause sound? How exactly is it that particles of light and electron waves in QED make H_2O molecules appear to us to be a fluid we call water? Why does water appear to be liquid when it is made of particles?

What exactly is the structure of iron when it is magnetized? Where are the electrons? How can they be non-local and have locality? How do electrons become soft and malleable in alkali metals? What is heat at the atomic level? The questions are endless and unanswered. They are about chemistry and electrons, yet they are unanswered at the atomic level. So how does any physicist ever have the audacity to say that their model describes all of chemistry? Quantum mechanics nor quantum electrodynamics explains all of chemistry. Yet there exists an explanation. The model simply has not been discovered. And when new models are discovered, they will all eventually be replaced or improved upon up to thousands of years into man's future. The atom is a complex system within systems within systems within the complexity of the universal system. The atom is also a complex system within the vastly more complex system of the living cell. Feynman says, "In these lectures I want to tell you about the part of physics that we know best, the interaction of light and electrons. Most of the phenomena you are familiar with involve the interaction of light and electrons—all of chemistry and biology, for example. The only phenomena that are not covered by this theory are phenomena of gravitation and nuclear phenomena; everything else is contained in this theory."[71,p.77] However, QED has not created any medical cures so how does it explain biology? How does it explain the function of DNA and genetics? It can't even explain where the power from a battery comes from except to say an exchange of virtual particles. Quantum theory is rather like a priest explaining, "It's God's will. It's a mystery." That a physicist says, "the way we describe Nature is generally incomprehensible to us" is in the end unacceptable.[71,p.77] One day we will be able to describe nature by some simple model. But as of today, we haven't "got it". Einstein was right to say so. Einstein in 1936:

"In spite of this [i.e. quantum mechanics], however, I believe that the theory is apt to beguile us into error in our search for a uniform basis for physics, because, in my belief, it is an *incomplete* representation of real things, although it is the only one which can be built out of the fundamental concepts of force and material points (quantum corrections to classical mechanics). The incompleteness of the representation leads necessarily to the statistical nature (incompleteness) of the laws."[62,p315]

Quantum mechanics does not rely on a physical model of the atom with causality, but relies on statistics. This is the problem. The statistics may closely approximate the spectrum, but the model according to Einstein is only statistical and therefore does not represent reality which is physical.

And it should behoove science to realize that one thousand years from now, we still won't "have it" either. A model that we can make may in fact explain the system of the universe, but it won't be the universe; it will always be just a model.

So what does QED teach us? It is basically telling us that it is wrong to think of the path of any body, particle, photon, or otherwise will not necessarily be that the body will take a straight line from point A to point B, but that a body will always take the quickest path from point A to point B and sometimes that path is a straight line and sometimes it isn't. If all the paths go in a circle, the object won't go anywhere. That's QED in a nutshell. An interesting slight variation on the shortest path between two objects is a straight line, but it doesn't solve the nature of the universe as it claims. Actually, Einstein already said in general relativity that the shortest path between two objects isn't a straight line. So QED is a minor clarification. The path an object takes will be the quickest. Figure out all the paths and find the quickest and that is what the object will take. Although the importance and ingenuity that Feynman and the other developers of QED exhibited cannot be underestimated since it gives a method, despite no underlying structural model, of mathematically calculating motion with predictability, it is incomplete in that it is statistical and again gives no account of a cause for the effects.

25 The Partial Reflection Of Light

Every wave has a different velocity between Planck barriers, but each photon has the same energy. Photons traveling through glass are encountering the same atoms arranged in molecules throughout the glass. The photons are forced through the glass according to their energy which is according to the speed of the photon accelerating and decelerating between the Planck barrier. The true undiscovered substructure of glass determines how much light is reflected by glass according to the velocity of the energy directed at it.

Therefore in Feynman's illustration of light directed at glass, each wave of light will reach a maximum reflection in glass of 16% at the lowest energy of any wave because the energy carried by each photon is exactly the same for every wavelength. This means at the lowest energy of every wave, the wave reflects back 16% of its photons. Therefore, the fluctuation of energy reflection in glass by any wavelength of light is 16% because the photon has equal energy that fluctuates according to velocity between Planck barriers and the photons are forced through the glass according to the varying velocity of the individual photon which represents its energy. The photon has different speeds between the Planck barrier and the atoms of glass absorb photons according to the increase in speed and the decrease in speed of the photon between Planck barriers. Each photon carries the same energy in a larger volume density or smaller volume density according to wavelength. At the highest velocity of each photon, all photons travel through the glass and are not reflected, therefore, there is 0% reflection. Therefore, the peaks of reflection are spaced closer together for blue than for red light. Other than the difference in spacing between peaks of reflection, the energy carried by the photon is the same and the atoms in the glass are the same, therefore, the reflection remains consistent despite the wavelength and only varies with wavelength.

Since the measurable energy drops to zero at the Planck barrier, the photon sounds like a particle in the photomultiplier with a distinct absence of sound where the measurable energy drops to zero. For each smaller wavelength the click is louder in the photomultiplier because the velocity is higher between the Planck barrier and therefore the apparent energy, the measurable energy, is higher. As with any two particle collisions, a photon colliding with the energy wave of an atom will impart energy to the wave and the wave will impart energy to the photon. The amount of energy transferred will depend upon the carrier energies of the wave and the particle. In any collision of particles which is elastic or a reflection, whatever excess energy is transferred in the collision from each particle to the other is immediately emitted by the particle that absorbed the excess energy in the direction in which the absorption took place. Therefore, a photon being reflected from an energy wave will emit energy to the wave and absorb energy from the wave according to the angle of the collision and then reemit the energy in the direction it was absorbed. Therefore the angle of incidence will be equal to the angle of reflection. The photons actually reflected from the glass from 0 to 16% will follow this rule.

26 On Liquids and Fluidity

Certainly it is not a stretch of the imagination to say that the difference between a gas, a liquid and a solid can be explained. A gas is explained by current scientific theory as atoms which are separated in distance further than in solids which gives them more latitude of movement. It should not be difficult then to conclude that solids contain atoms whose molecules are all bound to each other in the majority of cases in locked positions and that there is an interlocking of molecules into an arbitrary structure. However, this structure can be rearranged through crystal formation so that the structure is symmetrical and repeating in orientation. So then we have gas molecules that have great freedom of movement. Under this new theory, this means that no electron binding inside a single electron nucleus takes place in the connections between every molecule of the gas. The electrons between gas molecules are not bound to each other by sharing the nucleus of their energy waves. This keeps the molecules moving in random trajectories according to the state of the system being affected by free energy in the form of heat or electromagnetic waves. A solid on the other hand, has its molecules each bound to each other at least somewhere in the structure of the solid. At certain locations in solids, there is an electron nuclear bond between molecules that holds the solid in a structure by restricting movement. The pattern does not become a repeating orientation except in a crystal, even though the bonds between molecules may appear in the same regions of the molecule. For a liquid, there is a situation between a solid and a gas. In a liquid, the molecules form bonds in chains.

Current chemistry holds that water is unique because its oxygen atom has two lone pairs and two hydrogen atoms, meaning that the total number of bonds of a water molecule is up to four. Therefore, chemistry does not account for the fluidity of water. What happens in the macro world must have a cause in the atomic world.

If a liquid has the motion of a chain then it can take on the motion of a wave. The chain sequence can break under pressure because there are less bound electrons between the molecules of liquids than in the molecules of solids. Liquids have more freedom of movement than solids due to this sequence of molecules being bound mainly by one electron nucleus energy wave bond between molecules. However, the molecules hold together in a chain-like wave-like fashion so that their freedom of movement allows them to fill the volume of variously shaped containers while a solid cannot. When a drop of liquid begins to form from a faucet, the chain of molecules holds until the drop becomes so heavy that gravity breaks the electron nucleus binding the molecules and the drop falls. This is a simple theory and is made possible by the concept of energy waves around particles being able to bind other particles. Quantum mechanics does not allow for binding of particles in this fashion due to the Pauli Exclusion Principle and the very structure of the atom created in the Schrödinger wave mechanics which separates the electrons into always more and more distinct quantum states and the Uncertainty Principle which negates any locality for the electrons in which to bind. Chemistry is left floundering for an explanation under quantum mechanics. Where the electrons bind to create liquids and solids must be decided from known chemical data. The hydrogen molecule has already been discussed. Both electrons of hydrogen bind in a single nuclear energy wave of the electron. However, the hydrogen gas does not bind in this way with other molecules of hydrogen and the gas therefore has great freedom of movement. In one possible scenario, when hydrogen is combined with oxygen, the hydrogen molecule electrons may unbind and each hydrogen electron may bind with two electrons of an oxygen atom creating H_2O. This must be adjusted according to the empirically known binding strengths because atoms could still bind with weaker bonds simply by the electrons of one atom sharing the node of another atom's nuclear energy wave. Without two electrons sharing each other's energy wave, the electrons between two atoms that appear in nodes of two overlapping energy wave would necessarily take more energy to transition between shells and more ionization energy to leave the two energy wave of the overlapping atoms. Therefore, the strength of a bond between shared orbitals would depend on the combined energy required to transition from the node that has constructive interference between two atoms. This bond however is not as strong as the bond of two electrons sharing their energy wave and sharing the nodes of two atoms. In water, the chain sequence may go two hydrogen atoms then one oxygen atom then two hydrogen atoms and so on. Or it may take another form that resembles the twist of the DNA molecule, but at any rate, water undoubtedly creates a chain longer than simply four bonds in order for water to represent fluidity and have the ability to create waves. Because water would be merely countless chains of connecting atoms, it has a degree of freedom of movement less than gas but more than a solid. A propensity would exist for the chain of molecules to crumple up into the shape of the container or form waves in the ocean when the wind blows. Since there are less bonds than in a solid, the molecules would tend to break apart easily when pulled upon by external forces. The bonds break at certain points into water droplets on the ground. These in turn can be broken further by heat and thus reappears the gases that once formed the chains in the form of evaporation of the droplets. This simple model can be developed to explain many things about liquids. It should be developed further in order to explain the fact that water tends to form spheres. This probably has to do with the spherical shape of the energy waves of the atoms, but may also have to do with which electrons are bound between the molecules. This cannot be known because there is no knowledge to this day of the grouping of electrons in the atom in each shell. Cooper pairs show that electrons can pair in a shared nuclear energy wave, but do not tell us how many electrons can group in the same-shared energy waves in the atom. Since the electrons are in the nodes of the nuclear energy wave, then there is external pressure on the electron wave that would not exist in free electrons, therefore, it is possible that more might be able to bind into one electron energy while under the pressure of the nuclear energy wave. The early chemists Langmuir, Bury, Stoner, et al., appear to believe that eight electrons may group together in the atom in accordance with the chemical properties of the periodic table. They did not know how this occurred and neither did Bohr but he put these grouping into non-empirical suborbitals. If this occurs all in one shell or if this occurs all in one electron energy wave that is shared between all eight electrons cannot be known from empirical evidence, seen thus far, unless the naturally occurring doublets and triplets in the spectrum reveal their secret. If doublets are two electrons in a shared electron energy wave, then a triplet may be three in a single shared electron. As said before, the Zeeman effect shows how far a magnet may create waves on the surface of the energy wave and cause an electron to pop into slightly different places in the node of the nuclear wave because the node is rippled. The energy wave carries a charge and a magnet can ripple the waves.

A superfluid is undoubtedly a fluid with strong atomic-molecular bonds that form chains as does any fluid. Since the bonds are so strong in the chain links, the evaporation of the fluid, meaning the expanding of the molecules so that they should form a gas does not happen. Rather, the molecules expand and the superfluid should evaporate, but instead, the molecular chain of the fluid pulls the fluid up over the lip of the container and when the chain gets heavy enough it pulls the partially evaporated fluid back down on the outside of the container. The bonds of the fluid are so strong that what would normally evaporate stays connected to the fluid.

27 The Uncertainty Principle

The Uncertainty Principle developed by Werner Heisenberg in 1927 from his results on matrix mechanics in 1925 is attractive from a philosophical standpoint and a comforting principle when one uses it to deal a crushing blow to the extremist fatalistic determinism of Marquis de Laplace in order to argue "free will" for humans. But philosophy should not be taken into consideration at all in science. The only philosophy of science should be a rational cause for every effect. With the Uncertainty Principle, its philosophical appeal arises from a fatal mathematical flaw in its conception.

The Uncertainty Principle according to one account—that is undoubtedly an origin myth—was conceived from actual measurements. In a history of its development, first we examine the development of matrix mechanics:

"The basic idea of Heisenberg's paper was to get rid of the orbits in atoms and to arrive at new mechanical equations by working backwards from the observed frequencies and intensities of the light emitted and absorbed by matter. Working with an actual atom proved too complicated at this point. So Heisenberg studied instead a charged ball on a spring, an oscillator, whose motion was not quite regular (anharmonic). Heisenberg looked first at the connection between the observable properties of the emitted light--its color (frequency) and the intensity-- and the motion of the charged ball according to the classical mechanics of Newton. [37]

"The unfamiliar rule may be expressed as follows. If two position variables can be expressed as Fourier series consisting of amplitudes A(n,k) and B(k,m), where n,m,k are integers, then multiplying two amplitudes together to obtain an intensity results in an infinite sum over all values of k:
$C(n,m) = \Sigma_k A(n,k) B(k,m)$. [37]

"This led to the puzzling result that the 'commutation law' in arithmetic is no longer necessarily valid. That is, A times B does not necessarily equal B times A in quantum mechanics. This was particularly important when Heisenberg obtained the quantum mechanical expression corresponding to the "quantum conditions" in the old quantum theory. If the momentum p and the position q of a particle could be represented by Fourier series, then a differential expression for the multiplication pq in the old quantum theory became a difference expression in which pq does not equal qp. Instead, there is the famous "commutation relation" for the quantization condition that is at the basis of quantum mechanics:
$\Sigma_k p(n,k) q(k,n) - q(n,k) p(k,n) = h / 2\pi i$, h being Planck's constant."[37]

Heisenberg did not immediately assume in 1925 with the above formulation of matrix mechanics that the fact that the position and momentum did not commute in his equations was fundamental to the universe. This account continues that the Uncertainty Principle was derived from this in 1927 after he had studied the results of measurements made by other physicists as is explained:

"Studying the papers of Dirac and Jordan, while in frequent correspondence with Wolfgang Pauli, Heisenberg discovered a problem in the way one could measure basic physical variables appearing in the equations. His analysis showed that uncertainties, or imprecisions, always turned up if one tried to measure the position and the momentum of a particle at the same time. (Similar uncertainties occurred when measuring the energy and the time variables of the particle simultaneously.) These uncertainties or imprecisions in the measurements were not the fault of the experimenter, said Heisenberg, they were inherent in quantum mechanics. Heisenberg presented his discovery and its consequences in a 14-page letter to Pauli in February 1927. The letter evolved into a published paper in which Heisenberg presented to the world for the first time what became known as the uncertainty principle."[38]

The deviation of h-bar or h/2pi that arose in matrix mechanics does not appear to disclose the standard deviation of the measuring instrument. No treatment of any scientific subject, experiment, or measurement is said to be accurate without disclosing the standard deviation of the measurement. It does not appear that Heisenberg allowed for spectroscope error when he was making measurements of spectral emission and absorption. No instrument is perfect. This does not mean that there is no uncertainty in position and momentum, but if there is, the standard deviation of the spectroscope needs to be subtracted from the uncertainty relations. The Fourier Transform spectroscope was invented in 1911 and is highly accurate, but not perfect. Indeed, the instrument that Heisenberg used is not disclosed. It would appear that without subtracting out instrument error or standard deviation of the instrument that the uncertainty relations are estimated at too high a deviation if indeed there is any deviation left at all after allowing for instrument error.

However, Heisenberg himself contradicts this source concerning the inception of the uncertainty principle because Heisenberg attributes the foundation of the Uncertainty Principle to another cause. This cause stems from Heisenberg's main concern in matrix mechanics to eliminate Bohr's use of periodic orbits for the electrons. It bears repeating what Heisenberg said of Bohr's model: "So I now have to point out the difficulties and the errors of this model. The worst difficulty was perhaps the following. The electron described a periodic motion in the model, defined by quantum conditions, and therefore it moved around the nucleus with a certain frequency. However this frequency never turned up in the observations. You could never see it. What you saw were different frequencies, which were determined by the energy differences in the transitions from one stationary state to another."[74] So the periodic orbit of the atom is not empirically observed. Heisenberg says that it had to be abandoned and he in fact did abandon it with matrix mechanics which applied point locations for the position of the electron without applying a periodic orbit to the electron. Heisenberg invented matrix mechanics based upon the proposition that the electron does not orbit at all, ever, but only oscillates. Heisenberg used anharmonic oscillations to describe the electron at a locality, a distinct locality, with respect to the nucleus. Heisenberg used anharmonic oscillations to describe the movement of electrons between transitions, because Einstein's paper[60] in 1917 had explained the transition probability of the atom. And the Compton effect had shown that the reason x-rays had a greater wavelength after being scattered by the atom was that the electron had absorbed some of the energy and transitioned to a higher state.[77] Thus the oscillation between electron transitions was likened to an anharmonic oscillator and then quantized into quantum jumps in order not to show a streak across the jump in the spectrum, but discrete spectral lines. Matrix mechanics then describes all the possible distinct localities that the electron can be positioned before transitioning through creating an infinite number of locations in the electron shell. However, matrix mechanics did not yet interpret this as a probability cloud when it was invented in 1925. Heisenberg first invented locality for the electron and then destroyed locality. But the reason for doing away with any periodic orbit for the electron was that Heisenberg knew that in classical mechanics an electron making revolutions around the nucleus would create an observable frequency which he states is not observed in the spectrum. Heisenberg discusses his development of matrix mechanics and says: "In so doing I had the feeling that I should renounce any description of electron pathways, indeed that I ought deliberately to repress such ideas... A substitute for the Bohr-Sommerfeld quantum condition still had to be found; for the latter, of course, employed the concept of electron pathways, which I had expressly forbidden myself... After my return to Göttingen, I showed the paper to Born, who found it interesting but somewhat disconcerting, inasmuch as the concept of electron pathways was totally eliminated."[74,p.45] According to one account: "Heisenberg's paper marked a radical departure from previous attempts to solve atomic problems by making use of observable quantities only. 'My entire meagre efforts go toward killing off and suitably replacing the concept of the orbital paths that one cannot observe,' he wrote in a letter dated 9 July 1925."[78] He meant all orbital paths as he himself said none could be observed.

Schrödinger had also located the electron in the atom in relation to the nucleus, but Schrödinger did not locate the electron as a particle, but as a wave with a distribution. Therefore, Schrödinger's electron location had intrinsic uncertainty because a wave is a disturbance that has a distribution by definition. Schrödinger's wave mechanics did not allow for electron orbits either. As Einstein put it speaking of Schrödinger's equation: "On the basis of this theory there was obtained a surprisingly good representation of an immense variety of facts which otherwise appeared entirely incomprehensible. But on one point, curiously enough, there was failure: it proved impossible to associate with these Schrödinger waves definite motions of the mass points—and that, after all, had been the original purpose of the whole construction."[62,p.332]

This lack of movement of the electron in its shell suited Heisenberg. Heisenberg said that the motionless state of an electron in a shell without a periodic orbit was necessary because the atomic spectrum showed no frequency in the shell state and only a frequency at the transitions. However, one cannot have a frequency of an electron at transitions if the electron is not motionless before the transition. This is what Heisenberg understood. If the electron can radiate when it transitions, then it can radiate when it moves, therefore, it must be stationary before it moves. Heisenberg explains the development of matrix mechanics: "By this time you see that the idea of an electronic orbit, connected with the discrete stationary state, had been practically abandoned. The concept of the discrete stationary state had, however, survived. This concept was necessary, and had its basis in the observations. But the electronic orbit could not be connected with observations and had therefore been abandoned, and what had remained were these matrices for the coordinates."[74,p.26]

Now we come to why Heisenberg introduced uncertainty into his electron particle that was completely motionless in its shell with respect to the nucleus and only oscillated between transitions which created an anharmonic frequency that was quantized. Heisenberg was clearly bothered by a peculiar problem. It was actually Einstein who first raised the problem to Heisenberg in 1926 upon their first meeting. Einstein had invited Heisenberg to his home for a discussion of matrix mechanics upon its introduction. As Heisenberg describes the discussion:

"On the way home, he questioned me about my background, my studies with Sommerfeld. But on arrival he at once began with a central question about the philosophical foundation of the new quantum mechanics. He pointed out to me that in my mathematical description the notion of 'electron path' did not occur at all, but that in a cloud-chamber the track of the electron can of course be observed directly. It seemed to him absurd to claim that there was indeed an electron path in the cloud-chamber, but none in the interior of the atom."[74,p.113]

Heisenberg was clearly bothered by this peculiar problem and mentioned that it was the source of "many discussions, difficult discussions". [74,p.28] It was the fact that apparently electrons actually traveled on paths through a cloud chamber with or without radiating that was the problem. Paths had been taken from the atom, but appeared in the cloud chamber. He had based matrix mechanics on the assumption that electrons were absolutely motionless and did not radiate in their shell because they had no periodic orbit, but electrons definitely radiated when they moved between shells, but that radiation could be quantized so as to create discrete lines in the atomic spectrum. Heisenberg explains this himself through his written record of a conversation with Einstein. He has Einstein saying that Heisenberg is trying to reconcile classical electrodynamics with the atom by not having the electron move in a periodic orbit. Einstein is quoted as saying of Heisenberg's matrix mechanics, "You are, in fact, assuming that your theory does not clash with the old description of radiation phenomena in the essential points."[91,p.64]

However, in the cloud chamber, here was an observable electron moving in a path in a cloud chamber. This argument that Einstein had pointed out was greatly distressing to Heisenberg and he states this himself many times. He says, "Furthermore, it was clear that sometimes there are non-stationary states. The simplest example of a non-stationary state was an electron moving through a cloud chamber. So the question really was, how to handle such a state, which can occur in nature. Can such a phenomenon as the path of the electron through a cloud chamber be described in the abstract language of matrix mechanics?"[74,p27] He continues: "One could, of course, try—and it was tried very soon—to see whether one could describe the path of the electron through a cloud chamber by means of Schrödinger's wave mechanics. It turned out that this was not possible. In its initial position, the electron could be represented by a wave packet. This wave packet would move along and thereby one got something like the path of the electron through the cloud chamber. But the difficulty was that this wave packet would become bigger and bigger, so that, if only the electron ran long enough, it might have a diameter of one centimeter or more. This is certainly not what we see in the experiments, and so this picture again had to be abandoned. In this situation, of course, we had many discussions, difficult discussions, because we all felt that the mathematical scheme of quantum or wave mechanics was already final. It could not be changed, and we would have to do all our calculations from this scheme. On the other hand, nobody knew how to represent in this scheme such a simple case as the path of an electron through a cloud chamber."[74p.28]

The Wilson cloud chamber had been invented in 1911. Heisenberg explains the conception and introduction of the Uncertainty Principle was "the result" of this dilemma that

Einstein had exposed of electrons having paths in cloud chambers but not in the atom. He needed to destroy the path in the cloud chamber, because as he says "we all felt that the mathematical scheme of quantum or wave mechanics was already final. It could not be changed, and we would have to do all our calculations from this scheme." [74p.28]

Does anyone find anything wrong with this reasoning? Is this scientific? Or is it rather that in 1925 when Heisenberg invented matrix mechanics, he was 23 years old without a professorship or a solid career prospect.

"There was more at stake than personal preferences, for jobs were now in the balance for the creators of matrix mechanics. Most of the young men who created matrix mechanics were ready to move into teaching positions as professors, and the older generation of theoretical physicists was beginning to vacate positions at German universities. Heisenberg's family was exerting pressure on the young man to capture one of the vacancies at the same time that his best work, matrix mechanics, seemed to be overshadowed by wave mechanics."[38]

How would it look if Heisenberg changed his matrix mechanics due to Einstein pointing out to him that electrons move in paths in a cloud chamber, something that he so obviously overlooked until then. So his final solution was to invent the Uncertainty Principle to destroy the electron path in the cloud chamber. His excuse was from something Einstein had said:

"I remembered Einstein telling me, 'It is always the theory which decides what can be observed.' And that meant, if it was taken seriously, that we should not ask, 'How can we represent the path of the electron in the cloud chamber?' We should ask instead, 'Is it not perhaps true that, in nature, only such situations occur as can be represented in quantum mechanics or wave mechanics?' Turning the question around, one saw at once that this path of an electron in a cloud chamber was not an infinitely thin line with well-defined positions and velocities; actually, the path in the cloud chamber was a sequence of points which were not too well-defined by the water droplets, and the velocities were not too well-defined either. So I simply asked the question, 'Well, if we want to know of a wave packet both its velocity and its position, what is the best accuracy we can obtain, starting from the principle that only such situations are found in nature as can be represented in the mathematical scheme of quantum mechanics?' That was a simple mathematical task and the result was the principle of uncertainty, which seemed to be compatible with the experimental situation. So finally one knew how to represent such a phenomenon as the path of an electron, but again at a very high price. For this interpretation meant that the wave packet representing the electron is changed at every point of observation, that is, at every water droplet in the cloud chamber. At every point we get new information about the state of the electron; therefore we have to replace the original wave packet by a new one, representing this new information."[74,p29]

Heisenberg explains that he could introduce a measure of uncertainty in position and velocity if it doesn't interfere with experimental results. In other words, if he chose an appropriately small length of uncertainty, then it wouldn't cause experimental difficulties, but it could eliminate the bothersome path of an electron in a cloud chamber by giving the electron ambiguity of position and velocity. He explains: "The right question should therefore be: Can quantum mechanics represent the fact that an electron finds itself approximately in a given place and that it moves approximately with a given velocity, and can we make these approximations so close that they do not cause experimental difficulties?"[91,p.78]

After all, Heisenberg said that quantum mechanics was already final and could not be changed. Therefore, a small manipulation of the data should not affect the experiments. Heisenberg is here admitting and confessing to the fact that he is adding a level of uncertainty that would be so small as not to affect experimental results in the laboratory. He says: "can we make these approximations so close that they do not cause experimental difficulties?" He is asking whether he could manipulate the math by introducing an uncertainty that wouldn't matter because it would be so close as to approximate experimental results.

So what "approximations so close" could Heisenberg use? What number would be small enough not to affect experimental results. Well, Heisenberg already had his matrix mechanics. When he invented matrix mechanics, he did not actually multiply the matrices together himself. He didn't know how to do so. He was merely assembling sets of numbers to represent the possible electron transitions. Heisenberg explains, "Only much later did I learn from Born that it was simply a matter here of multiplying matrices, a branch of mathematics that had hitherto remained unknown to me. It bothered me, that in this way of multiplying series, a x b was not necessarily equal to b x a."[74,p.45] However, now Heisenberg found a exceptional use for this difference in the delta of two "canonically conjugated variables" which is just a formal way of saying the product between two pairs of values one can experimentally observe such as velocity and position. Heisenberg took the Gaussian curve of the delta as his small deviation that wouldn't affect experimental values. Heisenberg says he took Planck's constant as his uncertainty deviation in various sources including his own. The product of the delta is the full Planck's constant and the difference is one-half Planck's constant. As Heisenberg explains, "the product of the uncertainties in the measured values of the position and momentum (i.e., the product of mass and velocity) cannot be smaller than Planck's constant."[74,p.78] Whatever the uncertainty relation introduced by Heisenberg, the point here cannot be overstated. The Uncertainty Principle was invented for one purpose and one purpose alone: to abolish the pathway of the electron in a cloud chamber because matrix mechanics had no periodic orbital pathway which was so inconveniently pointed out to Heisenberg by Einstein after Heisenberg had published and before Heisenberg had secured a secular position.

Heisenberg is quoted as saying: "especially because I myself have thought so much about these questions and only came to believe in the uncertainty relations after many pangs of conscience, though now I am entirely convinced" (Heisenberg, 1927b)[96]

So here is the Uncertainty Principle in its full glory. One first says that all electrons are identical, therefore, the path of an electron in the cloud chamber is not one electron but several. At each water droplet, you have an electron in a different state than the previous state i.e. in an initial state therefore it is a different electron i.e. not the same one that would spread in the Schrödinger equation and it is motionless and therefore has no path. Each electron is identified by its wavefunction—new wave function, new electron i.e. different electron. Heisenberg is saying, "At every point …we have to replace the original [electron] by a new one." (Word in brackets added to replace "electron" with "wave packet" due to the equivalent meaning as the Schrödinger wave packet represents the electron i.e. if it were the same electron its path would enlarge with time. The whole purpose of quantum mechanics is to establish a distinction for each electron in order not to violate the Pauli Exclusion Principle so that every wave packet of the Schrödinger equation explains a different electron position i.e. four different quantum numbers representing distinct states for each electron i.e. a different electron for each wavefunction.) In this way, by introducing the Uncertainty Principle, one understands that the apparent movement of one electron in a cloud chamber is just several electrons all standing still or stationary i.e. not spreading and therefore their path is abolished i.e. there is no pathway. No path in the cloud chamber, no path in matrix mechanics, no path in wave mechanics, no path in quantum mechanics, no orbit in the atom. What is more, without an electron path, what is the use of the de Broglie pilot wave? Evidently the mass of the electron is not moving and the pilot wave therefore has nowhere to pilot it.

One cannot disregard Heisenberg's explanation and say that quantum mechanics does not really mean this because the quantum mechanical equations do not give motion to the electrons in a periodic orbit. Neither matrix mechanics, nor wave mechanics, nor Dirac's equation give the electron a pathway or orbit. It was in 1972 that Heisenberg explained at a lecture that in a cloud chamber the wave packet of the electron, which represents the electron itself in the Schrödinger equation, was constantly being replaced by a new one so one cannot say that this is the old theory. It cannot be said to be an outdated theory because quantum mechanics cannot explain under its equations anything different than what Heisenberg explained. The fact is that there does not exist a quantum equation with an electron traveling in a periodic orbit nor one that unites the phenomena of particles in the atom, with the phenomena of particles in an accelerator, with the phenomena of particles in the macro world i.e. electricity, batteries, classical mechanics, etc. And the Uncertainty Principle was an attempt by Heisenberg to combine two areas: 1. an electron only radiated by oscillating between transitions in an atom but did not have a real orbital pathway, and 2. electrons in a cloud chamber appear to have a pathway. Heisenberg did this by destroying particle paths in quantum mechanics. The important point that cannot be overemphasized is that this means that quantum mechanics teaches that the same electron does not move in a cloud chamber, therefore, you cannot say that any particle in a cloud chamber is splitting, changing direction, accelerating, being affected by magnetism, or disintegrating because at every position there is a new particle. In fact, all of particle physics which is based on the tenets of quantum mechanics is in error. Under the laws of quantum mechanics, no particle has a pathway. Therefore, it is impossible to test particles in a particle accelerator. There is no CP symmetry violation of kaons because you cannot say that the original kaon is the same kaon as the one at the end of the accelerator. All particles in fact cannot be accelerated under quantum mechanics because they cannot move in a path. All the testing in accelerators and observations in cloud chambers is superfluous and unscientific and without merit under quantum mechanics because quantum mechanics states that all particles are replaced at every point by a new particle and therefore you cannot observe a single particle doing anything whatsoever. You can't observe decay, you can't observe particle lifetime, you can't observe particle movement under quantum mechanical law. Heisenberg abolished particle movement with the Uncertainty Principle. Particles cannot move in a path. They can only do quantum jumps in atoms.

Destroying the path of the electron was no new idea for Heisenberg. Back in 1920 before he met Niels Bohr, Heisenberg's great friend was, the physicist, Wolfgang Pauli. Pauli had asked Heisenberg, "Do you honestly believe that such things as electron orbits really exist inside the atom?" After Heisenberg's non-commital reply, "But what is the alternative?", Heisenberg quotes Pauli as saying, "Yes, perhaps. Niels Bohr claims that he can tell the electron orbits of every atom in the periodic system, and the two of us do not even believe in the existence of such orbits." [91,p.36] One might claim that Pauli was putting words into Heisenberg's mouth, but it is Heisenberg who is relating the details some 50 years later and Heisenberg does not disagree with Pauli. In fact, we see clearly that Heisenberg institutes a model of the atom for the very purpose of, as he says, "killing off" once and for all the electron orbits.[78]

I take a moment to deviate here from the argument to contradict two foundation myths: 1. Heisenberg did NOT invent matrix mechanics. 2. As shown above Heisenberg decided that atoms didn't have orbits when he was 19-years-old before university and before studying physics.

What do I mean that Heisenberg did "not" invent matrix mechanics. Just that. His first paper written in Heligoland, does not include the name "matrix" and does not include a set of numbers in "matrix format," that is, no series of numbers in brackets. Heisenberg's first 1925 paper is not taught or even shown in school. Even now, only the German paper can be bought online from Springer.[xlv] In fact, Heisenberg starts by saying that he is going to eliminate orbits because they are unseen, then he jumps around with mathematical conclusions with no logical progression steps. Weinburg describes Heisenberg's first paper as "the paper of a magician physicist...Perhaps we should not look too closely at Heisenberg's first paper..." [xlvi]

I say let's look: What we find is that Heisenberg used regular algebra using the Fourier transform. He made up the equation to try to get a version for a linear atomic model that would jump between shells in the algebra, but against all the rules of normal algebra, his formula did not commute meaning the math could not be multiplied both ways. But Heisenberg knew this was wrong using normal algebra and didn't like it, because he was no advanced mathematician. So his first paper makes little sense to most physicists who read it. It has NO matrices.

Heisenberg sent his paper to two colleagues who were older and were also advanced mathematicians, Max Born and Paul Jordan. "They" realized that you can't use normal algebra if the equation does not commute, so "they" changed Heisenberg's equation to a matrix equation because that type of mathematics does not commute. Heisenberg did not invent matrix mechanics. Born and Jordan did. They didn't even put Heisenberg's name on the paper they wrote in 1926 called "Zur Quantenmechanik", Z. Phys. 34, 858-888 (1925). That is because Heisenberg couldn't help at all with the math. Heisenberg didn't know what matrix mathematics was. Yet, we always hear the myth that Heisenberg was so brilliant, he created a new math. It should infuriate us, but I don't believe that anyone cares what the truth is in science anymore. But we should. Because real science, good science, empirical science, can help us save this planet. And the good science is ignored and persecuted, because of cliques of leaders in the bad science like the science we find in the foundations of Quantum Mechanics. Now, science in the atomic field has gone completely crazy and off the rails. If you have read everything about the foundations of Quantum Mechanics that I've written in this book so far, you can't come to any other conclusion. Nobel laureates saying we've been brainwashed by Bohr, myths about Heisenberg inventing matrix mechanics, foundation theories that violate universal laws, and later theories that are even crazier because of bad foundations. Yes, there is emotion here. I'm angry. I'm angry because the answer is staring us in the face in the spectra itself. The paper at the end of this book gives 60 tables of how if we take a realistic look at the atom and do not violate any laws, we find that a wave exists around every particle creating the nuclear strong force and binding energy, and explains every single experiment and hypothesis of quantum mechanics with this one simple assumption. It's incontrovertible because it is real data from the real atomic spectra, yet no one wants to quit their quantum mysticism religion because it is a good approximation. No one wants the real truth even when they say they want the truth. It's frustrating. Please. Please. Just look at the data and tell me if there is any other way to interpret it. Yes, it will prove that the Standard Model is wrong. But the Standard Model is a mess anyway with its cancellation of infinities and stranger models like Everett's Many Worlds theory where every fraction of a second different universes are splitting off. Has everyone gone crazy? Well, you learned in school that Heisenberg invented Matrix Mechanics, so that must be true, right? It was Born and Jordan who wrote the equation that $qp - pq = i\hbar$. Not Heisenberg. Heisenberg wrote some kind of weird equation to try to simulate allowable spectral amplitudes and Born and Jordan made it into real math, but Heisenberg definitely did not.[xlvii] No wonder Einstein wrote to his friend Besso that Matrix Mechanics is a "veritable witches' brew where infinite matrices take the place of Cartesian coordinates," going on to say that it would be therefore hard to disprove.

So under quantum mechanics there are three equations to describe the atom (Schrödinger equation, Born/Jordan/Heisenberg matrix mechanics, Heisenberg Uncertainty equation), *none* which give the electron a periodic orbital pathway. Yet this is completely ignored by physicists who insist "when it suits them" that particles have pathways although this is forbidden by quantum mechanics whose rules they insist that they follow rigorously. Then you have particle physics where Cronin and Fitch win the Nobel Prize in 1964 due to their discovery that a neutral short kaon travels a pathway in a particle accelerator for a longer period than it should. If pathways are allowed for particles, then where is the basis for quantum mechanics? Where are the formulae for the pathways in quantum mechanics? Because as of now, there are none and never have been. Future generations of quantum physicists say they must play by the quantum mechanical rules and ignore the major one i.e. there are no pathways in the equations. This is the foundation of quantum mechanics. It is the foundation of the Uncertainty Principle, but it is ignored. Quantum mechanics teaches that the particles do not travel paths. Under the Uncertainty Principle every measurement made in a cloud chamber or a particle accelerator is creating a new particle with a fresh new wave packet that has not yet spread out. So what rules exactly are physicists following these days? Why even study the quantum mechanical equations? What do they represent? They certainly do not represent particles in an accelerator undergoing synchrotron radiation. And if the Uncertainty Principle was created solely for the purpose of proving that a particle did not have a pathway in the cloud chamber as Heisenberg himself admits, then why is the Uncertainty Principle considered valid when obviously science has ignored the sole purpose that it was created for and applied it to completely different experimental situations in which it was never meant to apply? Then science gives out Nobel prizes to those who violate the Uncertainty Principle by showing that particles in an accelerator do have a pathway and that the same particle that entered the accelerator has to be the same particles that decayed too slowly. It is not only unscientific, it is patently ridiculous.

Einstein predicted that doing away with orbital pathways in Heisenberg's matrix mechanics would lead to problems, but Einstein could never have predicted that science itself would just pick and choose the quantum mechanical rules at their pleasure. Heisenberg quotes Einstein as saying of the elimination of orbital pathways:

"I have a strong suspicion that, precisely because of the problems we have just been discussing, your theory will one day get you into hot water. I should like to explain this in greater detail. When it comes to observation, you behave as if everything can be left as it was, that is, as if you could use the old descriptive language. In that case, however, you will also have to say: in a cloud chamber we can observe the path of the electrons. At the same time, you claim that there are no electron paths inside the atom. This is obvious nonsense, for you cannot possibly get rid of the path simply by restricting the space in which the electron moves." [91,p.66] Although Heisenberg countered with the Uncertainty Principle, there is still no reconciliation between the path in a cloud chamber and the non-existence of an orbit in the atom. It has been pushed under the rug.

Based upon Heisenberg's admitted lame excuse for inventing the Uncertainty Principle, it must assuredly be untrue. It is often invoked to explain phenomena that cannot be understood, but such phenomena can be understood when one understands the existence of an energy wave that extends beyond the ground state shells of the atom and the electrons themselves.

And Heisenberg was wrong anyway. If you know current physics, then you know that the Uncertainty Principle used Planck's full constant originally because Heisenberg thought it was small enough not to affect measurement. He was wrong. Physics has had to make the number smaller than Planck's constant because now we measure so well that it does affect measurement. But does science care? No. They stick to this ridiculous rule anyway to try to explain all the things that Quantum Mechanics cannot explain. There are so many things that cannot be explained, so science says, it must be the Uncertainty Principle. Science can't explain these things because their model is wrong and based on poor foundations.

However, under this new model of the atom, moving particles are the same particle and they do not change into new wave packets with time. They behave the same way and under the same law in the atom, in a cloud chamber, in an electrical circuit, in a battery and in a cathode ray tube. They do have a pathway and they either radiate or do not radiate, depending upon the radiation that is captured in their energy wave and only excess energy is ever allowed to escape. Radiation is excess energy in the wave of each particle, therefore, particles radiate when they accelerate or decelerate, but not when their motion is constant in a balanced system. The law of nuclear equilibrium keeps the mass to energy wave ratio constant. The energy wave is a dynamic system. Therefore, there is only one particle in the cloud chamber that is moving and it does not radiate more than its mass to energy wave ratio because of nuclear equilibrium, not because the electrons are being replaced along the path by new electrons as Heisenberg had to assume to make sense of a model of the atom that was inherently flawed. In actuality then, according to Heisenberg himself, the Uncertainty Principle originally had nothing whatsoever to do with measurement such as Heisenberg's gamma ray microscope "thought experiment". Measurement was an excuse to introduce the Uncertainty Principle in order to justify matrix mechanics and land Heisenberg a good secular position. But Heisenberg knew the Uncertainty Principle had nothing to do with measurement, but rather it was the fact that an electron in a cloud chamber has a path and his matrix mechanics did not. So Heisenberg had to invent several electrons in different states to produce the effect of a moving single electron in order to explain how this could be reconciled with either matrix mechanics or wave mechanics. In order to do this he had to introduce an uncertainty in position and momentum to get to the explanation that several stationary electrons produce a path. Bottom line: The Uncertainty Principle is a sham. Heisenberg admits it is a sham by openly stating that the reason he invented the Uncertainty Principle was because his "version" of quantum mechanics i.e. matrix mechanics did not have a pathway for the electron and he needed to eliminate the pathway of the electron in the cloud chamber. Heisenberg further admits that Schrödinger's wave mechanics formula doesn't help because the electron as explained under it would grow too large in a cloud chamber. Einstein had challenged Heisenberg's lack of foresight and Heisenberg fearing to lose his ground, his reputation, and fearing loss of secular work cheated. He invented a complete fabrication based upon nothing, no evidence whatsoever. Heisenberg admits that he had many "pangs of conscience" about inventing it. [96] Why? Because it was a lie, a fraud, a cheat, to uphold his theory that he had already said to the world "could not be changed" and "the mathematical scheme of quantum or wave mechanics was already final". [74p.28] He could not embarrass himself by changing his matrix mechanics. Jobs were on the line. His future was at stake at 23 years old. The worst of it is that he didn't admit to the world how or why he invented the Uncertainty Principle in 1927. Instead, he invented some excuse about measurement and used an illustration which is called "Heisenberg's microscope" to explain the introduction of the Uncertainty Principle as if it had anything at all to do with measurement. It didn't! It was invented to hide the fact that matrix mechanics was flawed and didn't hold water. Schrödinger's wave mechanics had already dealt a crushing blow to matrix mechanics the previous year in 1926. That was only saved because Schrödinger pretended they were equivalent. They weren't equivalent at all in 1926 as will be illuminated clearly in the next few pages. Modern quantum mechanics based on Heisenberg's and Schrödinger's equations is a sham, a falsehood, a mockery of science, a tissue of lies, based upon the ambitions of men.

However, in this new theory being introduced in these pages, the true measure of uncertainty is the fact that matter is actually the subatomic particle and its energy wave that appears to create uncertainty in certain experiments when one is not aware of an energy wave extending from the particle. The wave carries a portion of the mass of the particle under Einstein's equation and a portion of its charge. Trying to measure solely the position of the mass of the particle is not going to give an exact measurement. Rather it is the position of the particle and its energy wave that describe the position of the particle. Therefore, the wave around the particle causes diffraction and an uncertainty of position, but not if one recognizes the wave itself as part of the particle's matter.

Tunneling is not caused by the uncertainty principle, but by the fact that a particle is surrounded by an energy wave. The space between a tip in a scanning tunneling microscope and the specimen is filled with particle energy waves.

"A very fine needle is brought very close to a sample surface, for example a crystal surface whose structure is to be examined. When very close, electrons might jump across the gap from sample to tip; especially if an electrical potential is applied (e.g. by connecting the tip to a battery and the sample to earth). The jump across the gap is explained within quantum mechanical theory by the phenomenon of tunneling. The electrons tunnel through the vacuum despite the classical, non-quantum mechanical theory predicting that they do not have the energy to surmount the obstacle provided by the vacuum. The tunneling electrons amount to an electrical current that can be measured with great precision. Quantum theory predicts that the tunneling current is very sensitive to the distance between tip and sample: proportional to the inverse of the distance squared."[51]

The energy of the wave surrounding a particle drops off in an inverse square law, because all energy waves emanating from a sphere fill larger and larger spheres whose surfaces are proportional to the square of the radius. Therefore, the energy wave between two spherical particles is proportional to the inverse of the distance squared as well. Physicists believe that there is a vacuum or space between the tip and the sample of the STM, but in actuality there is an energy wave emanating from the particles of both the tip and the specimen.

There is a population of individuals who cling to the Uncertainty Principle because they think it implies the only kind of universe that includes randomness instead of determinism. Personally I am not interested in philosophy. However, humans tend to turn observation and empirical evidence into a philosophy. I am speaking of this in a general sense in that human observation of the real world has created all philosophies and religions that have ever been in human history. This appears to be an evolutionary survival mechanism so I cannot question it. But science appears to have concluded that if classical physics is correct that means the universe is controlled by determinism and that if the current QM Uncertainty Principle is correct, the universe is indeterministic. Personally, I think that more complex systems develop random behavior which creates an indeterministic future for complex systems. I believe biology is a complex system and contains inbuilt randomness, as in i.e. genetic recombination, and that dispensing with the Uncertainty Principle does not change the randomness of complex systems. The point being that it doesn't matter if we eliminate the Uncertainty Principle because it was invented for the sole purpose of making it seem like atoms were impenetrable under normal circumstances so we don't fall through or see through all atoms. Personally, I believe that everyone is entitled to their own beliefs and philosophy but that this should not affect the proofs of the observations of scientific experiments. The philosophy should not dictate the interpretation of the experiment. The experiments have been done. The interpretation is incorrect.

In conclusion regarding philosophical arguments of determinism versus indeterminism, in the end, science should try to explain reality. Whatever the philosophic implication of the reality is should not be considered. Leave that to philosophers. That is not the science. And in the end, even if this new theory were to be embraced by the scientific community in the future, it does not prove determinism. No scientific theory yet stands the test of time as measured in centuries. Science is an evolving study of nature. Humans have only had the ability to write down their thoughts for a few thousand years. Compared to the amount of time that Homo sapiens have existed, we have just begun to learn. That is the beauty of science. There is always more to learn. Nature is complex and science is in its infancy.

The Uncertainty Principle is often given for the reason of Quantum Entanglement, but in reality, they are unrelated. Quantum Entanglement was actually discovered by Einstein when he was disputing the superposition of the Schroedinger wave equation. In fact, Einstein's arguments with Quantum Mechanics often led it to its foundation postulates.

Lasers are not based on the atomic models of Heisenberg and Schroedinger. Lasers are based on Einstein's photon theory of light.

Quantum Entanglement was Schroedinger's name for Einstein's argument against the changes at a distance described by Schroedinger's wave equation. Einstein with his two assistants wrote a paper called EPR theory which presented the superposition problem as untenable. Then Schroedinger realizing his equation did allow for correlations at a distance coined the term Quantum Entanglement in German.

In actuality, Quantum Entanglement is a result of the wave state of the particle being seen together at the same time as the particle state. They are inseparable. Since measures the photon of a laser beam shot in different directions and measures the polarization of the photon while forgetting that the laser beam is also a wave.

As seen from this theory of the atom, the quantum discontinuous states are not really discontinuous at all. They are merely only allowed discrete energies because the particles are restricted by the nuclear waves in which they travel.

As seen in this new model, when a particle moves, its wave front is moving. With massless particles like the photon, they are pure wave with energy pulses. The energy packets or quanta pulse so they act like particles. However, the wave form is never lost, not even in a laser.

So to understand entanglement, we can imagine the photons from the laser moving in a wavefront in the same way that we might consider a wave on a lake. If a prow of a boat hits the wave at wave normal, the wavefront splits to either side of the boat. But the two wave fronts remain in phase with respect to each other no matter how far apart they travel. If you measure the wave at a particular time on one side of the boat, you'll get a crest at the same time you measure the wave on the other side of the boat. If you measure a trough on one side of the boat, you will measure a trough on the other side of the boat as long as nothing interferes with one of the waves. This is entanglement. A photon is both a particle and a wave at the same time, so the photon keeps the wave property when a photon is split.

Many times when people speak of the great advances of Quantum Mechanics, they believe this applies to the success of the Heisenberg/Schroedinger model of the atom. However, most of the time this applies to the Einstein quantum mechanics of light that describe the photon.

28 Real Problems for Quantum Mechanics

We have seen that the quantum mechanical equations do not allow for moving electrons in periodic orbits. In fact, they do not allow electrons to move and radiate at the same time. This is generally ignored by particle physics and nuclear physics when convenient. The fact is that the branches of science that are supposed to rely on quantum mechanics to set the rules often ignore those very rules. There is no equation that has the electron particle itself orbiting or moving. Matrix mechanics absolutely does not. Schrödinger wave mechanics absolutely does not. The only thing one gets with Schrödinger's wave equation is a spreading of the wave packet of the electron with time.

Einstein explains it thus: "The difficulty appeared insurmountable, until it was overcome by Born in a way as simple as it was unexpected. The de Broglie-Schrödinger waves were not to be interpreted as a mathematical description of how an event actually takes place in time and space, though, of course, they have reference to such an event. Rather they are a mathematical description of what we can actually know about the system. They serve only to make statistical statements and predictions of the results of all measurements which we can carry out upon the system."[62,p332]

As another author puts it: "Schrödinger initially thought that the particle was genuinely smeared over space and that p(x; t) represented a literal mass density. Such an interpretation is rapidly seen to be inconsistent with physical reality, however. In practice, wavefunction solutions are often found to spread very rapidly in space whereas electrons, when observed, always seem to be point-like. Such an interpretation would not be consistent with the observation of particle tracks in cloud chambers, for example."[80]

Heisenberg explains further: "Born had made a first step by calculating from Schrödinger's theory the probability for collision processes; he had introduced the notion that the square of the wave function was not a charge density, as Schrödinger had believed; that it meant the probability of finding the electron at a given place."[74,p28]

So Schrödinger's stationary electron wave packet that did not describe the particle, only the de Broglie wave packet, was not moving in a periodic orbit, however, Born showed that squaring the wavefunction created a probability cloud for the wave packet.

As Heisenberg explains it:
"Then came the transformation theory of Dirac and Jordan. In this scheme, one could transform from $\psi(v)$ to, for instance, $\psi(\rho)$, and it was natural to assume that the square $|\psi(\rho)|^2$, would be the probability of finding the electron with momentum ρ. So gradually one acquired the notion that the square of the wave function, which, by the way, was not a wave function in three-dimensional space, but in configuration space, meant the probability of something."[74] This however was not the intent of Schrödinger in producing wave mechanics as Heisenberg points out: "Schrödinger at that time regarded his waves as true three-dimensional matter-waves—

comparable, say, to electromagnetic waves—and wanted to eliminate entirely the discontinuous features of quantum theory, especially the so-called quantum jumps."[74,p.51] Schrödinger's wave mechanics of 1926 is based on de Broglie's matter-wave, but if Schrödinger's wave mechanics does not describe "three-dimensional matter-waves", then what did Davisson-Germer find in their experiment? Furthermore, where is the electron mass in the Schrödinger equation? Schrödinger had depicted the mass of the electron in his equation as spread throughout the standing wave and surrounding the nucleus as Heisenberg explains: "Schrödinger thought that the wave picture of the atom—with continuous matter spread out around the nucleus, according to its wave function—could replace the older models of quantum theory."

The problem with this is as Einstein pointed out. "...It proved impossible to associate with these Schrödinger waves definite motions of the mass points..."[62,p332]

The point is that without movement of the electron particles in periodic orbits in the atom, there is no magnetic quantum number. There is no orbital momentum. There is no magnetic dipole moment. So what is quantum mechanics describing? A set of contradictions that need to be ignored and rules that need to be broken and mathematics that can only be calculated for hydrogen though in theory should be able to be calculated for heavier atoms although this has not been proven to be the case. One doesn't even use Schrödinger's equation or any other quantum mechanical equation to describe atoms. One must use the empirical data of Ritz's combination principle to describe atomic transitions. And ionization energies in atoms were described through Moseley's x-ray scattering experiments.

What exactly was going down in the 1920s in physics? We have the strange introduction of Planck's constant into the atom due to it being the proper "length" while there was no wave at all thought to be in the atom. We have Bohr introducing angular momentum of 1/2h due to Planck saying that 1/2h would not radiate so Bohr introduces half revolution orbits. And then Bohr has a ground state orbit in which the electron continues to make full orbits permanently. We have the strange interpretation of the Davisson-Germer experiment that defied the Bragg law. We have Heisenberg erasing the electron path from the cloud chamber because it doesn't fit his matrix mechanics. We have Schrödinger inventing wave mechanics with the intention of disproving quantum jumps and giving a picture of the atom where the electron was not a particle, but its mass was spread out over the de Broglie wave. We have Bohr and Max Born reinterpreting Schrödinger's wave mechanics against his will and still saying that it applies even though its creation was in direct conflict with quantum mechanics Copenhagen interpretation. Schrödinger's wave mechanics was devised to describe something very different than what it was twisted into:

"But Schrödinger, whose sympathies lay with the older classical physics, made more far-reaching claims for his wave mechanics. In effect, he claimed that it represented the reality of the interior of the atom, that not particles but standing matter waves resided there, that the atom was thereby recovered for the classical physics of continuous process and absolute determinism. In Bohr's atom electrons navigated stationary states in quantum jumps that resulted in emission of photon of light. Schrödinger offered, instead, multiple waves of matter that produced light by the process known as constructive interference, the waves adding their peaks of amplitude together."[82,p.128]

Bohr hounded Schrödinger day and night even when he was ill sitting on his bed and arguing that the quantum jumps were necessary and Schrödinger's equation could not explain how light could be emitted according to Planck's law of energy emission. One source describes Bohr's discussion with Schrodinger as Bohr insisting "fanatically and with almost terrifying relentlessness on complete clarity in all arguments."[82,p.129] It is surprising that Bohr insisted upon complete clarity when he had introduced complete ambiguity into quantum mechanics and complete acceptance of contradictions as being "wholeness". In fact Schrödinger's argument with Bohr on this occasion consisted of the basic argument accusing Bohr of not having clear explanations, of not having clarity of relationships, and therefore why should not Schrödinger's interpretation not be fully clear. Schrödinger responded to Bohr:

"I don't for a moment claim that all these relationships have been fully explained. But then you, too, have so far failed to discover a satisfactory physical interpretation of quantum mechanics. There is no reason why the application of thermodynamics to the theory of material waves should not yield a satisfactory explanation of Planck's formula as well—an explanation that will admittedly look somewhat different from all previous ones."[91,p.75]

In other words, Schrödinger argued that Bohr himself had not shown clarity in all arguments in quantum mechanics, so why should Schrödinger? And furthermore Schrödinger had pointed out on this occasion Bohr's use of contradicting physics in quantum mechanics.[91,p.74] But Bohr would not allow Schrödinger to assume that a solution could be found to Planck's formula through Schrödinger's interpretation. He replied, "No, there is no hope of that at all." Ironically, Bohr's interpretation did not possess the clarity that Bohr was fiercely, fanatically and terrifyingly demanding of Schrödinger. Bohr's interpretation by his own admission on many separate occasions was "poetic", "abstract", "symbolic", and demanded a "renunciation" of physical description. Is this clarity? Yet, Bohr obsessively insisted upon clarity from Schrödinger when Heisenberg's matrix theory of quantum mechanics, which Bohr was here endorsing, was the farthest thing from clarity that can be imagined. Matrix mechanics had removed periodic pathway orbits from the atom, described discontinuous quantum jumps, and had no "real" correspondence to classical mechanics as known in the modern world. Bohr explains:

"…that Heisenberg should find a stepping stone for the development of formalism of quantum mechanics, from which all reference to classical pictures beyond the asymptotic correspondence was completely eliminated. Through the work of Bohr, Heisenberg, and Jordan as well as Dirac this bold and ingenious conception was soon given a general formulation in which the classical kinematic and dynamical variables are replaced by symbolic operators obeying a non-commutative algebra involving Planck's constant."[90,p.88]

Schrödinger wondered rightly why Bohr could be so cryptic as to describe quantum mechanics as "symbolic" and yet Schrödinger was being forced to be unequivocally precise, unambiguous, and crystal-clear particularly "un-symbolic" in his explanations. But Bohr wasn't playing fair here, he was playing for keeps, and he was out to win a debate which takes the ability to manipulate words and that was something that he excelled at.

After Bohr's unrelenting arguments with him, "Schrödinger approached desperation. 'If one has to go on with these damned quantum jumps,' he exploded, 'then I'm sorry that I ever started to work on atomic theory.'" [82,p.129]

The circumstances here are amazing to think about. Bohr had invited Schrödinger to his home and then had argued with him and bullied him into a state of illness. As Heisenberg later recalled:

"Bohr's discussions with Schrödinger began at the railway station and were continued daily from early morning until late at night. Schrödinger stayed in Bohr's house so that nothing would interrupt the conversations. And although Bohr was normally most considerate and friendly in his dealings with people, he now struck me as an almost remorseless fanatic; one who was not prepared to make the least concession or grant that he could ever be mistaken. It is hardly possible to convey just how passionate the discussions were, just how deeply rooted the convictions of each; a fact that marked their every utterance."[xlviii]

This was not an isolated incidence with Bohr. It was well known that Bohr was relentless when someone did not agree with him. In the book The Age of Entanglement, Louisa Gilder, relates that Bohr put Kramers into the hospital because Kramers had agreed with Einstein about the photon.[xlix]

Both Heisenberg and Schrödinger were anxious that their equations be given precedence. However, Schrödinger realized that he hadn't included the line intensities in the spectrum and Heisenberg realized that no one liked working with his unusual matrix mathematics.

"He [Schrödinger] also argued for the superiority of wave mechanics over matrix mechanics. This provoked an angry reaction, especially from Heisenberg, who insisted on the existence of discontinuous quantum jumps rather than a theory based on continuous waves."[38]

Schrödinger said of Heisenberg's matrix mechanics:
"I knew of [Heisenberg's] theory, of course, but I felt discouraged, not to say repelled, by the methods of transcendental algebra, which appeared difficult to me, and by the lack of visualizability." [38]

And Heisenberg said of Schrödinger's wave mechanics:
"The more I think about the physical portion of Schrödinger's theory, the more repulsive I find it...What Schrödinger writes about the visualizability of his theory 'is probably not quite right,' in other words it's crap." [38]

Because Schrödinger realized that he couldn't think of a good explanation for what Bohr had said about his theory not explaining Planck's law and probably because he realized he had forgotten to include spectral line intensities, it behooved Schrödinger to write a paper saying that he had proven the two theories, matrix mechanics and wave mechanics, as mathematically equivalent. In this way, neither Schrödinger nor Heisenberg would lose out on the glory. But was this the truth?

In a paper entitled "The Equivalence Myth of Quantum Mechanics", it is proven that when Schrödinger made his assertion in 1926 that "at the time matrix mechanics and wave mechanics were neither mathematically nor empirically equivalent. That they were is the Equivalence Myth....That matrix mechanics and wave mechanics were ontologically distinct, in the sense of making conflicting assertions concerning atomic reality, was obvious from the very beginning and recognised by all the players."[87] Another example of a tissue of lies and a mockery of science in the name of ambitious men seeking a name for themselves and careers.

"...the nearly simultaneous emergence of two theories and their alleged almost unanimous acceptance by the physical elite barely 2 years later (at the Solvay Conference of October 1927 in Brussels) constitute a rare sequence of events in the history of physics..."[87] The motives of all physicists involved in quantum mechanics may be safely called into question during the 1920s which calls into question the very foundation of quantum mechanics and its validity as a scientific theory of the atom. Will Science take as long as the Catholic Church took to apologize about Galileo before they will stand up and admit that modern quantum mechanics is based on a general conspiracy and cover-up?

Of course, Schrödinger's equation and its expansion by Dirac to include spin and relativity are wonderful "mathematical" achievements, but do they really describe the atom? Dirac seems to have thought that was *not* important. He is said to have said of Schrödinger's failure to include relativity:

"And back in the 1960s, Paul Dirac famously asserted that: 'It is more important to have beauty in one's equations than to have them fit experiment.' ... Dirac's statement at the start of this essay on the importance of 'beauty in one's equations' was intended for Erwin Schrödinger. In Schrödinger's first attempt to concoct his famous wave equation, he looked for one that agreed with relativity theory. The equation he came up with, however, was not supported by experiment. Eventually he produced the Schrödinger equation, which was not beautiful, but did at least fit the data. Dirac thought that Schrödinger should have ignored the data and persevered in his pursuit of a beautiful equation."[81]

Dirac said: "I think that there is a moral to this story, namely that it is more important to have beauty in one's equations that to have them fit experiment. If Schrödinger had been more confident of his work, he could have published it some months earlier, and he could have published a more accurate equation. It seems that if one is working from the point of view of getting beauty in one's equations, and if one has really a sound insight, one is on a sure line of progress. If there is not complete agreement between the results of one's work and experiment, one should not allow oneself to be too discouraged, because the discrepancy may well be due to minor features that are not properly taken into account and that will get cleared up with further development of the theory.
Scientific American, May 1963."

Unfortunately, because the Greeks thought it was more important that the universe have beauty, the idea of circular orbits held sway for 2,000 years. Even Copernicus did not allow himself to believe that the orbits could be anything but perfect circles although he put the sun at the center. Thus the beauty of an equation can be the undoing of science. Some will argue that

Dirac's beautiful equation predicted antimatter. But did it? Is antimatter really antimatter? Dirac's equation predicted negative energy. Heisenberg called it the saddest chapter in theoretical physics. The question is whether a positron really has negative energy. The answer is no. A positron merely has opposite charge. It comes back to the vast number of particles in the Standard Model. If you predict a particle, you are likely to find one closely matching its description. There is no antimatter. There are only particles with opposite charge that combine together when they collide. Negative energy has not been proven. This is irrelevant to the discussion of the barrier in the atom between the electrons and the nucleus, but one must question whether or not these beautiful equations based as they are on weak foundations, false assumptions, random ideas, and arbitrary rules which have been shown in this treatment to include theories that ignore evidence, suborbitals that do not coincide with frequencies of the shells, introduction of Planck's constant before a wave was known, electrons without pathways creating magnetic moment, and all manner of irrelevant theory forced into a formula with known elements of spectral analysis and known elements of special relativity can really be sound.

Another author ascribes Schrödinger's failure and Dirac's comment not to the inability to incorporate relativity, but to the inability to describe helium.[73]

In this new theoretical model of the atom, the magnetic moment comes from the movement of the energy wave of each particle.

The Schrödinger wave equation can only be solved for one electron. Or at least until supercomputers it could not be solved for more than one electron at a time. This questions its real usability as a model of the atom.

One might say that quantum mechanics has been very successful. Didn't it produce the atomic bomb? The answer to that is an emphatic, "No, it did not." Classical mechanics produced the bomb. It was literally the work of Ernest Rutherford's scattering experiments that produced the atomic bomb. You don't have to know anything about the atomic spectrum, the atomic transitions, or electron shells or suborbitals as described by the quantum mechanical equations to create the bomb. The equations were all created strictly to describe electrons. The atomic bomb was created by experiments in nuclear physics i.e. the nucleus. Ernest Rutherford by 1911, i.e. two years before Bohr's theory, had shown that the nucleus contained the mass at the center of the atom.[4] This discovery by Rutherfold led him to eventually discover that the weight of helium was four times the weight of hydrogen although the charge on helium was only twice the charge on hydrogen. Rutherford postulated in 1920 the existence of a massive particle called the neutron which was discovered by his co-worker James Chadwick in 1932.[83,84] A Hungarian theoretical physicist, Leo Szilard, took notice of Rutherford's work.

"Sometime during the 1920s, a new wave of research caught Szilard's attention: nuclear physics, the study of the nucleus of the atom, where most of its mass—and therefore its energy—is concentrated....The nuclei of some light atoms could be shattered by bombarding them with atomic particles; that much the great British experimental physicist Ernest Rutherford had already demonstrated. Rutherford used one nucleus to bombard another, but since both nuclei were strongly positively charged, the bombarded nucleus repelled most attacks....The neutron, a particle with nearly the same mass as the positively charged proton that until 1932 was the sole certain component of the atomic nucleus, had no electric charge, which meant it could pass through the surrounding electrical barrier and enter into the nucleus. The neutron would open the atomic nucleus to examination. It might even be a way to force the nucleus to give up some of its enormous energy…. 'As the light changed to green and I crossed the street,' Szilard recalls, 'it…suddenly occurred to me that if we could find an element which is split by neutrons and which would emit two neutrons when it absorbs one neutron, such an element, if assembled in sufficiently large mass, could sustain a nuclear chain reaction.'"[82]

The progressive investigations of radiation by Madame Curie, the Joliot-Curies, and Enrico Fermi as well as others, and Fermi's nuclear bombardment, and Frederick Soddy's discovery of isotopes was unfortunately all that was primarily needed to develop the bomb. One didn't use quantum mechanics to create the atomic bomb, but rather it was based on the classical scattering experiments of Rutherford and Chadwick and the finding of an element that reacted to neutron bombardment. To find the solution to the bomb, Enrico Fermi irradiated every element from hydrogen to uranium until he found the one that would chain-react. The details of the bomb were based upon experiment, not upon quantum mechanics formula which thought the nucleus to be a solid sphere even after the discovery of the neutron so that Bohr had great difficultly accepting the fission of uranium until he realized the nucleus was not solid.[82] Not much information was known about the nucleus in the 1930s and 1940s because quantum mechanics dealt with electrons but that did not prevent the development of the bomb. In fact today, there is no "beautiful equation" that accurately describes the nucleus. The point is science does not yet have a complete picture of the atom as Einstein never gave up pointing out until his death.[74]

Another real failure of quantum mechanics is described in this explanation of particles in a particle accelerator.

"Portable sources of gamma-rays are useful for some X-ray studies, but most X-rays are generated from electron beams. Deceleration of a free electron results in the emission of a continuous spectrum of electromagnetic radiation (Bremsstrahlung: German for braking radiation) with a limiting frequency n = E/h, where E is the kinetic energy of the electron. The spectrum has a toroidal angular distribution at low electron velocity, but is compressed into a fine jet at high velocity, as is exploited in the continuous spectrum produced by the centripetal force in an SSR. Interactions between electrons and atoms in a conventional Xray tube, an EMP, or an EM, produce a continuous spectrum from deceleration during angular deflection and a characteristic spectrum from electron transitions between quantized energy states in atoms ionized by emission of one or more electrons."[79]

The problem is as was known to Heisenberg. Quantum mechanics is specific. It has hard fast rules. However, continuous emission by an electron is a complete collapse of the quantum mechanical system. Everyone knows this but ignores it outright. Why? Because there is nothing else except quantum mechanics. No one has come up with another viable solution so they stick with a failed theory because it measures parts of the atomic world.

But this is a blatant contradiction. How can quantum mechanics explain continuous emission? This is not allowed. Yet, when an electron is decelerated, there is an "emission of a continuous spectrum of electromagnetic radiation". Why does a free electron refuse to obey the rules of quantum mechanics? The answer is simple. The rules of quantum mechanics do not apply. The electron may emit across the continuous spectrum when it is not in the atom because it is not making transitions between nodes of the waves of the energy wave the proton or waves of the nucleons. Therefore it emits at will. Whatever excess energy is given to the electron, it will emit in any amount. It does not need to be balanced according to the energy needed to remain inside a particular node of the atomic nucleus. The fact that there is a limiting frequency of emission is wholly and completely explained by the law of nuclear equilibrium. No particle may emit more energy than its excess energy and the ratio between its mass and energy wave must remain the same.

29 Inertia and motion

The universe strives to maintain all systems in balance. This is why in any field of science based on physics whether chemistry, biology, astronomy, medicine, etc. one will find the use of equilibrium, balance, symmetry, and reactions. Entropy is nothing more than an attempt toward balance, sometimes called sameness. Entropy is the universe's natural law of balance, attempts to achieve the same temperature, the same energy, the same charge. Even the human body tries to maintain balance sometimes at the expense of health, sometimes to its benefit. (That however is off subject.) The point of balance is that even large bodies such as planets or comets or anything acting under inertia seems to have a certain amount of intrinsic energy to mass ratio. It is different from the atom, but analogous. Newton said any body or object such as a planet will continue in its motion unless acted upon by another force. The body is said to be moving under inertia. Inertia is then a balance or equilibrium of the system of the mass of the body and the energy it is carrying. Mass carries energy. Therefore, whatever energy a body is carrying in space compared to its mass creates its velocity which is called inertia. Only absorption of energy will change the course of the body, because there is an unbalance on the system. The body emits the excess energy and continues on its new course in opposition to the direction that it has emitted the energy. This is seen in rocket fuel expelled in space. But the same principle applies to a baseball thrown into space (where there is no other gravitational field to complicate the system—all theories must start without the complexity of several variants in the system). It carries the energy of the throw in its mass and will retain that energy with its mass until it is forced to absorb or emit that energy. If another baseball is thrown into space so that its path will intercept the path of the baseball and it will impact the baseball, when both objects collide there is an exchange of energy with each absorbing and emitting. Both baseballs absorb energy from the direction of the path of the other ball and both baseballs emit the energy absorbed in the direction of the incoming ball. Therefore both balls will change direction and velocity according to the amount of energy emitted and absorbed. This is just a small variation on Newton's laws of motion. Instead of opposing forces, we have forces that do not oppose but rather absorb the energy and re-emit the energy in the same direction that it came. Where the energy carried by each ball is uneven, the ball with the most energy makes the ball with the lesser energy absorb more energy than the other. That energy is emitted in a more forceful way and it increases the velocity or carrier energy of the slower ball. The equations result in the end to be exactly the same as Newton's; the analogy is different—more like the atom. Inertia is mass that carries excess energy and that energy does not change unless more is absorbed or emitted. Gravity is the natural mass to energy ratio. An object may absorb or emit more or less energy but the natural mass to energy ratio remains the same so that gravity always remains the same according to the mass. This is not a complete answer yet, at least, not without the mathematics. This does not explain gravitational attraction, the gravitational constant or the perihelion of Mercury. However, neither does general relativity explain the tides nor the separation of solar systems and galaxies, nor the plane of the ecliptic of solar systems and galaxies. This is merely a way to view the macro world in a similar way to the atomic world.

If we carry the analogy of gravity completely to this model of the atom, we have an interesting picture. Parallax of a star's light occurs not exactly because spacetime is curved. But rather the gravitational field around each celestial body is a series of consecutive spheres of energy that deflect electromagnetic waves that the system does not need to maintain balance or that the system is not forced to absorb through intense energy of the wave. Therefore, one has a system of energy about the sun that is stronger near with each consecutive sphere of energy that approaches the sun's surface. A portion of light from a star that is slightly behind the sun that is not absorbed by the sun will travel the path of least resistance around the spherical energy field of the sun. That means that the light not absorbed curves around the sun's energy field. This does not mean however that spacetime itself is curved, but that the energy fields around each celestial body form concentric spheres of energy. Einstein's field equations would be applicable but incomplete, because they neither account for an actual force pressing upon the celestial body in order to create a spherical pressure nor does spacetime curvature account for the force of the spherical fields of nearby celestial bodies affecting each other such as the moon's spherical energy field's affect on the earth and because they do away with the gravitational force, there is no accounting for an inertial path having a one-way direction when directed toward the center of a body. But this should remind us that models of the universe are strictly models and not reality. The more accurately that the model defines reality, the more useful the model is, but that does not make the model real anymore than a futuristic exact robotic replica of a human can ever be a human. To make the point even clearer: if one sends this replica to aliens in order to show what humans are like, then the aliens may think they know how humans behave, what humans look like, and what they can expect from humans, but the aliens will not really know what a human is.

When one thinks of the main differences between Newtonian gravity and Einsteinian gravitation, one thinks of the differences in the model i.e. especially of the curvature of spacetime. But if one considers that one of the main differences is that Newton explained the two-body concept of gravity and in his concept he imagined that the mass of each body could be pushed spherically toward its center until it created a point mass, yet the force of gravity would remain the same between the two bodies and this did not explore the effect of the density of a single body. This is where Einstein differs. Einstein's theory, aside from its conceptualization, is showing how a single body changes its gravitational field in its own space. Einstein showed that density enhances a single body's gravitational field. Therefore, solutions to the Einstein equations show gravity wells that become deeper as the mass is condensed for the single body. If one draws an analogy between this model of the atom and gravity, one can see that if the mass of the atom i.e. the nucleons and electrons are drawn closer together, then their energy wave fields expose more of the high energy waves near the particle. In this way, the pressure increases for the atom when the atom becomes denser.

Einstein's great idea was not that gravity was equivalent to acceleration because that just gives you Newton. As the distance increases the energy is greater in either direction depending upon the mass. Einstein's original idea that set him apart from Newton was his solution to describing gravity around a sphere in terms of acceleration. His idealization or visualization of curved forces added another factor to Newton's force between two bodies being dependent upon the masses. (Although technically it has been said that Einstein eliminated gravitational "force", but did he? Einstein's field equations describe a "gravity well", but what makes bodies fall into the gravity well if not a force?) Another Einstein novelty was incorporating the additional factor of density by curving spacetime. Two objects of the same mass have two equal sized gravitational fields—if you will, energy wave fields. Therefore, the size of the field depends upon the mass, but the strength of the field at any particular place near one body varies with the density. This was the new factor incorporated by Einstein. If one compresses the atoms, one gets closer to its higher energy waves and the pressure increases with density. For two bodies with equal mass, the energy fields are exactly the same with the radius from the center, but the force one feels near a single body depends on how much of the atomic energy wave field is exposed. The denser the body, the more exposure there is to the higher energy waves, therefore, the more pressure there is for a denser body. The denser the body, the closer the electrons are to the nucleus of each atom in the body and the more the strong force is exposed. Also, more the incoming waves in the energy field are exposed causing a greater incoming photoelectric effect upon each atom of the body. In a gaseous nebula in space of the same mass as a neutron star, the atoms in the nebula do not feel the same pressure as the atoms of a neutron star because the atoms of the nebula are not in the higher energy waves of the field because the field is not dense enough in the nebula. The two bodies, the nebula and the neutron star attract each other with the same force according to the distance between them, because their masses create equal sized fields that fall off with equal energy. However, density plays a key role. Stars and planets form only when enough mass becomes dense enough, in the higher energy parts of the field, to pull the atoms toward each other. Einstein is equivalent to Newton when it comes to mass at a distance between two bodies, but different when it comes to the density of a single body. This is the main difference between the two approaches no matter what the model was that was used to arrive at the equations.

The problem with the macro world energy fields and the atomic world energy fields is whether to describe them as the same field extending out from the atom into space itself or as a secondary field. It seems certain that a secondary field is created from the primary atomic field. This displays itself in the magnetosphere of the earth. It seems that the atomic energy fields create around their exteriors—possibly in a honeycomb effect—a secondary lower energy field. This creates many macro world questions: Does the existence of a primary atomic energy field versus a secondary macro world energy field explain the Landé g-factor? (Or is the g-factor evidence of the two fields in the hydrogen atom, the proton field and the electron field, both singly charged fields, constantly revolving from source to field horizon?) Is gravity a secondary energy wave field that creates a weaker field around macro world objects? If so, does gravity work to hold the nucleus of the solar system together and keep other solar systems separate from each other? Is this true for galaxies? The waves of the secondary field may be submicroscopic or even of flat amplitude which would account for the energy not being measurable but the effects being seen. Macro world objects would not fit into the nodes of such a wave whether submicroscopic or flat so there is no observable macro world moving from node to node although the atoms of the macro object would be doing so. What causes solar systems and galaxies to exist on the plane of the ecliptic? Is it the reversing of the magnetic poles of the sun every few years that keeps the solar system on the plane of the ecliptic since the poles of the sun are perpendicular to it?

If gravity is actually what science calls the strong force yet is a secondary effect thus weaker due to the strong force being shielded by the electrons, then the universe is held together by a single force. Energy in a loop presses against mass and free energy radiates. Electromagnetic waves push electrons off of metal in the photoelectric effect. Therefore, electromagnetic waves could push electrons and nucleons together into a massive planet if they looped back to that planet. The more energetic pressure of the electromagnetic waves would be pressing against the individual nucleons and held at a distance from the macro world by the electrons. The particle wave is atomic and reaches toward the tip of a scanning microscope, but not wholly into the macro world. Therefore macro world effects would be a secondary field. Gravity and light would be forms of the same energy in different configurations and strengths, and the universe would be filled with both.

The theory of entropy given by science is that the universe will run down if it continues to expand, that is, the universe will reach absolute zero and die. However, $E=mc^2$ and the conservation of energy and matter means that energy never dies. Even if the entire universe were to turn into energy, the energy would move at the speed of light because that is how it is released and inertia keeps it moving at that speed. As long as the universe contains radiation, it contains energy at lightspeed. Energy released at gamma ray velocity has a tendency to create matter. Matter has a tendency to release excess energy. No one can stop pure energy from traveling at light speed therefore no one can kill the universe. On the other hand, if all energy slows down by becoming matter, there is still an energy wave field about each particle causing pressure on the particle. That pressure is enough at near absolute zero to cause an implosion in an imbalanced magnetic field i.e. a sweep across the Feshbach resonance. That implosion releases the bound energy and the mass is lost to energy. Oppositely charged particles tend to assemble to maintain balance. Therefore, atoms never die, but have a tendency to form to maintain balance. Energy never dies and cannot slow down. Absolute zero is no heat, but that is because there are no electron transitions creating heat, but the atom still exists at absolute zero so the energy wave field is still moving fast enough to keep the particles held together.

30 Heat

It might be speculated that some of the excess absorbed energy does not show in the emission spectrum because the wavelength of the excess energy re-emitted is too low in intensity to measure. However, it appears more likely that the absorption of excessive energy in electron transitions of less than a quantum must result in the excess being trapped by the Planck barrier as heat held by the atom and the loss of less than a quantum of light must remain unobserved or is observed only as an indistinct blurred spectral line. Any excess energy for that n-shell or node causes the electron to higher oscillation and the excess energy is released as heat through contact with neighboring atoms. This does not affect the continuous spectrum background of absorption spectrometry due to the partial quantum of light being unable to either remove light or replace light into the spectrum. A partial quantum of light is incapable of movement across the Planck barrier.

The oscillation of an electron in the same shell is caused by the absorption of too much energy in the form of half a quantum or less than a quantum, but not sufficient energy to pop through the nuclear wave field antinode to the next outer orbital. However, this excess energy can be at any frequency according to the amount of excess energy that was absorbed causing the oscillation of electrons at that frequency. If the excess energy is great enough to form an electromagnetic wave, the energy from excess absorption is then emitted as the exact frequency in excess of the difference needed between inner and outer electron shells. Electrons can absorb any amount of excess energy. The excessive energy absorbed causes the electron to vibrate at the frequency absorbed and this in turn emits the excess energy as a wave of the same frequency. The light of this excess electrical energy is not seen in the absorption spectrum because the exact excessive frequency of light absorbed is re-emitted at the speed of light and the absorption spectrum only appears to show the exact wavelength of the energy that is required to be absorbed for the transition of an electron to a higher electron shell. As individual atoms are absorbing and re-emitting the excessive energy at different rates, the spectrum appears to show a continuous spectrum with only discrete absorption wavelengths for each element.

If the necessary energy needed to be absorbed by an electron to pop outward to the next outer orbital is less than a photon or whole quantum then the above equation is inaccurate and the excess energy would not be emitted as a wavelength of light. Rather that energy would need to be released in another form such as heat. This release of energy is necessary so that when the electron transitions to a lower state, it only emits the correct wavelength. Under the law of nuclear equilibrium, the excess energy would be released immediately if possible.

Max Planck postulated that energy may be absorbed in any amount but only be emitted in discrete amounts.[48] The amount of excess energy absorbed by an electron determines its frequency of oscillation and the frequency of the wave it will emit. When humans create electrical power, they cause the electrons to oscillate with controlled amounts of excess energy, thus creating waves of a single frequency according to the imbalance of the system.

When the excess energy of the electron is in the form of electrical energy, it cannot be transmitted as a wave until it reaches a minimum energy so that it can penetrate the Planck barrier and can be transmitted as a wave. This excess energy lower than Planck's constant is probably the cause of heat that like mass moves at speeds less than the speed of light although heat is massless slow energy and mass is massive slow energy. Heat is trapped energy in the atom. Heat is Planck's 1/2h i.e. energy too weak to emit. Heat moves from atom to atom by direct contact of the energy wave fields. Accumulation of heat energy to a full quantum results in formation of far infrared waves. The absorption of the partial quantum results in a larger atom, however, the frequencies of the energy wave fields of all the particles in the atom each increase in energy by an equal amount. Therefore, all factors of the equation increase equally so that the transition energies of electrons remains constant and each element retains its unique spectrum no matter how large the atom increases by absorption of a partial quantum.

In order to get electrons to transition, one uses heat because heat moves slowly from atom to atom. Heat takes time to build enough energy in an electron energy field for a transition. In a general way, the electron absorbs 1/2h and then waits to absorb another 1/2h before it can transition to the next node. This is only general in the sense that h is the correct amount of energy, but alone does not describe the frequency of the energy. A full quantum of energy is described by h, but the energy varies with the frequency.

Quantum mechanics gives the intensity of the spectral lines in matrix mechanics and incorporates the spacing of the spectral lines, but does not give them physical meaning other than light emission from electron shell transitions. Under this new model of the atom, the diffusion seen in some spectral lines can be explained as being due to excess energy in the form of heat that does not release at the speed of light and distorts the emission lines.

31 Heat and Work

James Prescott Joule proved that heat is the mechanical equivalent of work. But what is work? Work is said to be the transfer of energy by a force to an object. But what is happening at the atomic level then? Work is when one energy system of atoms described by its energy wave fields or even one atom's energy wave field transfers its excess energy to another atom's energy wave field or system of atoms. Work is the transfer in the form of excess electromagnetic radiation transferring from one atom to another atom. Inertia is caused by excess energy being retained by the system and therefore the system maintains its motion due to the excess electromagnetic energy held in its atomic energy wave fields. Work is however defined as the transfer of energy from one system to another. When an atom or system of atoms retains its initial energy to mass ratio, no mechanical work is done. However, even when no mechanical work is done, the attempt to do work releases energy from one system even if it does not transfer enough energy to do the work. An example is when one pushes very hard on a heavy stone and is unable to move it. Energy has passed from the person's atoms making up the person's cells to the atoms in the stone. The stone either deflects or reflects the energy. The person has lost energy from the energy fields of the atoms making up his body even though the stone has not moved therefore no mechanical work has been done. This is because the atoms of the person have been unable to force the stone to absorb the energy from those atoms. If a person is strong enough, then they can push their atoms with greater force upon the stone causing the stone to absorb the energy and reemit that absorbed energy in the direction in which it was absorbed, thereby, moving the stone. When a person is not forcing the stone to absorb energy, that energy from the person is being deflected by the stone so that the person is losing the energy from their own atoms although no mechanical work is being done. The person is losing their own energy which is released into the environment by pushing on the stone without the stone absorbing the energy from the person's atoms. On the atomic level, work means transferring energy from one atom to another atom in another system. This energy is stored in the matter's energy wave field and emitted when there is a full quantum of energy to reemit in the direction of absorption or when there is a system of lesser energy that the matter can emit into.

The energy wave field around a particle forms a loop back to the particle therefore it does no mechanical work. It takes in as much energy as it emits. Therefore, its presence cannot be felt in the macro world as doing any mechanical work. It causes no excess energy between particles or atoms to be exchanged because under the law of nuclear equilibrium the mass to energy ratio is preserved for the particle. Therefore, the energy wave field of the particle is dark energy. The only way it can be detected is from its effects. It is the cause of the effects.

When we examine heat, which is electromagnetic free radiation that cannot break through the Planck barrier being too small a quantity of energy, i.e. less than Planck's quantum, we see that heat travels only through atom to atom contact. When heat is applied to a container containing a liquid or a gas, the container is usually a solid. Heat placed under the container passes atom to atom through the solid container and then to the liquid or gas. If the heat is passed from the solid container to a gas atom, the gas atom cannot throw off the heat since the heat is below the Planck barrier so instead the gas atom carries the heat in its energy wave field. This causes the atom to accelerate. As the atom increases in velocity, it is able to collide with another atom and pass the heat energy onto the other atom. Atoms of gas at the bottom of the container are continually coming into contact with the hot solid container and are continually gaining the heat energy which makes them accelerate and pass on the energy. When an atom has been passed heat energy i.e. less than a quantum, at times, the atom will a second time receive more heat energy before it is able to pass on its previously obtained heat energy and in this way heat will accumulate in the atom. Once heat has accumulated so that it creates a quantum, then the heat is released in the infrared region as an electromagnetic radiation. However, as long as the source keeps heating the container there is both below the Planck barrier energy exchange by contact and infrared radiation being formed. The gas atoms carrying the heat energy at high velocities also hit the solid container which will also absorb the heat energy from the gas atoms. However, the atoms in a solid are all connected to each other by molecular bonds and cannot accelerate as can the gas atoms. The solid is forced to remain solid and cannot under normal heat energy throw off its atoms into the surrounding air outside the container. Only the gas atoms outside the container that are in contact with the container can receive the heat energy. So it is very difficult for heat to pass through a solid as the atoms in a solid do not have the freedom of movement to pass on the heat energy through contact with other atoms. The heat of the solid container in the form of excess energy will however cause the electrons in the atoms to transition outwards and therefore the solid container will enlarge as Isaac Newton explained about a metal rod in summer versus winter. The mass of an atom that is heated has more volume, due to the extra energy it is carrying, causing the electrons to transition outward. Even before the outward transition, trapped heat causes the energy wave field of the atom to expand which causes the atoms to distance themselves from each other. Therefore hot objects generally tend to expand. This means that essentially conduction and convection are exactly the same thing at the atomic level. Conduction is the direct contact transfer of energy from one atom to another. Convection is exactly the same. In convection atoms are sped up or in motion so that they contact faster and therefore they transfer heat energy or conduction energy faster through agitation of the atoms. Heat is transfer of partial quantum energy through direct contact. When heat reaches the quantum level, it transfers through light waves in the form of infrared heat radiation.

32 More on Properties of the Wave

Science often speaks of matter carrying energy, but it doesn't describe how the same amount of mass could carry more energy than another mass of the same amount. The energy carrier is the energy wave field surrounding the particle. This is the holder, the receptacle, the storage container of excess energy for the mass.

It occurs that there must be an explanation as to why protons emit only x-rays and gamma rays and electrons emit all sorts of radiation especially in particle accelerators. The explanation that suggests itself is that protons while able to absorb any amount of energy as does the electron, however, cannot emit energy except in higher photon packets. This is inferred from the fact that the electron emits more types of radiation more frequently and the proton does not during particle acceleration.

Therefore, the proton must be made of waves that consist primarily of gamma rays and x-rays in its energy wave field. A transition between nodes of protons therefore takes a large amount of energy. Therefore a proton must keep absorbing energy in greater amounts before radiation is possible between transition levels. On the other hand, since the electron is made up of lower energy waves in its wave field, it can emit much more frequently than a proton due to the ease of transitions between its nodes.

It would then appear that protons in a particle accelerator are all in each other's energy wave fields since they are only radiating in transition wavelengths. On the other hand, because electrons can radiate across the spectrum, they too may or may not always be in each other's wave fields. If they are in each other's wave fields in a particle accelerator, then they emit more quickly because the energy transitions between levels becomes very weak at the extremity of their fields and possibly even flat at some places so that the emission is across the entire spectrum. If they are able to separate from other electron energy wave fields, then continuous emission can be expected.

It would be an erroneous temptation to infer that if the protons in a particle accelerator are transitioning in the energy wave fields of other protons which appears the case since they only transition at high emission levels, then all atomic nuclei must overlap. The problem with this visualization is that gas molecules would be tightly bonded to each other's nuclei which is not true.

The answer appears to lie in the difference in nature between electrons and protons that displays itself in the particle accelerator. The electrons appear unable to hold onto energy. There is immediate release. In a completely free electron that does not overlap, the spectrum would be continuous. On the other hand, protons appear to hold onto energy and be unable to release it except when overlapping in another proton's energy field, therefore, the releases are proton shell transitions which occur at the gamma ray and x-ray energies.

The role of the electron in the atom would take on the character of a direction indicator and the nucleons would be the energy carriers. That is their energy wave fields would create these roles. In a collision, the electron would utilize the energy to transition as far as possible then reemit the energy in the direction it was absorbed. On the other hand, the proton would absorb the energy and transition in the great sphere of the nucleus and hold onto the excess energy while maintaining the same velocity. Inertia could be described as the electron expelling energy to give the object its direction and the proton or nucleons holding the energy to give the object its velocity. Proton velocity energy would be unable to leave the object unless it encountered the mass of another object.

Has science been able to calculate the distance between individual particles while they emit radiation while definitely ruling out the possibility of neutral particles interfering? Can science control the release of individual particles into a vacuum and study the radiation of individual particles at great distances from other particles while they are accelerating? The proximity to the walls of the container must be taken into consideration as well so that the particle is not transitioning through the energy waves of the container's atoms.

For the macro world, there is an interesting correspondence. A larger body must absorb more energy before it will accelerate and holds the energy longer than a smaller body. Therefore the larger a body is, the higher energy and the shorter the wavelengths are, in its energy wave field and the longer it takes to emit because more energy must be absorbed and the emission across the Planck barrier waits for a sufficiently large amount of energy in order to be able to penetrate it.

This suggests an unfathomed deeper structure to the atom. It would appear that all subatomic particles that we can measure have energy wave fields that overlap each other. There may be no such thing as a free electron in a particle accelerator or a free proton. Each proton is transitioning with other protons between shells in the accelerator and this creates the necessity of only emitting high energy waves for the proton because it takes so much more energy to make the transition. The energy must build in the protons energy wave field to a strength great enough to make the transition.

Naturally under this model an alpha particle has a four-fold nuclear field and double charge of a proton and therefore has more difficulty colliding with the nucleus of an atom because of the strength of both colliding energy fields i.e. that of the alpha particle and that of the nucleus while a neutron which does not have to overcome the Coulomb force of the nuclear proton field and itself has a less energetic field than a four nucleon alpha particle has can make collisions with the nucleus more energetically.

Electrons on the other hand, although they too probably exist in each other's energy wave fields in the accelerator, they transition between nodes of significantly less energy than protons which means they transition between each other's electron energy wave fields more frequently or, if the energy fields of electrons may be separate, then they are not transitioning at all which causes continuous radiation since the electron may radiate any energy absorbed while not making a transition.

This raises the question of the radius of the energy wave fields. If the energy wave fields of all particles have the same radius, but simply lesser or greater energy waves, then if the protons are transitioning in an accelerator because their energy wave fields are overlapping, then the electrons must also be transitioning in the accelerator between other electrons.

However, if the radius of the energy wave field of the electron is smaller than the radius of the proton wave field then what accounts for atomic neutrality when there is one electron to one proton? If the charge of each particle extends into the energy wave field and only becomes neutral where the particles overlap, in this scenario of waves of different radii for different particles, then the positively charged fields would always extend further than the negatively charged fields and the universe would have a positive charge which is empirically shown not to be the case. Therefore, the radius of the energy field of the electron must be the same radius as the field of the proton in order to hold equal extent of charge. The energy fields must rather have different wavelengths in the fields of smaller particles versus larger.

The electrons do not appear to emit or absorb as much energy moving in the nuclear wave field of the atom as the protons or nucleons emit or absorb moving between nodes. It may be that the energy wave field of the electron only has less intense waves of greater wavelength and therefore electrons pass with greater ease through other electron energy wave fields emitting and absorbing at a great variety of frequencies. Whereas because the energy wave field of the proton is equal in radius to the electron, the intensity i.e. density of the waves is greater and it therefore takes a greater amount of absorbed energy before there is a transition and an emission. However, due to the electron and proton fields being drawn together by the Coulomb force, the electron passes with greater ease through transitions in the proton or nucleon energy wave field even though the intensity is great. Therefore, the emission and absorption is less for an electron passing through the proton field than for a proton passing through a proton field, because the proton instead of getting assistance in transitions by the Coulomb force gets greater resistance.

There is the further question of just how far the energy field of each particle extends. If the energy wave fields of human atoms extend into other humans or if humans are intrinsically connected to all things living and nonliving, it would have amazing and bizarre philosophical consequences without end.

34 The Implications for Gravity and the Structure of the Universe.

If the atom is modeled as having a wave emanating from the particle that has energy surrounding the particle which protects the particle from intrusion, but can also be penetrated so that other particles can be grouped together in the great sphere of their combined particle waves, then the universe can also be modeled in this manner.

If we imagine, the particle wave to surround the particle and the nuclear wave to surround the nucleons, we can imagine the atomic waves to surround the molecules by grouping atoms with a lessened pressure due to a secondary great sphere around the molecules. If we then imagine the molecules being grouped by tertiary great spheres. This would mean that all mass has both a repellent and attractive nature, but the effects of this would be very weak the larger the great sphere.

Given enough mass, the effects would again be significant. This is how I picture the universe. The clumping of galaxies and the clumping of galaxy clusters is due to the gravitational effect as described by General Relativity, but also by an energy wave field sphere around galaxies creating clusters.

This is just a model, but so was particle physics. In particle physics, the forces were modeled as particles in fields to try to describe their interaction. However, if the particle is modeled as particles surrounded with waves, and the atom is modeled as nucleons surrounded with the particle waves, and molecules as the residual particle waves surrounding the atoms, then by extension, the energy wave created by clumps of matter could surround planets, then solar systems, then galaxies, then galaxy clusters. This would unite the very small with the very large. If this can be described mathematically, then this would be an excellent model for the universe.

It would describe the repulsive effects of matter on a large scale so that the universe is expanding. It would describe the attractive effects of matter within the great spheres created by the energy field. This model could describe dark matter and dark energy.

35 Energy

It would appear from this initial concept that the E for energy in $E=mc^2$ is the universal E. There is one energy and Maxwell discovered it. It is not Dalton's atom, but it is Maxwell's electromagnetic wave that is the absolute one fundamental substance from which everything else is formed. All particles and all matter are forms of this one fundamental electromagnetic energy in a different form. Free infinite electromagnetic energy can be slowed and confined by the Planck barrier to become matter and it itself becomes bound to the matter it forms. All forms of energy come back in the end to electromagnetic free infinite unbound energy. Heat is this energy in Planck's $1/2h\nu$ which we call zero point fluctuations, therefore, heat is a form of slow light i.e. light that cannot break the Planck barrier. It travels slowly through direct contact moving from hotter to cooler systems that will accept it. Due to the law of nuclear equilibrium, atoms will not generally accept more heat or light energy than their mass to energy field ratio. Excess energy releases from the atomic energy field as soon as possible. Therefore, heat always moves from hotter to cooler systems due to this universal atomic ratio. Mechanical energy is also electromagnetic energy transfer at its most fundamental. Battery energy and electrical energy are basically electromagnetic energy. Motion is a form of transfer of electromagnetic energy between atoms. In the final analysis, it may be found that chemical energy and hydropower are all transfers of the electromagnetic energy bound in the atom or the free electromagnetic energy of the infinite wave.

Sound itself appears to be independent of a transfer of electromagnetic radiation or trapped electromagnetic radiation below the Planck barrier i.e. heat. Sound appears to be merely the collision of two atomic energy wave fields creating a shock wave. So aside from whatever electromagnetic energy transfers that may take place in the collision there is a shock wave that vibrates each atom's energy wave field and moves out into space at the frequency of the vibration passing the energy from atom to atom as vibrational energy. This wave is not going fast enough to hit the Planck barrier, but the wave is caused instead by the vibration of the shock wave which gives it its characteristic pitch when it hits the human ear. Just as not all light is visible to the human eye, not all atomic energy wave field collisions produce vibrations detectable to the human ear. So sound appears to be a form of energy and independent of true electromagnetic energy transfer, but a secondary effect of such transfer. Since it is a shock wave that locally vibrates from atomic energy wave field to atomic energy wave field and is not electromagnetic free radiation, it can only move from atom to atom and not in a vacuum. It is not true energy as in $E=mc^2$ but a secondary energy of vibrational motion. Sound energy cannot transform into matter. However, sound is kinetic energy. Kinetic energy is not truly energy but is movement. Movement or motion has a cause. The cause is the substance we refer to as energy in the formula $E=mc^2$. Sound is the collision of atoms which can cause atoms to transfer contact energy such as heat. Heat is already the E in the formula, but in too small a quantity to move at light speed. Sound appears to be motion caused by collisions of energy wave fields that are made of the substance E in the formula. Just as heat has a mechanical equivalent and can do work, so too sound can be converted to do work. Kinetic energy is an effect of true energy that is a substance that has volume and moves in waves whether in a loop of changing frequency about a mass or as free waves of radiation unchanging in frequency. Potential energy is energy stored in the energy wave field of atoms. Potential energy is the cause.

In the end, there would be one thing in the universe and it is Maxwell's energy in its various forms of energy and matter. This may not be the case, but is an interesting visualization.

36 Time

A Scientific Definition of Time

I would here like to propose a scientific definitive meaning for the word "time."

Time is our measure of change. The universe is in constant change. Humans from the beginning of awareness experience that change and it is disorienting, therefore, we find things that change consistently and we measure the inconsistent changes in our frame of reference against the consistent changes in nature. In other words, we originally measured our frame of reference, i.e. changes in our movements on the ground, with the consistent changes of the heavenly bodies. In this way, we could measure the changes and we utilized the word "time" for the measure of such changes. Our measure of changes is only as accurate as the consistency of natural changes which we use as a yardstick.

Our memory creates our perception of time. Our brain gets accustomed to the amount of changes that occur in our reference frame against the consistent changes that we use to measure inconsistent changes and we call this our internal "clock." If we are either consciously or unconsciously aware of changes in our body, we can experience changes, i.e. time, even in isolation. The more changes that our brain can imprint during our measuring frame of reference, the slower the measure of changes appear. We experience a slowing of time. Young people's brains can imprint many changes during a cycle of measurement and time seems to pass more slowly to them as there are many changes being measured and imprinted over the same measuring cycle. Older people's brains do not imprint as many changes during the cycle and therefore time appears to pass more quickly to them as less things appear to have changed. This can be deduced from the number of neurons firing from babyhood to old age.[1]

Our abstract reasoning helps us to visualize that changes have happened before and will happen again, therefore, we can extend our measurements into changes that we did not see but can visualize as having happened and we call it the "past." And we extend our measurements into future changes that we can visualize will happen and we call it the "future." Because change cannot be undone, we measure changes that have occurred, and reversing the direction of change only creates new changes so the measure of such changes is new. Trying to reverse a change only causes a new change and can only be measured against consistent changes (timepieces) that continue to change consistently in the same direction, therefore, we say time only goes forward.

In relativity theory[li], time slows when change slows respective to the observer. If atoms are slower to transition when at relativistic speeds, then in that reference frame, the measure of change will be slowed.

37 In Summary

Heisenberg said of Niels Bohr:
"Bohr had dared to publish ideas that later turned out to be right, even though he couldn't prove them at the time. Others have done a lot by rational methods and good mathematics. But the two things together, I think, are too much for one man."[39]

Unfortunately, science is very reluctant today to publish any new ideas like the one that Bohr had. It is nearly impossible if not *completely* impossible to publish ideas in a reputable journal without proof, as Bohr had done in 1913.

The modern model of quantum mechanics has the following features and has evolved thus:
- The uncertainty principle based on the path of an electron in a cloud chamber
- Pauli's exclusion principle proved originally by spin which is non-existent in current theory
- Quantum entanglement
- No barrier exists between the nucleus and electrons
- Planck's constant is introduced due to its length
- Electron orbitals are discrete due to a mathematical formula
- Electrons quantum leap between orbitals, momentarily disappearing in the process
- Electrons "know" what wavelength to emit before jumping inward or outward
- Charged particles emit no radiation except quantized radiation between stationary shells
- Electrons do not have orbits or pathways but are truly stationary
- Electrons form standing waves in circular orbits
- Electrons in suborbitals are not in circular orbits
- Wave-particle duality is detected in wavefunction collapse and does not obey the Bragg law
- Hypothetical particles are theoretically invented
- The nucleon-nucleon interaction is mediated by such particles
- The weak nuclear force is mediated by such particles
- When an electron is not being measured it is in a probability cloud

Many of these properties may appear to be empirically correct, but the foundation is flawed. You may well ask after reading this book, how could quantum mechanics have been invented in such an unscientific fashion? How could it be true that the ambitions of men were able to rule the day? How could it be true that evidence could have been manipulated and ignored? How would it be possible for a consensus of scientists to make such an error? The answer is a question: Do you know how many physicists there were in 1927 to create a consensus for quantum mechanics? The answer: 29. There were twenty-nine attendees at the October 1927 Fifth Solvay International

Conference on Electrons and Photons, where the world's most notable physicists met to discuss the newly formulated quantum theory. (Wikipedia: Solvay Conference) And guess what? Today with the hundreds of thousands of physicists having to form a consensus, no one would have accepted it. No one. Not on the meager evidence that existed in 1927. Not on the random conflicting formulae of Heisenberg and Schrödinger. Yes, contradictory equations of absolutely different formulations and representations. That was quantum mechanics in 1927 that was accepted by 29 men of whom most deferred to the great three, Bohr, Einstein and Heisenberg. De Broglie still young as a physicist at 35 would of course uphold the new theory as his reputation would be destroyed if he didn't. Schrödinger was bulldozed into keeping his mouth shut or he wouldn't keep his position. Heisenberg had done away with Einstein's objection that an electron has a path in a cloud chamber so he could secure a secular position being only 25. Einstein was not involved in the formulation of Quantum Mechanics and only reviewed the final work. He was dissatisfied, but was not fully involved and therefore could not object with conviction or authority.

However, let us contrast the above random ideas that form modern Quantum Mechanics with this simple idea:

New model:
- There is an energy field around all particles that creates a barrier
- The law of nuclear equilibrium causes all nuclear interactions

Reality cannot be replaced with mathematics. Even if this new model presented here is incorrect on each and every of its details, it is yet correct in that some physical cause is making the electrons overcome the electrostatic attraction to the nucleus of the atom. That is science. Anything else is a product of "the gods". This new model of the atom tells us why Planck's constant has had a measure of success in describing the atom. The science of physics itself should tell us that there is a *physical* barrier between the electron and the nucleus, but it does not. Quantum mechanics merely tells us that electrons are not supposed to be in between the nucleus and electron orbitals, but does not give a cause. Causality should be primary in physics. Where there is no cause, there is no science.

This new model is merely in conceptual form. It is unproven and only suggests a basis for investigation. The theory presented here is not meant to, and in fact cannot, undermine experimental evidence but the experimental results should be reanalyzed from the perspective of this new paradigm shift for the model of the atom. It is not pretended that this model of the atom is complete. In some cases, it raises more questions than it answers, but the questions are good ones whose answers are truly meaningful. It needs to be mathematically calculated by those with access to available data and is meant to stimulate mathematical calculation from the data. Therefore it is a new beginning and starts with a new assumption, a new concept, that can open the way to new investigation and calculation. If this model is not a better approximation of reality, then perhaps it will inspire others to seek one. All theories and models begin with an assumption. Einstein assumed that gravitation and acceleration were equivalent. Bohr assumed that the atom was mostly empty space, the laws of electrodynamics did not apply to the atom, but somehow still held, and Planck's constant could be introduced to constrain electron orbits without knowing of a wave in the atom. Bohr assumed that an electron in the atom although moving does not produce an electrical current, yet Bohr knew that a moving electron in a wire produces electrical current for cities. How can the same electron produce electrical current in the macro world and not in the atom?

There is too much data for one person to develop a complete model of the atom without adjustments. Most ideas need developing. In fact, quantum mechanics has undergone adjustments. This new model needs adjusting to meet all data, but the basis is at least grounded in physicality and causality.

The problem with quantum mechanics is that it has developed along superstitious lines due to the philosophical influences and religious beliefs of its inventors. It is a magical model that explains how far the magician is levitating his assistant, but not how the magician is accomplishing this feat. It is a return to the mysticism of the ancient human past in which superstition rather than reason ruled the explanation of the natural world. Until man lets go of his need for the mysterious, the need for "atomysticism", and turns back again to the rational, then atomic science will remain in the dark ages. Humans will remain shrouded in the mysticism they have created and the rational science invented by the Greeks and passed on through the ages by Kepler, Galileo, Newton and Einstein will remain apart from quantum mechanics.

One cannot find a solution unless one identifies the problem. It is time to take away the power of the identical electron to decide exactly how much energy it will absorb or emit. It is time to unite the atomic world with the macro world. It would appear that, if this model is a better approximation of reality, the universe is not so strange, complicated, and cluttered as the Standard Model and quantum mechanics would make it appear. The observed particles in the Standard Model represent all the possibilities that the universe explores in maintaining nuclear equilibrium. It is hoped that this theory depicts the atom in a way that more closely approximates the real nature of the atom. If it does so, it will open up new vistas and new channels of exploration of the atom, thereby helping science in its continual quest to come to a better understanding of the natural world.

Einstein on quantum mechanics: "For the quantities which figure in its laws make no claim to describe physical reality itself, but only the *probabilities* of occurrence of a physical reality that we have in view. ... I am still inclined to the view that physicists will not content themselves with that sort of indirect description of the real, even if the theory can be adapted to the postulate of general relativity in a satisfactory manner. We shall then, I feel sure, have to return to the attempt to carry out the program which may be described properly as the Maxwellian—namely, the description of physical reality in terms of fields which satisfy partial differential equations without singularities. ...Some physicists, among them myself, cannot believe that we must abandon, actually and forever, the idea of direct representation of physical reality in space and time..."[62]

Einstein said in his own "obituary" at age 67 about theory: "A theory is the more impressive the greater the simplicity of its premises is, the more different kinds of things it relates, and the more extended is its area of applicability."[lii] The theory proposed in this book is simple i.e. a particle is surrounded by a wave. It has the ability to predict the concepts described by the Strong Force, electron quantum-like transitions, the Weak Force, Tunneling, Entanglement, the

Ionization levels of the elements, Van Der Waals forces, and with some mathematical ingenuity it can lead to a unification of the proton-wave extending through the Balmer-Rydberg formula into the infinity of space to the reason for gravity, dark matter, and dark energy. This one simple concept can eventually lead to a real true simple unification of all forces and complete understanding of the universe. Without even having the mathematics yet to prove these hypotheses, one can visualize them from this simple atomic picture of the atom. It is an exciting and awe-inspiring concept. I can only hope that the equations will be as easy to extract from the model as the model is easy to understand.

Adam Becker in his book on the history of Quantum Mechanics said this about Einstein:

"Einstein also wanted to find other ways to describe what was going on at the quantum level altogether. The Copenhagen interpretation and Bohr insisted on the necessity of using classical concepts, along with classical descriptions of measuring apparatuses. Bohm's ideas broke with both, but not as thoroughly as Einstein had hoped for. Einstein wanted a new way of looking at nature, a theory underlying quantum physics that would reveal some previously unknown truth, rather than a new way of interpreting the existing quantum theory."[liii]

With the paper that follows, I hope to give Einstein what he wished for.

38 Paper Proving New Theory

Before I place my paper here showing data from the spectrum proving that the theory is correct and does indeed produce the ionization levels directly from spectral data which current Quantum Mechanics cannot do, I want to discuss why current QM works.

I felt like clumping myself on the side of my head when I figured out why the Schrödinger equation works at all when it is only an estimate and obviously a guess completely out of the blue considering the history you've just read in this book. Then I read the book the Quantum Story-A History in 40 Minutes and although I knew this, it finally hit me when I read it again. De Broglie created a wave in the atom to hypothesize that particles behaved like waves. He created a pilot wave for the particle that is in a circular orbit around the nucleus. This wave travels faster than light and guides the electron in a circular orbit. This wave is a standing wave. Of course! It finally hit me. The evidence I found in the atomic spectra shows that the wave emanating from each particle is a standing wave! Then I was reminded by this book that this caused Schrödinger to create his equation beginning with a classical equation for a standing wave.[liv] Of course! Every proton in the universe is surrounding by a standing wave that follows the Balmer-Rydberg equation. And this is key, even protons in more complex atoms than Hydrogen follow the Balmer-Rydberg equation. This is proved in the following paper. So of course the Schrödinger equation works for atomic systems! The problem is that Quantum Mechanics accidentally made a very good guess and it turned out to be so close to the truth that we were stuck with this weird theory for a century. It isn't right because it couldn't be, but it is so close because it got the two most important things right. There is a standing wave in the atom and every atom carries the Balmer-Rydberg equation. Now let's look at the data from the spectrum. If you take the time to study this, it will astound you. Everything becomes clear.

To further research on this theory, there is the need to take each element to practically Bose-Einstein Condensate, and then, slowly warm the element while detecting its absorption spectrum up to the highest possible temperatures. This should produce all the Balmer-Rydberg lines in every element of the periodic table. For now, the NIST tables carry only a couple of lines for each electron series in carbon so that it begins to become difficult to find the Balmer-Rydberg formula. In further research, one should try to analyze other ions as well as higher elements.

American Based Research Journal

Article

New Atomic Model from the spectra of Hydrogen, Helium, Beryllium, Boron, Carbon, and Deuterium and their ions

Janeen A. Hunt [1*]

[1] Affiliations: APS, American Physical Society
[*] Correspondence: voyajer17@hotmail.com; Tel: +1-949-643-8070

Received: 27 July 2019; Accepted: 31 July 2019; Published: date

Copyright © 2019 Janeen Hunt. This is an open access article distributed under the Creative Commons Attribution License, which permits unrestricted use, distribution, and reproduction in any medium, provided the original work is properly cited.

Abstract: A cohesive unifying theory of the atom does not currently exist in Quantum Physics. In this research, the atomic spectra are allowed to determine the model for the atom based upon the finding of patterns of the Balmer-Rydberg formula in the first 20 ions and neutral atoms of the periodic table. From this data, the model postulates a standing wave of varying energy antinodes originating from the particles in the nucleus of each atom which is able to predict the ionization energies of these atoms. The transitions of the electrons in atoms are defined by the energies of each antinode represented by the difference in energy between each spectral line. The spectral patterns for H, He-I, He-II, Li-I, Li-II, Li-III, Be-I, Be-II, Be-III, Be-IV, B-I, B-II, B-III, B-IV, B-V, C-I, C-II, C-III, C-IV, and Deuterium are charted and the ionization energies are calculated from the data including general inferences this model predicts about the unification of atomic forces, electron transitions, heat, and electromagnetism. This model predicts that the nucleus of every atom is held together by energy in the form of a standing wave originating from the nucleus and surrounding it. This is the Sollism Theory of the atom.

Keywords: Atom, Atomic Model, Spectral Analysis, Quantum Mechanics, Atomic Theory, Unified Theory, Electron Transitions, Nuclear Force, Strong Force, Ionization Energy, Sollism Theory.

---

1. Introduction

Quantum Mechanics is a set of rules governing the atom and the subatomic particles. No single formula nor single rule describes a single atom with its electrons and nuclear forces. A study of the history of Quantum Mechanics shows that ad hoc observations were ascribed ad hoc rules such as the Pauli Exclusion Principle and the Uncertainty Principle. Although this set of rules and abstruse formulae based on observation hold up well in the real experimental world, they do not form a cohesive atomic model nor do they relate well to general relativity. The method used to construct this new model is to use the most observable and non-controversial property of the atomic elements, that is, the atomic spectrum to form a model of a single atom that is comprehensive and takes into account both the electron paths and the nuclear forces and can be deduced directly from the spectra of elements in a way that unites all the atomic forces and explains phenomena that remains unexplained by QM and also predicts the ionization of the elements solely by their spectral data.

Let us assume that causality and invariance rule in the macro and micro worlds. We can infer from this that a permeable, massless, neutrally charged barrier exists to overcome the Coulomb force of the proton and electron in the Hydrogen atom. This model unites the phenomena that an alpha particle with a kinetic energy of about 5 MeV may pass through the atom as in the Rutherford Gold Foil experiment[lv], and yet an electron of 20 to 200 eV may not pass through the atom as in Low Energy Electron Diffraction (LEED) experiments including the Davisson-Germer experiment[lvi]. Max Planck discovered in 1900 (elaborated by Einstein in 1905) that E=nhv where v is frequency and n is a whole number[lvii]. We can infer that there exists on the surface of every body of mass unradiated partial quanta as there is some energy that is not a whole number. To describe a partial quantum, the wavelength is less than a full quantum of energy necessary to radiate. Therefore, non-radiating massless, neutrally charged energy exists between the proton and the electron in the Hydrogen atom. This same energy separates each shell of the electron and exists in the form of a partial quanta standing wave. The electrons in atoms reside in the zero energy nodes of the proton standing wave. Let's examine the anomalies of the current theory and see how this assumption aids our understanding of the atom in decoding first the spectra of Hydrogen, Helium, and Lithium, and how this assumption causes us to arrive from theory to the known ionization energies. Then we will continue with the elements up to Carbon.

From the atomic spectra, we infer that all particles are held together by a standing partial-quanta wave of energy. Protons have an energy wave that is proportional to the Rydberg constant for the known transition series. The series of the spectral lines of Neutral Hydrogen are caused by different states of energy of the electron-particle with its standing wave, therefore, the Lyman, Balmer, Paschen, Brackett, Pfund, and Humphries series, etc., are electron transitions in Hydrogen atoms that have different energy states, meaning different amounts of energy stored in the partial quanta standing wave of the electron. The pattern for transitions was generally represented by Niels Bohr[lviii], but he did not analyze what the energy between transitions represented. Let us here take the difference in energy between spectral lines as literally energy between transitions.

We know that where n1 is Lyman ground state and n2 is an allowed transition line from Lyman ground state and E is energy, the spectrum tells us that if the distance of the transition between n1 and n2 is 1, then the energy for transition outward and inward is $E_{n2} - E_{n1} = 1 \text{ and } E_{n1} - E_{n2} = 1$. If we calculate the Coulomb force between the electron and proton into the equation, we get an inequality because in one direction the electron is aided by the attractive force and in the other direction the electron must overcome the attractive force and the distance is different between shells:

$$E_{n2}\left(k_e \frac{q_1 q_2}{r^2}\right) - E_{n1}\left(k_e \frac{q_1 q_2}{(r-1)^2}\right) \neq E_{n1}\left(k_e \frac{q_1 q_2}{r^2}\right) - E_{n2}\left(k_e \frac{q_1 q_2}{(r+1)^2}\right)$$

Therefore, electromagnetism, i.e. the Coulomb force, is not determining the energy of the electron transition nor the direction of the electron transition. We will assume that a spherical standing wave of energy determines the difference in transition energy. The absorption line is the same energy as the emission line in the spectra. Electromagnetism guides the direction of the transition where possible, but the hypothesized proton-wave restricts the possible transitions. Let us examine how visualizing a proton-wave in the Neutral Hydrogen atom explains why an electron must be more excited to get to ground state of the Lyman Series. An anomaly in the current model is that ground state, or state of least energy, is the Lyman series and it is in the ultraviolet which would indicate a more excited state than other H series. Let us question how the inner Lyman series has higher frequency energy.

2. Results

We will go through the spectral series of each of the first 20 ions and neutral atoms of the first six elements in the periodic table showing how there exists a repeating pattern and how to calculate the expected ionization energy directly from the spectra of the ion.

2.1. Hydrogen

Let's examine the spectra for neutral Hydrogen. For visual convenience, the energies will be rounded in the column so marked[lix].

Lyman Series	Larger	Next Shorter	Larger	Next Shorter	Transition	Rounded	Difference
Antinode Number	Wavelength	Wavelength	Λ in eV	Λ in eV	Energy eV	Energy eV	eV Between Series
101	121.566824	102.57222	10.1988704	12.0875251	1.8886547	1.89	
102	102.57222	97.25367	12.0875251	12.7485604	0.6610353	0.66	10.198
103	97.25367	94.9743	12.7485604	13.0545241	0.3059637	0.31	10.198
104	94.9743	93.78034	13.0545241	13.2207271	0.166203	0.17	10.198
105	93.78034	93.07482	13.2207271	13.3209421	0.100215	0.10	10.198
106	93.07482	92.62256	13.3209421	13.3859859	0.0650438	0.07	10.198
107	92.62256	92.3	13.3859859	13.433	0.0470141	0.05	10.198
108	92.3	92.1	13.433	13.462	0.029	0.03	
109	92.1	91.9	13.462	13.491	0.029	0.02	
110	91.9		13.491				

Balmer	Larger	Next Smaller	Larger	Smaller	Transition	Rounded	Difference
Antinode Number	Wavelength	Wavelength	Λ in eV	Λ in eV	Energy eV	Energy eV	eV Between Series
102	656.2711	486.1287	1.88922579	2.55044454	0.6612188	0.66	
103	486.1287	434.0462	2.55044454	2.85648	0.3060355	0.31	1.889
104	434.0462	410.174	2.85648	3.02272764	0.1662476	0.17	1.889
105	410.174	397.0072	3.02272764	3.12297683	0.1002492	0.10	1.889
106	397.0072	388.9049	3.12297683	3.18803977	0.0650629	0.07	1.889
107	388.9049	383.5384	3.18803977	3.23264708	0.0446073	0.04	1.889
108	383.5384		3.23264708			0.03	

Paschen	Larger	Next Smaller	Larger	Smaller	Transition	Rounded	Difference
Antinode Number	Wavelength	Wavelength	Λ in eV	Λ in eV	Energy eV	Energy eV	eV Between Series
103	1875.1031	1281.807	0.66121467	0.96726284	0.3060482	0.31	
104	1281.807	1093.81	0.96726284	1.13350974	0.1662469	0.17	0.661
105	1093.81	1004.94	1.13350974	1.23374956	0.1002398	0.10	0.661
106	1004.94	954.597	1.23374956	1.29881435	0.0650648	0.07	0.661
107	954.597	922.9	1.29881435	1.3434	0.0445856	0.04	0.661
108	922.9		1.3434			0.03	

Brackett	Larger	Next Smaller	Larger	Smaller	Transition	Rounded	Difference
Antinode Number	Wavelength	Wavelength	Λ in eV	Λ in eV	Energy eV	Energy eV	eV Between Series

	104	4051.16	2625.15	0.30604673	0.47229464	0.1662479	0.17	
	105	2625.15	2165.53	0.47229464	0.57253618	0.1002415	0.10	0.306
	106	2165.53	1944	0.57253618	0.6378	0.0652638	0.07	0.306
	107	1944	1817	0.6378	0.6824	0.0446	0.04	0.306
	108	1817		0.6824			0.03	0.306

Pfund Series Antinode Number	Larger Wavelength in nm	Next Smaller Wavelength in nm	Larger Λ in eV	Smaller Λ in eV	Transition Energy eV	Rounded Energy eV	Difference eV Between Series
105	7460	4654	0.1662	0.2664	0.1002	0.10	
106	4654	3741	0.2664	0.3314	0.065	0.07	0.166
107	3741	3297	0.3314	0.3761	0.0447	0.04	0.166
108	3297	3039	0.3761	0.408	0.0319	0.03	0.166
109	3039		0.408			0.02	

Humphreys Antinode Number	Larger Wavelength in nm	Next Smaller Wavelength in nm	Larger Λ in eV	Smaller Λ in eV	Transition Energy eV	Rounded Energy eV	Difference eV Between Series
106	12370	7503	0.1002	0.1653	0.0651	0.07	
107	7503	5908	0.1653	0.2098	0.0445	0.04	
108	5908	5129	0.2098	0.2417	0.0319	0.03	
109	5129	4673	0.2417	0.2653	0.0236	0.02	
110	4673		0.2653		0.0190	0.02	

Table 1. Table of Antinodes of Hydrogen

Please note how each series repeats the energy differences between the spectral lines as is shown in blue. Note how each series starts with a spectral line (shaded) whose energy is equal to the energy difference of the first and second lines of the previous series. Now let's use our first assumption that the nodes of the proton standing wave are where the electrons reside. We see from the repeating differences between energies in each series shaded in blue that the energies represent the same antinodes being traversed by the electron through each series, but the electrons are in different states of excitation between the antinodes that have the same energies. See the chart below (Table 2) which shows this.

Hydrogen		Proton-wave Energy in eV	Nodes may contain electrons in energy states per spectrum in eV (rounded). Note: For Hydrogen, the ground state energy is zero excess energy in the electron-wave.					
Number			Lyman	Balmer	Paschen	Brackett	Pfund	Humphrey
99N	Node	0	Ground					
100A	Antinode	10.20						
100N	Node	0	10.20	Ground				
101A	Antinode	1.89						
101N	Node	0	12.09	1.89	Ground			
102A	Antinode	0.66						
102N	Node	0	12.75	2.55	0.66	Ground		

103A	Antinode	0.31						
103N	Node	0	13.06	2.86	0.97	0.31	Ground	
104A	Antinode	0.17						
104N	Node	0	13.22	3.02	1.13	0.47	0.17	Ground
105A	Antinode	0.10						
105N	Node	0	13.32	3.12	1.23	0.57	3.27	0.1002
106A	Antinode	0.07						
106N	Node	0	13.39	3.19	1.30	0.63	0.33	0.1653
107A	Antinode	0.04						
107N	Node	0	13.43	3.23	1.34	0.68	0.38	0.2098
108A	Antinode	0.03						
108N	Node	0	13.46	3.26	1.37	0.98	0.41	0.2417

Table 2. Table showing excitation of electrons in Hydrogen

(Note: Since every particle is assumed to be surrounded by a standing wave of energy originating from the particle, the electron and the proton will be referred to as the electron-wave and the proton-wave.) We see from the spectrum of Hydrogen in the charts that each series, representing the excitation state of the electron-wave in Hydrogen, shows that the proton-wave maintains its antinode energy levels no matter what excitation state that the electron-wave is in. The antinode energy levels repeat with every series. Niels Bohr only allowed transitions per series as transitions from ground state and to ground state. However, intermediate transitions can be accomplished that would create allowable spectral lines. Therefore, an electron in the Lyman series will start to absorb light that is more and more infrared as its energy increases incrementally. Conversely, an electron slowly losing energy in the Lyman Series will be emitting infrared light as it goes to Lyman ground state. In fact, a graph of light wavelengths from the transitions of an electron would look like Planck's Radiation Law Spectral Graph due to the preponderance of infrared light emitted by incremental transitions.

To make this clear: An increase from Lyman ground state through the first 10.20 eV antinode is an increase in energy by 10.20 eV. An incremental increase from 10.20 eV to 12.09 eV produces an absorption of 1.89 eV. An incremental increase from 12.09 eV to 12.75 eV produces an absorption of .66 eV. As the electron transitions through the higher Lyman series, it is absorbing incremental energy changes. As a high-energy electron cools, it may be a Lyman electron and yet emit no Lyman lines unless the drop in temperature is large enough, but the cooling Lyman electron may emit spectral radiation of the lines associated with the lower energy series if the electron emits radiation slowly. The effect of the proton-wave antinodes causes most emission and absorption to occur in the infrared even if the electron is traveling through the higher Lyman energy series.

The spectrum in the tables agrees with what we already know, but by allowing the energy differences between lines to describe energy, a whole new picture of the atom emerges. For energy to be described at atomic scales, the energy must be contained in a wavelength that is of atomic proportions. Thus, we introduce Planck's partial quantum that does not radiate.

Taking the case of the Lyman Series excitation state of Hydrogen, the energy in the partial quantum antinode between ground state for the electron and the next node is 10.20 eV. To overcome this antinode and move from Lyman ground state, the electron must gain 10.20 eV of energy. The next energy difference is described by an antinode of 1.89 eV of energy which antinode is also seen in the Balmer series. The next energy difference is described by an antinode of 0.66 eV of energy which antinode is also seen in the Balmer and Paschen series. The next antinode in the Lyman series contains 0.31 eV of energy. This antinode is also seen in the Balmer, Paschen, and Brackett series. For all series, the partial quanta in the proton standing wave continues outward per this pattern, and we will examine how the energy of the antinodes continues inward toward the proton in a wave pattern. The electrons reach their ionization energy before they can transition inward through the higher energy antinodes of the proton-wave toward the nucleus.

The proton particle is emanating the standing wave in which the electron particle with its associated wave is transitioning. For each different excitation state of the electron-wave in the Hydrogen atom described by each series, the energy of the electron changes, but the proportion of the antinodes is always according to that described by the energy differences between the lines described by the Balmer-Rydberg formula. The spectra of Hydrogen defines the (rounded) antinodes as infinity to ... , 0.03 eV, 0.04 eV, 0.07 eV, 0.10 eV, 0.17 eV, 0.31 eV, 0.66 eV, 1.89 eV, and 10.20 eV.

Because we describe radiated light as a full quantum proportional to its wavelength, knowing the relative size of the atom, we can see that the partial quanta in the antinodes describes a wavelength for each related energy that is a small fraction of a full radiated quantum. Let us consider the wavelength which these antinodes maintain. Where a radiated energy of 0.6612eV would produce a wavelength of 1875 nm, an antinode of 0.6612 eV in a partial quanta wave must have a wavelength that is a fraction of the 0.11 nm size of the hydrogen atom[lx]. This relation between wavelength sizes in a partial quantum wave must be proportional in the same way that radiated light has a wavelength that is proportional to the energy of the light. So the larger the energy contained in the antinode, the smaller the wavelength of the antinode. So a partial quantum is described by **e=hc/jא** where **e** is energy and **j** is a constant that is a fraction of wavelength, **h** is Planck's constant, and א is the wavelength of the partial quantum. A non-rigorous maximum preliminary estimate of j would be .0000267 of א of radiated light although the spectrum leads us to believe the number is less than that.

Protons have an energy wave that is defined by the difference in the energy between the spectral lines. The series in the spectra are caused by different states of energy of the electron with its electron-wave in the Hydrogen atom as it transitions through the antinodes of the proton-wave. The energy of the electron describes the energy of the radiated light it produces. The Lyman Series is in the ultraviolet because the electrons must be in an energy state that will overcome the 10.20 eV antinode to transition inward closer to the nucleus.

2.1.1. Description of the Proton-Wave

The electron-wave stores excess energy equal to the energy required to maintain its position in the node of the Series that it describes, but the proton-wave antinodes maintain a consistent antinode energy in the spectra. Therefore, the electron-wave loses and gains the amount of energy in transitions equal to the antinode it passes through while the proton-wave remains a stable energy. The spectrum tells us that each series can be described by a difference in the total overall energy of the electron as it passes through the same energy antinodes. As we know when the electron is in a higher energy state, the electron is oscillating faster according to the frequency associated with the energy. So, a Lyman series electron will be vibrating more rapidly in a node than a Balmer series electron in the same node. The ground state for each series is a different node. Only higher energy electrons can penetrate further towards the nucleus. A Paschen electron of 0.66 eV, a Balmer electron of 2.55 eV, and a Lyman electron of 12.75 eV, may occupy the same node according to the spectrum for Hydrogen.

Each series has a ground state because the antinode of the next series holds the electron in place until it gains enough energy to pass through an inner antinode. This is in accord with Einstein's original findings on specific heat, as well as Debye's findings, and those of current QM theory i.e. that in solids the temperature increases in jumps of energy. [lxi,lxii]

It is easy to see why the electrons do not normally transition into the nucleus as the energy gets higher with each antinode as the antinodes approach the nucleus. The proton-wave is the force that dominates the atom, and the electromagnetic force determines the probable direction of movement.

The following are data and charts of the Hydrogen proton-wave in Table 3, Chart 1 and 2.

Per antinode	Energy eV	
10	0.02360	Outer wave
9	0.03190	
8	0.04460	
7	0.06500	
6	0.10020	
5	0.16620	
4	0.30610	
3	0.66122	
2	1.88923	
1	10.19800	Inner wave

Table 3. Data for graph of Proton Wave

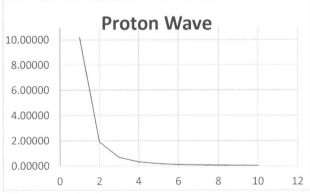

Chart 1. Graph of proton wave

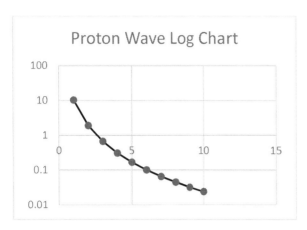

Chart 2. Graph of proton wave as a Log Chart

Spectral series energy levels between lines are shown in the chart. We can see that if the energy of the next antinode followed this curve that the next antinode would be much higher than 10.20 eV as the energy begins to increase exponentially to the nucleus. This would prevent the electrons from transitioning closer to the nucleus as an increase in energy would ionize the electron. Also, we can see why being in ground state of one series, the electron must be more excited to get into the higher energy ground state of an inner series of antinodes in the proton-wave toward the nucleus. The electrons normally transition through the antinodes that are less than or equal to 10.20 eV in Neutral Hydrogen.

Antinodes in eV	
10.19800	
1.88923	
0.66122	
0.30610	
0.16620	
0.10020	
0.06500	
0.04460	
0.03190	
0.02360	
0.01900	
0.01420	
0.00970	
13.52895	Subtotal
etc.	
≈13.59843	Rydberg Energy[lxiii]

Table 4. Table showing energy of antinodes calculating with a limit of infinity converges to the total for the Rydberg Energy

The Balmer equation which found the actual sequence between the lines was later transformed into the Balmer-Rydberg formula. It describes outer wave series beginning with the Lyman series, but does not describe the sequence extended to higher inner antinode energies as light is not abundantly radiated from inner nodes. We can see that the ionization energy described by the Rydberg energy is equal to the sum of all the antinodes from the Lyman series ground state to the limit described by the Balmer-Rydberg formula. Originally Rydberg wrote his equation as wavelengths divided by Rydberg constant. The Rydberg constant describes the energy of the wavelengths. The wavelengths divided by the entire energy of all the antinodes is then staggered per the Rydberg pattern formula which is the basis for the universal inverse square law. 13.6/n2. The inverse square law means that energy is spreading spherically over space. As the energy spreads over consecutive spheres, the energy is divided by the square of equal changes in the radius. It is the energy difference between the consecutive spheres that is shown to be the antinodes of the proton-wave. For every whole number n where Ry is Rydberg Energy, take the difference between each value obtained to arrive at the antinode energy for $\frac{Ry_\infty}{n^2}$ as shown in Table 5.

Sphere #	Total eV Ry	n^2	Ry/n^2	Difference
1	13.59840	1	13.5984	0.00000
2	13.59840	4	3.3996	10.19880
3	13.59840	9	1.510933333	1.88867
4	13.59840	16	0.8499	0.66103
5	13.59840	25	0.543936	0.30596
6	13.59840	36	0.377733333	0.16620
7	13.59840	49	0.277518367	0.10021
8	13.59840	64	0.212475	0.06504
9	13.59840	81	0.167881481	0.04459

Table 5. Sum of all energy differences converges to Rydberg constant

We know the Rydberg energy limit is theoretically calculated as 13.59843.... eV[lxiv] and this continues to infinity under the inverse square law. Therefore, for each proton in the universe, the approximately 13.6 eV of energy is spreading out across the universe to infinity. The first 13 antinodes account for 13.52895 eV of this energy which is concentrated in or near the atom forming matter.

However, we cannot consider Lyman ground state as a point in space where energy arbitrarily begins in the Neutral Hydrogen atom, because the center of the atom does not begin at Lyman ground state. Rather the Rydberg energy antinodes begin their inverse square dispersion from a point which may be considered as the surface of a sphere of energy in space around the nucleus from which energy has bled out. So far, we have established that in Hydrogen, at the surface of an energy sphere in space where the nucleus is the center of the sphere, the Rydberg Energy has bled out from Lyman ground state into the surrounding volume of space according to the inverse square law. However, this energy does not radiate away as it is partial quantum energy. It therefore forms a standing wave of partial quanta energy through which the electrons transition.

Having established the groundwork that energy exists between the electron transitions from Lyman ground state outward from the nucleus, we can then assume that energy exists between Lyman ground state and the nucleus. Because this energy does not radiate, it is also partial quantum energy. However, this energy has not propagated out from the nucleus in a manner conforming to the inverse square law.

Assuming **e=mc²** describes a relation between the proportion of mass and energy in a particle[lxv] and assuming further that for subatomic particles the mass is created by the energy surrounding it, we can estimate the inner antinodes of the standing proton-wave by retrofitting the data along the same curve from the nucleus to the known energy levels as in the data shown in Table 6 and Chart 3.

	Per cycle	Energy eV
Outer	15	.024
Antinode	14	.032
	13	.045
	12	.065
	11	.100
	10	.166
	9	.306
	8	.661
Inner	7	1.889
Antinode	6	10.198
Estimated	5	84.820
Estimated	4	1473.000
Estimated	3	59510.000
Estimated	2	5361000.000
Estimated	1	932850000.000
	Total energy	938272081.306
Energy of proton-wave: 938,272,081 eV		

Table 8. Estimate of entire proton wave energy through extending known curve

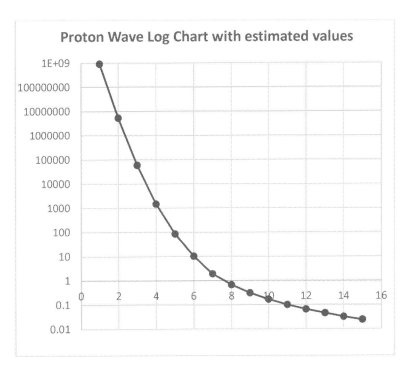

Chart 3. Log Chart of proton wave with estimated extension of Balmer-Rydberg curve

This would change the Bohr radius to much closer to the nucleus. This would indicate that most of the energy in the atom is right up against the nucleus. This steep inward curve of energy from the outer proton-wave to the proton-particle appears to follow the rules of special relativity. The inward curve changes from Isaac Newton to Einstein. No pattern in the relativistic change factor is readily apparent. The universe is very precise so these numbers must have a relation to one another. The graph shown in Chart 3 is an exponential decay curve from the nucleus and probably can be depicted mathematically by a negative e (Euler's number) exponential formula. Table 7 shows the estimated change factor.

Relativistic Change Factor	Inner Antinodes	% of c
8.317355249	84.820	0.992746
17.36611925	1473.001	0.998340700018
40.4006106	59510.139	0.99969361981
90.08548254	5360999.615	0.9999383868
174.0067302	932850013.784	0.99998348641
	938272081.359	

Table 7. Estimate of change in antinode energies as a percent of the speed of light

If we assume that the proton is spherical, beginning with the radius of the sphere of the proton particle, each antinode represents an increase in spherical energy which represents an increase in the radius of the proton—a proton being defined as including its proton-particle and its associated proton-wave.

2.1.2. Quantum Mechanical Effects Explained from Above Assumption of Sollism Theory

The objections to a new theory of the atom can be raised upon the grounds of the proofs of Quantum Mechanics in the form of quantum tunneling, quantum entanglement, the Uncertainty Principle, and the Pauli Exclusion Principle. Wave-Particle Duality is obviously explained in this new theory as the particle is a wave and a particle at the same time i.e. a particle surrounded by a wave. Because no new theory should be accepted unless it can encompass the phenomena of QM, they will be briefly addressed. In this model of the atom, tunneling is accounted for because the nucleon-waves create an atomic-wave that extends far beyond the particles that create matter. In current QM Theory, the Uncertainty Principle predicts phenomena such as the Casimir Effect. We can see this is the effect of the energy of the proton-waves resisting each other. Under this new theory, the Pauli Exclusion Principle is due to the resistance of the energy of particle-waves. Vacuum or Zero Point Energy is the energy of the proton-wave that fills space as can be deduced from the Balmer-Rydberg formula extending the energy of each single proton to infinity. In fact, the "vacuum catastrophe" shows that the Uncertainty Principle is inaccurate and need not be regarded as it is unnecessary and appears to give incorrect values[lxvi]. We know that many atomic observation devices input energy into the system, so that, inputting energy into a system changes its measurement. This is not a proof of the Uncertainty Principle. When we measure by absorption, we get exact measurements. Also, we can correct for the amount of energy we input into a system when observing the system. Therefore, the Uncertainty Principle is unnecessary and should be disregarded as it makes predictions like the Vacuum Catastrophe. In this model, any predictions made by the Uncertainty Principle can be attributed to the force of the proton-wave extending beyond the atom and extending into the nucleus of the atom. Since the universe is filled with proton-waves, there is, in fact, no true vacuum. In order to account for Quantum Entanglement which has been tested on photons, there needs to be a more precise theory for both light and the photon which will be attempted here and which seems to come naturally as a consequence of Sollism Theory.

So instead of many rules governing the atom, the amount of resistance of every point in space can theoretically be calculated as the amount of resistance caused by the energy being exerted by all the particle-waves in that space. The amount of pressure of every point in space can theoretically be calculated as the amount of pressure exerted on every particle by its particle-wave and by the energy-waves of every other particle in space. The proton-waves extending to infinity cause energy to surround all matter in great spheres where the nearer particles are to each other, the greater the energy surrounding them. Every particle is interior to wave-energy that exerts pressure and exterior to wave energy from other particles that exerts resistance.

We will see how theorizing energy between the spectral lines helps us decipher the spectra of the elements and gives us the ionization values directly from the spectra.

2.2. Helium

For two protons to fuse into Helium, or more precisely four nucleons to fuse into Helium, the nucleons must overcome their respective wave energies and become trapped in each other's particle-waves. We can see that by assuming causality for the separation of the electrons from the nucleus, we explain the Strong Force.

In Helium, there would be two protons, each with associated proton-waves, and two neutrons, each with associated neutron-waves. The neutron-waves have insufficient energy to create a stable particle. It is only by being enclosed in a proton-wave that the neutron-particle becomes stable. We know from Deuterium, shown later in this article, that the neutron-wave does contribute to the nucleon-wave energy, but as the contribution of energy is negligibly small, we will neglect the contribution of the neutron-wave in our evaluation of the proton-waves in the following elements. In Helium II we see that the two proton-waves are in superposition and being emitted from the nucleus. The waves are coherent and in phase. The energy is not linear superposition, but spherical superposition and the energy values are squared from Neutral Hydrogen to Helium II. The pattern of the antinodes of the proton-wave of Hydrogen repeats with four-fold energy in Helium II showing complete constructive interference as shown in Table 8 below[lxvii].

He-II Series 1 [lxviii] Antinode	Larger Wavelength	Next Shorter Wavelength	Larger λ in eV	Next Shorter λ in eV	Transition Energy eV	Rounded Energy eV	Difference eV Between Series
He II	30.37858	25.63177	40.8131087	48.3713879	7.5582792	7.558	
He II	25.63177	24.30266	48.3713879	51.0168141	2.6454262	2.645	40.81
He II	24.30266	23.73307	51.0168141	52.2412098	1.2243957	1.224	40.81
He II	23.73307	23.43472	52.2412098	52.9062984	0.6650886	0.665	40.81
He II	23.43472	23.25842	52.9062984	53.3073308	0.4010324	0.401	40.81
He II	23.25842	23.14541	53.3073308	53.5676097	0.2602789	0.260	40.81
He II	23.14541		53.5676097				

He-II Series 2 Antinode	Larger Wavelength	Next Shorter Wavelength	Larger λ in eV	Next Shorter λ in eV	Transition Energy eV	Rounded Energy eV	Difference eV Between Series
He II	164.0375	121.509	7.55829788	10.2037239	2.64542602	2.65	
He II	164.04897		7.55776942				Fine structure
He II	121.509	108.494	10.2037239	11.4277682	1.2240443	1.22	7.55
He II	108.494	102.527	111.4277682	12.0928564	0.6650882	0.67	7.56
He II	102.527	99.236	12.0928564	12.4938962	0.4010398	0.40	7.56
He II	99.236	97.211	12.4938962	12.7541563	0.2602601	0.26	7.56
He II	97.211	95.87	12.7541563	12.9325575	0.1784012	0.18	7.56
He II	95.87		12.9325575			-	

He-II Series 3 Antinode	Larger Wavelength	Next Shorter Wavelength	Larger λ in eV	Next Shorter λ in eV	Transition Energy eV	Rounded Energy eV	Difference eV Between Series
Calculated	468.7	320.31	2.645	3.870764	1.220764	1.22	
He II	320.31	273.33	3.8707636	4.53607101	0.66530741	0.665	2.65
He II	273.33	251.12	4.53607101	4.93725824	0.40118723	0.401	
He II	251.12	238.54	4.93725824	5.19763682	0.26037858	0.260	
He II	238.54		5.19763682			0.180	

He-II Series 4 Antinode	Larger Wavelength	Next Shorter Wavelength	Larger λ in eV	Next Shorter λ in eV	Transition Energy eV	Rounded Energy eV	Difference eV Between Series
Calculated	1012.9		1.22			0.665	

| | Calculated | | | | | | 0.401 | |

He-II Series 5	Larger	Next Shorter	Larger	Next Shorter	Transition	Rounded	Difference
Antinode	Wavelength	Wavelength	λ in eV	λ in eV	Energy eV	Energy eV	eV Between Series
He II	1863.68	3090.85	0.66526672			0.401	

He-II Series 6	Larger	Next Shorter	Larger	Next Shorter	Transition	Rounded	Difference
Antinode	Wavelength	Wavelength	λ in eV	λ in eV	Energy eV	Energy eV	eV Between Series
He II	3090.85		0.40113376	0.66526672	0.26413296		

Table 8. Table of the energies of the antinodes of neutral Helium

Next in Table 9 we see the proton-waves for Helium II with two proton waves in superposition and various ground states.

HELIUM II		Antinode eV	May contain electrons in energy states in eV (rounded)						
Number			Series H2-1	Series H2-2	Series H2-3	Series H2-4	Series H2-5	Series H2-6	Series H2-7
99N	Node	0	Ground						
100A	Antinode	40.81							
100N	Node	0	40.81	Ground					
101A	Antinode	7.56							
101N	Node	0	48.37	7.56	Ground				
102A	Antinode	2.65							
102N	Node	0	51.02	10.20	2.65	Ground			
103A	Antinode	1.22							
103N	Node	0	52.24	11.43	3.87	1.22	Ground		
104A	Antinode	0.67							
104N	Node	0	52.91	12.09	4.54	1.89	0.67	Ground	
105A	Antinode	0.40							
105N	Node	0	53.31	12.49	4.94	2.29	1.07	0.4	Ground
106A	Antinode	0.26							
106N	Node	0	53.57	12.75	5.20	2.55	1.33	0.66	0.26
107A	Antinode	0.18							
107N	Node	0	53.75	12.93	5.38	2.73	1.51	0.84	0.44

Table 9. Table showing excitation of electrons in Helium II

When we examine the proton-wave in Helium I, it gives us unexpected values which were not found by Niels Bohr nor any quantum theorist until now. In Helium II, the proton-waves are in phase, and the constructive interference causes the single electron to transition at a four-fold energy level relative to Neutral Hydrogen. In Helium I, the spectrum infers that although the proton-waves are still coherent, the proton-waves are arrayed as to allow independent transition for each electron so that the two electrons in Helium I transition each on their own proton-wave. Therefore, the very same antinode energies of the proton-waves in Helium I equal those of Neutral Hydrogen. The addition of an electron in Helium I polarizes one proton-wave to one electron rather than the polarization of two proton-waves to one electron in Helium II. In Helium I, the transition energy changes because the ground state rather than being zero energy changes to a higher energy level. The mathematical relation or equation that creates the new ground state energy causing the increase in transition energy is unclear, but we can derive the ionization energy from the non-zero ground state energy.

Note the repeating pattern shown in blue of the antinodes in the spectrum of Helium I showing that they are on average equal to the antinodes of Neutral Hydrogen. It should also be noted that for each series, the two electrons in Helium I are paired for each energy level as we shall see from the spectrum. The pattern found in Table 10 below is completely novel in the history of science.

Helium I

He I-1A Antinode	Larger Wavelength	Next Shorter Wavelength	Larger λ in eV	Next Shorter λ in eV	Transition Energy eV	Rounded Energy eV	Difference eV Between
He I	58.433436	53.702992	21.2180623	23.0870616	1.8689993	1.869	20.631349
He I	53.702992	52.221309	23.0870616	23.7421143	0.6550527	0.655	20.6151201
He I	52.221309	51.561684	23.7421143	24.0458455	0.3037312	0.304	20.6149288
He I	51.561684	51.209856	24.0458455	24.2110481	0.1652026	0.165	20.6148333
He I	51.209856	50.999829	24.2110481	24.3107538	0.0997057	0.100	20.6147844
He I	50.999829	50.864338	24.3107538	24.3755121	0.0647583	0.065	
He I	50.864338	50.771809	24.3755121	24.4199352	0.0444231	0.044	
He I	50.771809	50.705802	24.4199352	24.4517242	0.031789	0.032	
He I	50.705802	50.657057	24.4517242	24.475253	0.0235288	0.024	

He I-1B Antinode	Larger Wavelength	Next Shorter Wavelength	Larger λ in eV	Next Shorter λ in eV	Transition Energy eV	Rounded Energy eV	Difference eV Between
He I	2113.203	501.56783	0.58671329	2.47194151	1.8852282	1.885	
He I	501.56783	396.4729	2.47194151	3.12718546	0.655244	0.655	0.36179094
He I	396.4729	361.364	3.12718546	3.43101219	0.3038267	0.304	0.35440265
He I	361.364	344.759	3.43101219	3.59626373	0.1652515	0.165	0.34986754
He I	344.759	335.455	3.59626373	3.69600777	0.099744	0.100	0.35026411

He I-2A Antinode	Larger Wavelength	Next Shorter Wavelength	Larger λ in eV	Next Shorter λ in eV	Transition Energy eV	Rounded Energy eV	Difference eV Between
He I	587.562	447.148	2.11015057	2.77278281	0.6626322	0.663	0.25358208
He I	447.148	402.3973	2.77278281	3.08114465	0.3083618	0.308	0.25376225
He I	402.3973	381.9607	3.08114465	3.24599962	0.164855	0.165	0.25556565
He I	381.9607	370.5	3.24599962	3.34640833	0.1004087	0.100	
He I	370.5	363.423	3.34640833	3.41157353	0.0651652	0.065	1.9444857
He I	363.423	358.727	3.41157353	3.45623354	0.04466	0.0447	2.00026445

He I-2B Antinode	Larger Wavelength	Next Shorter Wavelength	Larger λ in eV	Next Shorter λ in eV	Transition Energy eV	Rounded Energy eV	Difference eV Between
He I	667.815	492.193	1.85656849	2.51902056	0.6624521	0.662	
He I	492.193	438.793	2.51902056	2.825579	0.3065584	0.307	1.85548204
He I	438.793	414.3761	2.825579	2.9920748	0.1664958	0.166	1.85579828

He I-3A Antinode	Larger Wavelength	Next Shorter Wavelength	Larger λ in eV	Next Shorter λ in eV	Transition Energy eV	Rounded Energy eV	Difference eV Between
He I	1868.534	1278.479	0.66353852	0.96978072	0.3062422	0.306	
He I	1278.479	1091.305	0.96978072	1.13611161	0.1663309	0.166	
He I	1091.305	1002.773	1.13611161	1.2364157	0.1003041	0.100	0.1002424
He I	1002.773	952.617	1.2364157	1.30151392	0.0650982	0.065	0.10072556
He I	952.617	921.034	1.30151392	1.34614388	0.04463	0.0446	

He I-3B Antinode	Larger Wavelength	Next Shorter Wavelength	Larger λ in eV	Next Shorter λ in eV	Transition Energy eV	Rounded Energy eV	Difference eV Between
He I	1196.912	1091.71	1.03586921	1.13569014	0.0998209	0.100	
He I	1091.71	1031.154	1.13569014	1.20238518	0.066695	0.0667	

He I-3C Antinode	Larger Wavelength	Next Shorter Wavelength	Larger λ in eV	Next Shorter λ in eV	Transition Energy eV	Rounded Energy eV	Difference eV Between
He I	1700.247	1196.912	0.72921421	1.03586921	0.306655	0.307	
He I	1196.912	1031.154	1.03586921	1.20238518	0.166516	0.167	0.06610366
He I	1031.154	951.66	1.20238518	1.30282274	0.1004376	0.100	0.06669504
He I	951.66	906.327	1.30282274	1.36798781	0.0651651	0.065	0.06682981

He I-3D Antinode	Larger Wavelength	Next Shorter Wavelength	Larger λ in eV	Next Shorter λ in eV	Transition Energy eV	Rounded Energy eV	Difference eV Between
He I	1869.723	1278.499	0.66311656	0.96976555	0.306649	0.307	
He I	1278.499	1091.71	0.96976555	1.13569014	0.1659246	0.166	
He I	1091.71	1003.116	1.13569014	1.23599293	0.1003028	0.100	
He I	1003.116	952.927	1.23599293	1.30109052	0.0650976	0.065	
He I	952.927	921.034	1.30109052	1.34614388	0.0450534	0.045	

Table 10. Table of Helium I

A comparison of Hydrogen and Helium antinodes is shown below in Table 11.

Hydrogen	Helium I	Ratio
10.20		N/A
1.89	1.87	1.0
0.66	0.66	1.0
0.306	0.304	1.0
0.166	0.165	1.0

Hydrogen	Helium II	Ratio
10.198	40.81	4.0
1.889	7.558	4.0
0.665	2.645	4.0
0.306	1.224	4.0
0.166	0.665	4.0

0.100	0.100	1.0		0.100	0.401	4.0
0.065	0.065	1.0		0.065	0.260	4.0
0.044	0.044	1.0		0.044	0.178	4.0
0.032	0.032	1.0		0.032		

Table 11. Comparison of Hydrogen and Helium I and a comparison of Hydrogen and Helium II.

Helium I antinodes are grouped below in Table 12. To arrive at the ground state energy, subtract the energy of the line above ground state from the antinode between ground state. I have shown the proton-wave energies as equal to Hydrogen for convenience. The variation from Hydrogen proton-waves to Helium proton-waves is small. And that variation is due to the distortion of the proton-waves by the increase in the size of the nucleus which decreases the overall energy and with an additional small increase in energy by the wave energy of the neutrons. In other words, the small differences in the energy of the antinodes of the Helium I atom's proton-waves when compared to the Neutral Hydrogen atom are due to the larger radius of the nucleus causing the great sphere of energy in the first antinode around the nucleus to change the energy levels of the entire proton-wave. Also, the cumulative effect of the neutron-waves is affecting the energy of the nuclear antinodes. But the proton-wave of Hydrogen according to the difference in energy between the lines in the Balmer-Rydberg equation is still clearly seen as the ratio is clearly 1:1 shown in Table 11.

Helium I		Energy in eV	May contain electrons in energy states in eV (rounded) Note: For Helium I, the ground state energy is 11.02 eV excess energy in electron-wave.							
Number			He I-1	He I-1	He I-2	He I-2	He I-3	He I-3	He I-3	He I-3
99A	Antinode	≈ 85 eV								
99N	Node	0	11.02							
100A	Antinode	10.20								
100N	Node	0	21.22	0.58	0.22					
101A	Antinode	1.89								
101N	Node	0	23.09	2.47	2.11	1.86	Ground	0.07		
102A	Antinode	0.66								
102N	Node	0	23.74	3.13	2.77	2.52	0.66	0.73		
103A	Antinode	0.31								
103N	Node	0	24.05	3.43	3.08	2.83	0.97	1.04	0.87	0.97
104A	Antinode	0.17								
104N	Node	0	24.21	3.59	3.246		1.14	1.20	1.04	1.14
105A	Antinode	0.10								
105N	Node	0	24.31		3.346		1.24	1.30	1.14	1.24
106A	Antinode	0.07								
106N	Node	0	24.38		3.412		1.30	1.37		1.30

Table 12. Table showing excitation of electrons in Helium I

The ability to calculate ionization energy directly from data without experimentation in atoms with more than one electron has until now not been possible. However, in this model, the ionization energy is seen by examining the electron in its ground state of the highest energy series for most atoms of an element. The ground state electron in Helium I holds an excess partial quantum of energy. We can analyze the electron energies from Helium II to Helium I and see how the electrons are paired in the Table 13 below.

	He II Energy in eV	He I Energy in eV	Nullius in verba, but the word of the stars.													
			He I	He I				He I	He I				He I			
	He II	He I	Series He II-1	He I-1	He I-2	Sub-total He I-1	%of He II	Series He II-1	He I-1	He I-2	Sub-total He I-2	%of He II	Series He II-3	He I-3	He I-3	Sub-total He I-3	%of He II
Anti-n	≈ 340	≈ 85															
Node	0	0	Ground	11.02													
Anti-n	40.81	10.20															
Node	0	0	40.81	21.22	0.58	21.79	0.53	Ground	0.22								
Anti-n	7.56	1.89															
Node	0	0	48.37	23.09	2.47	25.55	0.53	7.56	2.11	1.86	3.97	0.52	Ground	0.07	Ground		
Anti-n	2.65	0.66															
Node	0	0	51.02	23.74	3.13	26.87	0.53	10.20	2.77	2.52	5.29	0.52	2.65	0.73	0.66	1.38	0.52
Anti-n	1.22	0.31															
Node	0	0	52.24	24.05	3.43	27.48	0.53	11.43	3.08	2.83	5.91	0.52	3.87	1.04	0.97	2.00	0.52
Anti-n	0.67	0.17															
Node	0	0	52.91	24.21	3.59	27.80	0.53	12.09	3.25	3.00	6.24	0.52	4.54	1.20	1.14	2.34	0.52
Anti-n	0.40	0.10															
Node	0	0	53.31	24.31	3.70	28.01	0.53	12.49	3.35	3.10	6.44	0.52	4.94	1.30	1.24	2.54	0.51
Anti-n	0.26	0.07															
Node	0	0	53.57	24.38	3.77			12.75	3.41	3.17	6.58	0.516	5.20	1.30	1.30	2.60	0.50

Table 13. Table pairing the electrons in Helium I and showing the ground states.

Because the same antinodes appear in Helium I as in Neutral Hydrogen, the ionization energy should be the same as Neutral Hydrogen except that Helium I has a ground state that is not equal to zero, but 11.02 eV. Therefore, the 11.02 eV must be added to the Rydberg energy of Neutral Hydrogen to arrive at the ionization energy of Helium I as shown in Table 14 below which calculates the ionization energy directly from the spectra:

Hydrogen	Helium II
Antinodes in eV	Antinodes in eV
10.1980	40.7920
1.8892	7.5569
0.6612	2.6449
0.3061	1.2244
0.1662	0.6648
0.1002	0.4008
0.0650	0.2600
0.0446	0.1784
0.0319	0.1276
0.0236	0.0944

Helium I
Antinodes in eV
10.1980
1.8892
0.6612
0.3061
0.1662
0.1002
0.0650
0.0446
0.0319
0.0236

	0.0190	0.0760		0.0190	
	0.0142	0.0568		0.0142	
	0.0097	0.0388		13.5193	Subtotal
	13.5290	54.1158	Subtotal	etc.	Add all antinodes
	etc.	etc.	Add all antinodes	≈13.59	Rydberg Energy
NIST	≈13.5984	≈54.3936	H*4=Rydberg Energy	+ 11.02	Ground State energy
NIST	≈13.6057	≈54.4227	H*4=Rydberg Energy	≈24.61	Calculated Ionization Energy
		54.4177	NIST ionization	24.5874	NIST ionization

Table 14. Table showing Helium II and Helium I ionization calculation from collected spectral data.

As we have seen previously, when Helium I loses its electron, the proton-waves polarize toward one electron and change formation to the state of Helium II and, therefore, the proton-waves arrange themselves in spherical superposition and the energy for ionization changes to that of Helium II.

We can infer that the chemical properties of elements are not only dependent on the number of protons, but are dependent upon the structural pattern that the proton-waves take in the atom.

2.3. Lithium

As we know from current theory, Lithium III follows the one electron polarization pattern of Helium II. In Lithium III, the three proton-waves are in spherical superposition paired to one electron and the overall energy is nine times that of Neutral Hydrogen. [lxix] Table 15 below shows the series for Lithium III.

L3-1 Series	Larger	Next Shorter	Larger	Next Shorter	Transition	Rounded	Difference
Antinode Number	Wavelength	Wavelength	λ in eV	λ in eV	Energy eV	Energy eV	eV Between Series
Li III	13.5	11.39	91.8403177	108.853756	17.013438	17.013	
Li III	11.39	10.8	108.853756	114.800397	5.946641	5.947	91.85
Li III	10.8	10.55	114.800397	117.520785	2.720388	2.720	91.84
Li III	10.55	10.41	117.520785	119.101276	1.580491	1.580	91.85
Li III	10.41	10.34	119.101276	119.907571	0.806295	0.806	
Li III	10.34	10.29	119.907571	120.490212	0.582641	0.583	
Li III	10.29		120.490212			0.000	

L3-2 Series	Larger	Next Shorter	Larger	Next Shorter	Transition	Rounded	Difference
Antinode Number	Wavelength	Wavelength	ƛ in eV	ƛ in eV	Energy eV	Energy eV	eV Between Series
Li III	72.91	54	17.0051335	22.9600794	5.9549459	5.955	
Li III	54	48.3	22.9600794	25.669654	2.7095746	2.710	17.01
Li III	48.3	45.6	25.669654	27.1895677	1.5199137	1.520	16.96
Li III	45.6		27.1895677				16.99

L3-3 Series	Larger	Next Shorter	Larger	Next Shorter	Transition	Rounded	Difference
Antinode Number	Wavelength	Wavelength	ƛ in eV	ƛ in eV	Energy eV	Energy eV	eV Between Series
Calculated	208.4	142.4	5.95	8.70677169	2.7567717	2.757	
Li III	142.4	121.5	8.70677169	10.2044797	1.497708	1.498	
Li III	121.5		10.2044797			0.000	

Table 15. Table of Lithium III series energy of antinodes.

Lithium III		Energy in eV	May contain electrons in energy states in eV		
Number			Li III-1	Li III-2	Li III-3
99A	Antinode				
99N	Node	0	Ground		
100A	Antinode	91.782			
100N	Node	0	91.84	Ground	
101A	Antinode	17.001			
101N	Node	0	108.85	17.01	Ground
102A	Antinode	5.940			
102N	Node	0	114.80	22.96	5.96
103A	Antinode	2.754			
103N	Node	0	117.52	25.67	8.71

| 104A | Antinode | 1.494 | | | |
| 104N | Node | 0 | 119.101 | 27.19 | 10.20 |

Table 16. Table showing excitation of electrons in Lithium III

The proton-wave antinode energy levels of Neutral Hydrogen appear in Lithium II and Lithium I. In Lithium I, the electrons travel through lower energy antinodes when compared to Hydrogen thereby extending the size of the atom as far as matter is concerned.

In Lithium II, we see an interesting pattern arise in Table 17. In Lithium II, we have a Helium II superposition of two proton-waves upon which only one electron travels i.e. a single electron polarizes two proton-waves toward itself. And in the same Lithium II atom, we have a Hydrogen-type proton-wave upon which the other electron travels. This other electron has polarized one proton-wave to itself. The Lithium II nucleon waves take the form of two proton-waves in spherical superposition and a single proton-wave in the manner of Hydrogen emanating from the nucleus. We will see from nature that the atom has coherent waves despite the number of nucleons and that the electron is plane-polarizing the wave or waves toward itself.

Lithium II

L2-1 Series	Larger	Next Shorter	Larger	Next Shorter	Transition	Rounded	Difference
Antinode	Wavelength	Wavelength	λ in eV	λ in eV	Energy eV	Energy eV	eV Between Series
Li II	19.928	17.8014	62.2161927	69.6486955	7.4325028	7.433	He II-type
Li II	17.8014	17.1575	69.6486955	72.2625259	2.6138304	2.614	
Li II	17.1575	16.874	72.2625259	73.4766083	1.214	1.214	61.3086767
Li II	16.874	16.721	73.4766083	74.1489318	0.6723235	0.672	61.2959555

L2-3 Series	Larger	Next Shorter	Larger	Next Shorter	Transition	Rounded	
Antinode	Wavelength	Wavelength	λ in eV	λ in eV	Energy eV	Energy eV	
Li II	113.188	101.788	10.9538492	12.1806528	1.2268036	1.227	He II-type
Li II	101.788	96.5	12.1806528	12.8481273	0.6674745	0.667	
Li II	96.5	93.6	12.8481273	13.2461996	0.3980723	0.398	
Li II	93.6	91.75	13.2461996	13.5132892	0.2670896	0.267	
Li II	91.75	90.55	13.5132892	13.692372	0.1790828	0.179	

L2-2 Series	Larger	Next Shorter	Larger	Next Shorter	Transition	Rounded	
Antinode	Wavelength	Wavelength	λ in eV	λ in eV	Energy eV	Energy eV	
Li II	96.5	80	12.8481273	15.4980536	2.6499263	2.650	He II-type

L2-4 Series						Rounded	
Antinode						Energy eV	
Li II	273.047	238.3199	4.54077242	5.20243709	0.66166467	0.662	H-type
Li II	238.3199	224.921	5.20243709	5.51235451	0.30991742	0.310	H-type
Li II	224.921	218.3	5.51235451	5.67954324	0.16718873	0.167	H-type

Table 17. Table showing energy of antinodes of Lithium II series.

We can see by putting the values in a table of antinodes (Table 18) how we arrive at the ionization energy, because the electron in ground state has 21.40 eV of energy.

Lithium II		Energy in eV	May contain electrons in energy states in eV				
Number		First Nucleon Wave	Li II-1	Li II-2	Li II-3	Second Nucleon Wave	Li II-4
99A	Antinode					≈ 85 eV	
99N	Node	0	21.40			0	
100A	Antinode	40.81				10.19	
100N	Node	0	62.22			0	
101A	Antinode	7.56				1.89	
101N	Node	0	69.65	12.85		0	4.54
102A	Antinode	2.65				0.66	
102N	Node	0	72.26	15.49	10.95	0	5.20
103A	Antinode	1.22				0.31	
103N	Node	0	73.48		12.18	0	5.51
104A	Antinode	0.67				0.17	
104N	Node	0			12.85	0	
105A	Antinode	0.40				0.10	
105N	Node	0			13.25	0	
106A	Antinode	0.26				0.066	
106N	Node	0			13.51	0	

Table 18. Table showing excitation of electrons in Lithium II

Whereas, Lithium I takes the formation of three separate Hydrogen-like proton-waves as seen in Table 19.

LI-1 Series	Larger	Next Shorter	Larger	Next Shorter	Transition	Rounded
Antinode	Wavelength	Wavelength	λ in eV	λ in eV	Energy eV	Energy eV
Li I	323.2633	274.12	3.8354007	4.52299828	0.68759758	0.688
Li I	274.12	256.231	4.52299828	4.83877551	0.31577723	0.316
Li I	256.231	247.506	4.83877551	5.00935043	0.17057492	0.171
Li I	247.506	242.543	5.00935043	5.11185352	0.10250309	0.103
Li I	242.543	239.439	5.11185352	5.17812173	0.06626821	0.066
Li I	239.439	237.354	5.17812173	5.22360815	0.04548642	0.045
Li I	237.354	235.893	5.22360815	5.25596049	0.03235234	0.032
Li I	235.893	234.822	5.25596049	5.27993241	0.02397192	0.024

Li I	234.822	234.015	5.27993241	5.29814024	0.01820783	0.018
Li I	234.015	233.394	5.29814024	5.3122372	0.01409696	0.014
Li I	233.394	232.902	5.3122372	5.32345917	0.01122197	0.011
Li I	232.902	232.511	5.32345917	5.33241132	0.00895215	0.009

LI-2 Series Antinode	Larger Wavelength	Next Shorter Wavelength	Larger λ in eV	Next Shorter λ in eV	Transition Energy eV	Rounded Energy eV
Li I	391.535	379.472	3.16662441	3.26728794	0.10066353	0.101
Li I	379.472	371.87	3.26728794	3.334	0.06671206	0.067
Li I	371.87	366.2	3.334	3.38	0.046	0.046

Table 19. Table showing antinode energy of Lithium I

We can arrive at the ionization energy shown in table 20.

Lithium I Number		Energy eV	electron energy eV Li I-1	Li I-1
99A	Antinode	≈ 85 eV		
99N	Node	0		
100A	Antinode	10.198		
100N	Node	0	1.95	
101A	Antinode	1.89		
101N	Node	0	3.84	
102A	Antinode	0.66		
102N	Node	0	4.50	
103A	Antinode	0.31		
103N	Node	0	4.84	
104A	Antinode	0.17		
104N	Node	0	5.01	3.166
105A	Antinode	0.10		
105N	Node	0	5.11	3.267
106A	Antinode	0.066		
106N	Node	0	5.18	3.334
107A	Antinode	0.045		
107N	Node	0	5.22	
108A	Antinode	0.032		
108N	Node	0	5.26	

Lithium I	
13.598	Rydberg Energy
-10.198	Less antinode
3.40	Subtotal
+1.95	Ground state energy
5.35	Calc. Ionization Energy

Table 20. Table showing excitation of electrons in Lithium I and calculation of ionization energy from spectral lines.

Lithium II

Antinodes in eV

Lithium I

Antinodes in eV

40.7920		1.8892	
7.5569		0.6612	
2.6449		0.3061	
1.2244		0.1662	
0.6648		0.1002	
0.4008		0.0650	
0.2600		0.0446	
0.1784		0.0319	
0.1276		0.0236	
0.0944		0.0190	
0.0760		0.0142	
0.0568		0.0097	
54.0770	Subtotal	3.3310	Subtotal
etc.	Add all antinodes	13.59	Rydberg Energy
54.39360	H*4=Rydberg Energy	-10.19	Less antinode
+21.40	Ground State energy	3.4	Subtotal
75.79360	Calculated Ionization energy	+1.95	Ground state energy
75.64	NIST ionization	5.35	Calculated Ionization Energy
		5.391714	NIST ionization

Table 21. Tables showing energy in antinodes of Lithium II, Lithium I, and calculation of ionization from spectral lines.

We can describe the laws of Hydrogen, Helium, and Lithium III as:

For Neutral Hydrogen: The excess energy of the electron stored in its electron-wave is equal to the antinodes of the proton-wave that the electron transitions through and at ground state the electron-wave has zero excess energy.

For Helium I: The energy of the antinodes is approximately equal to the proton-wave of Neutral Hydrogen, but the ground state energy is equal to 11.02 eV and therefore the transition energies are higher than Neutral Hydrogen. The two electrons share the energy of the node.

For Helium II: The nucleon-waves are coherent and in spherical superposition and the electron polarizes the two proton-waves toward itself so that the energy of the antinodes is four times that of the antinodes of the proton-wave in Neutral Hydrogen.

For Lithium III: The nucleon-waves are in spherical superposition and the energy of the antinodes is nine times that of the antinodes of the proton-wave in Neutral Hydrogen.

Table 22 shows the pattern that the whole periodic table follows as regards to formation of the proton-waves in each element according to the number of electrons of that element.

Element	Formation of nucleon waves	Electrons

Hydrogen	(1) H-type proton-wave	1
Helium II	(2) waves form (1) polarized proton-wave	1
Helium I	(2) H-type proton-waves	2
Lithium III	(3) waves form (1) polarized proton-wave	1
Lithium II	(1) Helium II wave and (1) Hydrogen wave	2
Lithium I	(3) H-type proton-waves	3

Table 22. Comparison of types of proton waves with respect to other ions and neutral atoms.

We can deduce from this emerging pattern that the rule for all atoms of all elements is that each electron pairs with one proton-wave creating a neutral electron-proton pair until the final electron pairs with any remaining proton-waves which makes the remaining waves into a single coherent proton-wave polarized toward that remaining electron. Ions contain at least one unequal electron-proton pairing and therefore have a higher ionization energy, because the unequal electron-proton pairing forms coherent proton-waves polarized to a single electron so that the proton-waves are in spherical superposition with associated higher energies.

The sections for Beryllium, Boron and Carbon follow[lxx]. The multi-polarized wave is easy to detect and a single series for each of these ions is shown in the sections below, but the Hydrogen-type wave is hypothesized in the "Formation of Nucleon Waves" table 23 for these higher elements, as it becomes increasingly more difficult to detect as there are not enough spectral lines in the NIST tables for each ion and neutral atom as the atoms become larger, and the Hydrogen-type wave where surmised for these higher elements and is shown in brackets in Table 23. The tables from here to Carbon speak for themselves by incorporating the energies of the Hydrogen proton-wave along with the energies of the previous elemental ions. The tables showing the ionization calculations for each element and its ions is also presented.

Element	Formation of Nucleon Waves	Electrons
Hydrogen	(1) H-type proton-wave	1
Helium II	(2) waves form (1) polarized proton-wave	1
Helium I	(2) H-type proton-waves	2
Lithium III	(3) waves form (1) polarized proton-wave	1
Lithium II	(1) Helium II wave and (1) Hydrogen wave	2
Lithium I	(3) H-type proton-waves	3
Beryllium IV	(4) waves form (1) polarized proton-wave	1
Beryllium III	(1) Lithium III wave [and (1) Hydrogen wave]	2
Beryllium II	(1) Helium II wave [and (2) Hydrogen]	3
Beryllium I	(4) H-type proton-waves	4
Boron V	(5) waves form (1) polarized proton-wave	1
Boron IV	(1) Be IV proton-wave [and (1) H wave]	2
Boron III	(1) Li III proton-wave [and (2) H waves]	3
Boron II	(1) He II proton-wave [and (3) H waves]	4
Boron I	(5) H-type proton-waves	5

Carbon VI	[(6) waves to (1) polarized proton-wave]	1
Carbon V	(1) B V proton-wave [and (1) H wave]	2
Carbon IV	(1) Be IV proton wave [and (2) H type waves]	3
Carbon III	(1) Li III proton-wave [and (3) H type waves]	4
Carbon II	(1) He II proton-wave [and (4) H type waves]	5
Carbon I	(6) H-type proton-waves	6

Table 23. Configuration of proton waves for each ion and neutral atom.

2.4 Beryllium

Beryllium-IV Antinode	Larger Wavelength	Next Shorter Wavelength	Larger λ in eV	Next Shorter λ in eV	Transition Energy eV	Rounded Energy eV	Difference eV Between
Be IV	8.932	7.593	138.8093	163.2878	24.47855	24.479	
Be IV	7.593	6.406	163.2878	193.5442	30.25642	30.256	163.168
Be IV	6.406	6.074	193.5442	204.1232	10.57897	10.579	163.168
Be IV	6.074	5.857	204.1232	209.030	4.9067	4.907	163.168
Calculated	5.931	5.857	209.030	211.6859	2.656	2.656	163.168
Be IV	5.857	5.813	211.6859	213.2882	1.602302	1.602	163.168
Be IV	5.813		213.2882				

Table 24. Spectral sequence of Beryllium-IV (Shaded yellow: spectral line is calculated and hypothesized.)

Beryllium-IV Number		Energy in eV	eV Be-IV
98A	Antinode		
98N	Node	0	Ground
99A	Antinode	163.168	
99N	Node	0	163.28
100A	Antinode	30.224	
100N	Node	0	193.54
101A	Antinode	10.560	
101N	Node	0	204.12
102A	Antinode	4.896	
102N	Node	0	209.03
103A	Antinode	2.656	
103N	Node	0	211.686

Table 25. Table showing energy of antinodes and excitation of electrons in Beryllium-IV

Beryllium-III Antinode	Larger Wavelength	Next Shorter Wavelength	Larger λ in eV	Next Shorter λ in eV	Transition Energy eV	Rounded eV	
Be III	10.025	8.831	123.6752	140.3968	16.72158	16.722	Li III type
Be III	8.831	8.476	140.3968	146.2771	5.880235	5.880	

Be III	8.476	8.32	146.2771	149.0197	2.742695	2.743	
Be III	8.32	8.238	149.0197	150.5031	1.483323	1.483	
Be III	8.238	8.189	150.5031	151.4036	0.900556	0.901	

Table 26. Spectral sequence of Beryllium-III

Beryllium-III		Energy in eV	eV
Number		Lithium III type wave	Be-III
98A	Antinode		
98N	Node	0	31.89
99A	Antinode	91.782	
99N	Node	0	123.68
100A	Antinode	17.001	
100N	Node	0	140.40
101A	Antinode	5.940	
101N	Node	0	146.28
102A	Antinode	2.754	
102N	Node	0	149.02
103A	Antinode	1.494	
103N	Node	0	150.50

Table 27. Table showing energy of antinodes and excitation of electrons in Beryllium-III

Beryllium IV		Beryllium III	
Antinode in eV		Antinode in eV	
163.168		91.7820	
30.224		17.0031	
10.560		5.9510	
4.896		2.7549	
2.656		1.4958	
1.600		0.9018	
1.040		0.5850	
0.704		0.4014	
0.512		0.2871	
0.384		0.2124	
0.288		0.1710	
0.224		0.1278	
0.160		0.0873	
216.4160	Subtotal	121.7606	Subtotal

	etc.	Add all antinodes
	217.5744	Rydberg Energy H*16
	217.719	NIST ionization

	etc.	Add all antinodes
	122.3856	Rydberg Energy H*9
	31.89	Ground state energy
	154.2788	Calculated Ionization Energy
	153.896	NIST ionization

Table 28. Calculation of ionization from the spectrum and comparison to NIST.

Beryllium-II	Larger	Next Shorter	Larger	Next Shorter	Transition	Rounded	
Antinode	Wavelength	Wavelength	λ in eV	λ in eV	Energy eV	Energy eV	
Be II	151.24	114.3039	8.19779457	10.8469	2.64912	2.649	He II
Be II	114.3	102.6926	10.8469115	12.0734	1.22644	1.226	type
Be II	102.69	97.3266	12.0733557	12.739	0.66565	0.666	
Be II	97.327	94.3559	12.7390075	13.1401	0.40107	0.401	
Be II	94.356	92.5246	13.1400822	13.4002	0.26008	0.260	

Table 29. Spectral sequence of Beryllium-II

Beryllium-II		Energy in eV	eV
Number		Helium II type wave	Be-II
98A	Antinode		
98N	Node	0	
99A	Antinode	40.810	
99N	Node	0	0.64
100A	Antinode	7.558	
100N	Node	0	8.19779
101A	Antinode	2.645	
101N	Node	0	10.8469
102A	Antinode	1.224	
102N	Node	0	12.0734
103A	Antinode	0.665	
103N	Node	0	12.739
104A	Antinode	0.401	
104N	Node	0	13.1401

Table 30. Table showing energy of antinodes and excitation of electrons in Beryllium-II

Beryllium I	Larger	Next Shorter	Larger	Next Shorter	Transition	Rounded	
Antinode	Wavelength	Wavelength	λ in eV	λ in eV	Energy eV	Energy eV	
Be I	166.1478	149.1762	7.462297	8.311274	0.848977	0.849	H-type
Be I	149.1762	142.6117	8.311274	8.693847	0.382573	0.383	
Be I	142.6117	139.39	8.693847	8.894786	0.20094	0.201	
Be I	139.39	137.56	8.894786	9.013116	0.11833	0.118	

Be I	137.56	136.43	9.013116	9.087769	0.074652	0.075
Be I	136.43	135.65	9.087769	9.140024	0.052256	0.052
Be I	135.65	135.13	9.140024	9.175196	0.035172	0.035
Be I	135.13	134.74	9.175196	9.201754	0.026557	0.027
Be I	134.74	134.47	9.201754	9.22023	0.018476	0.018
Be I	134.47	134.22	9.22023	9.237403	0.017174	0.017

Table 31. Spectral sequence of Beryllium-I

Beryllium-I		Energy in eV	eV
Number			Be-I
98A	Antinode		
98N	Node	0	
99A	Antinode	10.198	
99N	Node	0	5.57
100A	Antinode	1.889	
100N	Node	0	7.46
101A	Antinode	0.660	
101N	Node	0	8.31
102A	Antinode	0.306	
102N	Node	0	8.69
103A	Antinode	0.166	
103N	Node	0	8.89
104A	Antinode	0.100	
104N	Node	0	9.01
105A	Antinode	0.065	

Table 32. Table showing energy of antinodes and excitation of electrons in Beryllium-I

Beryllium II	Beryllium I
Antinode in eV	Antinode in eV
40.7920	1.8892
7.5569	0.6612
2.6449	0.3061
1.2244	0.1662
0.6648	0.1002
0.4008	0.0650
0.2600	0.0446
0.1784	0.0319
0.1276	0.0236
0.0944	0.0190

0.0760			0.0142	
0.0568			0.0097	
0.0388			3.3310	Subtotal
54.116	Subtotal		13.598	Rydberg Energy
etc.	Add all antinodes		-10.198	Less antinode
54.394	H*4=Rydberg Energy		3.4	Subtotal
-40.792	Less antinode		5.571	Ground state energy
13.602	Subtotal		8.971	Calculated Ionization Energy
0.64	Ground state energy		9.323	NIST ionization
14.241	Calculated Ionization Energy			
18.211	NIST ionization			

Table 33. Calculation of ionization from the spectrum and comparison to NIST.

2.5 Boron

The following is an analysis of Boron.

Boron-V	Larger	Next Shorter	Larger	Next Shorter	Transition	Rounded	Difference
Antinode	Wavelength	Wavelength	ʎ in eV	ʎ in eV	Energy eV	Energy eV	eV Between
B V	4.859	4.098	255.164496	302.54863	47.384134	47.384	
B V	4.098	3.887	302.54863	318.972032	16.423402	16.423	255.3
B V	3.887	3.794	318.972032	326.790798	7.818766	7.819	255.2
B V	3.794	3.748	326.790798	330.801571	4.010773	4.011	255.3
B V	3.748	3.721	330.801571	333.201905	2.400334	2.400	255.2
B V	3.721	3.703	333.201905	334.821574	1.619669	1.620	

B V	26.24	19.44	47.2501634	63.7779984	16.527835	16.528	
B V	19.44	17.35	63.7779984	71.4607659	7.6827675	7.683	47.24
B V	17.35	16.4	71.4607659	75.6002615	4.1394956	4.139	47.26
B V	16.4	15.87	75.6002615	78.1250339	2.5247724	2.525	47.25

B V	74.97	51.23	16.5378723	24.2015281	7.6636558	7.664	
B V	51.23	43.74	24.2015281	28.345777	4.1442489	4.144	

Table 35. Spectral sequence of Boron-V

Boron V		Energy in eV	May contain electrons in energy states in eV		
Number			B V-1	B V-2	B V-3

99A	Antinode				
99N	Node	0	Ground		
100A	Antinode	255			
100N	Node	0	255.16	Ground	
101A	Antinode	47.2			
101N	Node	0	302.55	47.25	Ground
102A	Antinode	16.5			
102N	Node	0	318.97	63.78	16.53
103A	Antinode	7.7			
103N	Node	0	326.79	71.46	24.20
104A	Antinode	4.2			
104N	Node	0	330.8016	75.60	
105A	Antinode	2.5			
105N	Node	0	333.202	78.13	
106A	Antinode	1.63			
106N	Node	0	334.822		

Table 36. Table showing energy of antinodes and excitation of electrons in Boron-V

Boron-IV Antinode	Larger Wavelength	Next Shorter Wavelength	Larger Λ in eV	Next Shorter Λ in eV	Transition Energy eV	Rounded Energy eV	Difference eV Between
B IV	4.8437	4.335	255.970495	286.007909	30.037414	30.037	
B IV	6.03144	5.26853	205.563561	235.330213	29.766652	29.767	Be IV
B IV	5.26853	5.04347	235.330213	245.831597	10.501384	10.501	
B IV	5.04347	5.019	245.831597	250.702011	4.870414	4.870	
B IV	25.989	24.602	47.7065023	50.3960771	2.6895748	2.690	
B IV	24.602	23.771	50.3960771	52.1578515	1.7617744	1.762	
B IV	23.771	23.308	52.1578515	53.1939372	1.0360857	1.036	
B IV	23.308	23	53.1939372	53.9062734	0.7123362	0.712	

Table 37. Spectral sequence of Boron-IV

Boron IV		Energy in eV	May contain electrons in energy states in eV		
Number			B IV-1	B IV-2	B IV-3
99A	Antinode				
99N	Node	0	42.40		
100A	Antinode	163			

100N	Node	0	205.56	
101A	Antinode	30.2		
101N	Node	0	235.33	
102A	Antinode	10.6		
102N	Node	0	245.83	
103A	Antinode	4.9		
103N	Node	0	250.70	47.70
104A	Antinode	2.7		
104N	Node	0		50.40
105A	Antinode	1.6		
105N	Node	0		52.16
106A	Antinode	1.04		
106N	Node	0		53.19
107A	Antinode	0.70		
107N	Node	0		53.90

Table 38. Table showing energy of antinodes and excitation of electrons in Boron-IV

Ionization

B IV

217.44	Rydberg * 16
42.4	Ground state
259.84	Calculated
259.3715	NIST

Table 39. Calculation of ionization from the spectrum and comparison to NIST

Boron-III Antinode	Larger Wavelength	Next Shorter Wavelength	Larger λ in eV	Next Shorter λ in eV	Transition Energy eV	Rounded Energy eV	Difference eV Between
B III	36.569	33.83	33.9042437	36.6492547	2.745011	2.745	Li III
B III	33.83	32.433	36.6492547	38.2278632	1.5786085	1.579	
B III	32.433	31.654	38.2278632	39.168645	0.9407818	0.941	

B III	39.841	38.014	31.1198084	32.615465	1.4956566	1.496	

B III	43.4627	42.129	28.5266283	29.4297108	0.9030825	0.903	Li III
B III	42.129	41.305	29.4297108	30.0168088	0.587098	0.587	
B III	41.305	40.76	30.0168088	30.4181621	0.4013533	0.401	

B III	49.34	47.542	25.128583	26.0789257	0.9503427	0.950	Li III
B III	47.542	46.557	26.0789257	26.630674	0.5517483	0.552	
B III	46.557	45.8729	26.630674	27.0278157	0.3971417	0.397	

Table 39. Spectral sequence of Boron-III

Boron III		Energy in eV	May contain electrons in energy states in eV		
Number			B III-1	B III-2	B III-3
99A	Antinode				
99N	Node	0			
100A	Antinode	91.8			
100N	Node	0	10.96		
101A	Antinode	17.0			
101N	Node	0	27.96		
102A	Antinode	5.94			
102N	Node	0	33.90		
103A	Antinode	2.75			
103N	Node	0	36.65		
104A	Antinode	1.49			
104N	Node	0	38.23	28.43	25.13
105A	Antinode	0.900			
105N	Node	0		29.43	26.08
106A	Antinode	0.585			
106N	Node	0		30.02	26.63

Table 40. Table showing energy of antinodes and excitation of electrons in Boron-III

B III Ionization

122.39	Rydberg
-91.782	less antinode
30.60	Subtotal
10.96	Ground state
41.56	Calculated ionization
37.93058	NIST

Table 40. Calculation of ionization from the spectrum and comparison to NIST

Boron-II	Larger	Next Shorter	Larger	Next Shorter	Transition	Rounded	Difference
Antinode	Wavelength	Wavelength	λ in eV	λ in eV	Energy eV	Energy eV	eV Between

B II	51.13	49.78	24.2488615	24.9064742	0.6576127	0.658	He II
B II	49.78	48.946	24.9064742	25.3308603	0.4243861	0.424	
B II	48.946	48.4	25.3308603	25.6166175	0.2857572	0.286	
B II	48.4	48.035	25.6166175	25.8112686	0.1946511	0.195	
B II	48.035	47.79	25.8112686	25.9435925	0.1323239	0.132	
B II	47.79	47.604	25.9435925	26.0449602	0.1013677	0.101	
B II	47.604	47.457	26.0449602	26.1256356	0.0806754	0.081	
B II	47.457	47.352	26.1256356	26.1835675	0.0579319	0.058	

B II	77.08	68.722	16.085162	18.0414465	1.9562845	1.956	H-type
B II	68.722	66.3066	18.0414465	18.6986557	0.6572092	0.657	
B II	66.3066	65.064	18.6986557	19.0557649	0.3571092	0.357	
B II	65.064	64.56	19.0557649	19.2045274	0.1487625	0.149	
B II	64.56	64.158	19.2045274	19.3248587	0.1203313	0.120	

B II	108.2073	88.2543	11.4580466	14.0485425	2.5904959	2.590	He II
B II	88.2543	80.915	14.0485425	15.3227991	1.2742566	1.274	
B II	80.915	77.529	15.3227991	15.9920067	0.6692076	0.669	

Table 41. Spectral sequence of Boron-II

Boron II		Energy in eV	May contain electrons in energy states in eV			Energy in eV	
Number		He II-type	B II-1	B II-2	B II-3	H-type	B II-4
99A	Antinode					0	5.89
99N	Node	0				10.20	
100A	Antinode	40.81				0	16.09
100N	Node	0		3.90		1.89	
101A	Antinode	7.558				0	18.04
101N	Node	0		11.46		0.66	
102A	Antinode	2.645				0	18.70
102N	Node	0		14.05		0.31	
103A	Antinode	1.224				0	19.06
103N	Node	0	24.24	15.32		0.17	
104A	Antinode	0.665				0	19.20
104N	Node	0	24.91			0.10	
105A	Antinode	0.401				0	
105N	Node	0	25.33			0.065	
106A	Antinode	0.26				0	
106N	Node	0	25.62			0.044	

107A	Antinode	0.178			0	
107N	Node	0	25.81		0.032	
108A	Antinode	0.128			0	
108N	Node	0	25.94		0.024	
108A	Antinode	0.096			0	
108N	Node	0	26.04		0.018	

Table 42. Table showing energy of antinodes and excitation of electrons in Boron-II

B II Ionization

54.39360	Rydberg
-40.81	less antinode
-7.558	
-2.645	
3.38060	subtotal eV
24.24	Ground state
27.62060	Calculated

25.15483 NIST

Table 43. Calculation of ionization from the spectrum and comparison to NIST

Boron-I Antinode	Larger Wavelength	Next Shorter Wavelength	Larger λ in eV	Next Shorter λ in eV	Transition Energy eV	Rounded Energy eV	Difference eV Between
B I	249.7722	181.7843	4.96390026	6.82041457	1.85651431	1.857	H-type
B I	182.64	166.685	6.78845975	7.43824752	0.64978777	0.650	H-type
B I	181.7843	166.7272	6.82041457	7.43636484	0.61595027	0.616	
B I	624.466	594.263	1.98544722	2.08635619	0.10090897	0.101	H-type
B I	594.263	573.1943	2.08635619	2.16304364	0.07668745	0.077	
B I	573.1943	563.3069	2.16304364	2.2010103	0.03796666	0.038	
B I	564.4321	556.3185	2.19662256	2.2286591	0.03203654	0.032	
B I	556.3185	550.4565	2.2286591	2.25239285	0.02373375	0.024	

Table 44. Spectral sequence of Boron-I

Boron I		Energy in eV	May contain electrons in energy states in eV		
Number			B I-1	B I-2	B I-3
99A	Antinode				

99N	Node	0				
100A	Antinode	10.198				
100N	Node	0	4.96	4.90		
101A	Antinode	1.889				
101N	Node	0	6.82	6.79		
102A	Antinode	0.66				
102N	Node	0	7.48			
103A	Antinode	0.306				
103N	Node	0				
104A	Antinode	0.166				
104N	Node	0			1.985	
105A	Antinode	0.100				
105N	Node	0			2.086	
106A	Antinode	0.065				
106N	Node	0			2.163	
107A	Antinode	0.044				
107N	Node	0			2.196	
108A	Antinode	0.032				
108N	Node	0			2.229	

Table 45. Table showing energy of antinodes and excitation of electrons in Boron-I

B I Ionization

13.590	Rydberg
-10.198	Less antinode
3.392	Subtotal eV
4.96390026	Ground state
8.356	Calculated ionization
8.298019	NIST

B I

Table 46. Calculation of ionization from the spectrum and comparison to NIST

2.6 Carbon

The difference of the size of the nucleus becomes more pronounced in Carbon affecting the energy of the antinodes of the hydrogen-type proton-wave and the other superpositioned proton-waves. However, the pattern is still very clear.

Carbon I	Larger	Next Shorter	Larger	Next Shorter	Transition	Rounded
Antinode	Wavelength	Wavelength	ʌ in eV	ʌ in eV	Energy eV	Energy eV
C I	127.73	119.339	9.7069	10.389	0.682	0.682
C I	119.34	115.804	10.389	10.706	0.317	0.317

Carbon I	Larger	Next Shorter	Larger	Next Shorter	Transition	Rounded
Antinode	Wavelength	Wavelength	λ in eV	λ in eV	Energy eV	Energy eV
C I	193.09	148.176	6.4211	8.3674	1.946	1.946
C I	148.18	136.416	8.3674	9.0887	0.721	0.721
C I	136.42	131.136	9.0887	9.4546	0.366	0.366

Carbon I	Larger	Next Shorter	Larger	Next Shorter	Transition	Rounded
Antinode	Wavelength	Wavelength	λ in eV	λ in eV	Energy eV	Energy eV
C I	600.11	505.217	2.066	2.4541	0.388	0.388
C I	505.22	477.175	2.4541	2.5983	0.144	0.144

Carbon I	Larger	Next Shorter	Larger	Next Shorter	Transition	Rounded
Antinode	Wavelength	Wavelength	λ in eV	λ in eV	Energy eV	Energy eV
C I	786.09	658.761	1.5772	1.8821	0.305	0.305
C I	658.76	601.484	1.8821	2.0613	0.179	0.179

Carbon I	Larger	Next Shorter	Larger	Next Shorter	Transition	Rounded
Antinode	Wavelength	Wavelength	λ in eV	λ in eV	Energy eV	Energy eV
C I	1814	1261.41	0.6835	0.9829	0.299	0.299
C I	1261.4	1068.31	0.9829	1.1606	0.178	0.178
C I	1068.3	965.844	1.1606	1.2837	0.123	0.123
C I	965.84	910.573	1.2837	1.3616	0.078	0.078

Table 47. Spectral sequence of Carbon-I

Carbon I		Energy in eV	May contain electrons in energy states in eV				
Number			C I-1	C I-2	C I-3	C I-4	C I-5
99A	Antinode						
99N	Node	0					
100A	Antinode	10.198					
100N	Node	0	7.82	6.48			
101A	Antinode	1.889					
101N	Node	0	9.71	8.37			
102A	Antinode	0.660					
102N	Node	0	10.39	9.09	2.07	1.58	0.68
103A	Antinode	0.306					
103N	Node	0	10.70	9.45	2.45	1.88	0.98
104A	Antinode	0.166					
104N	Node	0				1.16	
105A	Antinode	0.101					
105N	Node	0				1.28	

106A	Antinode	0.065					
106N	Node	0					
107A	Antinode	0.045					
107N	Node	0					
108A	Antinode	0.032					
108N	Node	0					

Table 48. Table showing energy of antinodes and excitation of electrons in Carbon-I

Carbon I

13.59	Ryberg Energy
-10.20	less antinode
3.39	Subtotal
7.82	Ground state
11.21	Calculated ionization energy
11.26	NIST Ionization

Table 49. Calculation of ionization from the spectrum and comparison to NIST

Carbon II Antinode	Larger Wavelength	Next Shorter Wavelength	Larger λ in eV	Next Shorter λ in eV	Transition Energy eV	Rounded Energy eV	Difference eV Between
C II	90.448	90.4142	13.7078	20.8369	7.129136	7.129	11.99447

Carbon II Antinode	Larger Wavelength	Next Shorter Wavelength	Larger λ in eV	Next Shorter λ in eV	Transition Energy eV	Rounded eV	Difference
C II	723.642	723.132	1.71334	9.2823	7.568961	7.569	He II type
C II	133.5708	103.7018	9.2823	11.9559	2.67356	2.674	
C II	103.7018	90.448	11.9559	13.7078	1.751952	1.752	
C II	90.3624	85.8559	13.7208	14.441	0.720193	0.720	

Table 50. Spectral sequence of Carbon-II

Carbon II		Energy in eV	eV	
Number			C II-1	C II-2
99A	Antinode	≈ 340 eV		
99N	Node	0		
100A	Antinode	40.81		
100N	Node	0	13.71	1.71
101A	Antinode	7.56		
101N	Node	0	20.84	9.28
102A	Antinode	2.65		
102N	Node	0		11.96
103A	Antinode	1.22		

103N	Node	0		13.71
104A	Antinode	0.67		
104N	Node	0		14.441

Table 51. Table showing energy of antinodes and excitation of electrons in Carbon-II

Carbon II

54.39	Rydberg *4
-40.81	less antinode
13.58	Subtotal
13.71	Ground state
27.29	Calculated ionization
24.3845	NIST ionization

Table 52. Calculation of ionization from the spectrum and comparison to NIST

Carbon III	Larger	Next Shorter	Larger	Next Shorter	Transition	Rounded	
Antinode	Wavelength	Wavelength	λ in eV	λ in eV	Energy eV	eV	
C III	45.0734	38.6203	27.5072	33.3569	5.849717	5.850	Li III type

Carbon III	Larger	Next Shorter	Larger	Next Shorter	Transition	Rounded
Antinode	Wavelength	Wavelength	λ in eV	λ in eV	Energy eV	eV
C III	57.4281	45.0734	21.5895	27.5072	5.917722	5.918

Carbon III	Larger	Next Shorter	Larger	Next Shorter	Transition	Rounded
Antinode	Wavelength	Wavelength	λ in eV	λ in eV	Energy eV	eV
C III	272.59	117.637	4.54839	10.5396	5.991192	5.991

Carbon III	Larger	Next Shorter	Larger	Next Shorter	Transition	Rounded
Antinode	Wavelength	Wavelength	λ in eV	λ in eV	Energy eV	eV
C III	466.586	298.211	2.65727	4.15761	1.500339	1.500

Table 53. Spectral sequence of Carbon-III

Carbon III		Energy in eV	eV		
Number			C III-1	C III-2	C III-3
99A	Antinode				
99N	Node	0			
100A	Antinode	91.8			
100N	Node	0	10.51	4.59	
101A	Antinode	17.0			
101N	Node	0	27.51	21.59	
102A	Antinode	5.94			
102N	Node	0			

103A	Antinode	2.75			
103N	Node	0			2.66

Table 54. Table showing energy of antinodes and excitation of electrons in Carbon-III

Carbon III

122.3856	Rydberg *9
-91.782	less antinode
30.6036	subtotal
10.51	ground state
41.1136	Calculated
47.88778	NIST

Table 55. Calculation of ionization from the spectrum and comparison to NIST

Carbon IV	Larger	Next Shorter	Larger	Next Shorter	Transition	Rounded	Difference
Antinode	Wavelength	Wavelength	ʌ in eV	ʌ in eV	Energy eV	Energy eV	eV Between Series
C IV	41.971	31.246	29.5405	39.6801	10.1396	10.140	
C IV	41.952		29.55388			0.000	
C IV	38.418		32.27248			0.000	
C IV	38.403		32.28509			0.000	
C IV							
C IV	31.246	24.491	39.6801	50.62449	10.94439	10.944	
C IV	31.242		39.68518			0.000	
C IV	28.923		42.86707			0.000	
C IV	28.914		42.88041			0.000	
C IV	24.491		50.62449				

Table 56. Spectral sequence of Carbon-IV

Carbon IV		Energy in eV	eV	
Number			C IV-1	C IV-2
99A	Antinode			
99N	Node	0		
100A	Antinode	163.17		
100N	Node	0		9.46
101A	Antinode	30.22		
101N	Node	0	29.5405	39.6801
102A	Antinode	10.5		
102N	Node	0		
103A	Antinode	4.89		
103N	Node	0		

104A	Antinode	2.66	
104N	Node	0	
105A	Antinode	1.61	
105N	Node	0	

Table 57. Table showing energy of antinodes and excitation of electrons in Carbon-IV

217.719	Rydberg * 16
-163.17	antinode
54.549	Subtotal
9.46	Ground state
64.009	Calculated ionization
64.49351	NIST

Table 58. Calculation of ionization from the spectrum and comparison to NIST

Carbon V has too few spectral lines to recognize a series from the NIST webpages available in 2017 and Carbon VI has no spectral lines listed on the NIST webpages.

2.7 Deuterium

The neutron-wave, which can be deduced from an examination of Deuterium, increases the energy of the proton-wave transitions. The counter-effect of multiple nucleons enlarging the great sphere of energy in the first antinode around the nucleus lessens the overall energy of the proton-wave as a portion of the proton-wave is inside the nucleus.

Deuterium

Deuterium	Larger	Next Shorter	Larger	Next Shorter	Transition	Rounded	Difference
Antinode	Wavelength	Wavelength	λ in eV	λ in eV	Energy eV	Energy eV	eV Between Series
D	121.5338	102.5427	10.20164	12.091	1.88923	1.889	
Calculated	102.5427	97.23506	12.091	12.752	0.66122	0.661	10.20113
Calculated	97.23506	94.94848	12.752	13.05807	0.305982	0.306	10.20095
D	94.94848	93.75484	13.05807	13.22432	0.166249	0.166	10.20079
D	93.75484	93.0495	13.22432	13.32457	0.100244	0.100	10.20075
D	93.0495	92.28992	13.32457	13.38967	0.0651	0.065	10.20072
Calculated	92.59709	92.28992	13.38967	13.43423	0.044566	0.045	10.20074
D	92.28992	92.07125	13.43423	13.46614	0.031906	0.032	10.20068
D	92.07125	91.91013	13.46614	13.48975	0.023607	0.024	10.20067
D	91.91013	91.78796	13.48975	13.5077	0.017955	0.018	10.20067

Deuterium	Larger	Next Shorter	Larger	Next Shorter	Transition	Rounded	Difference
Antinode	Wavelength	Wavelength	λ in eV	λ in eV	Energy eV	Energy eV	eV Between Series

D	656.0925	486.0039	1.88974	2.551099	0.661359	0.661
D	485.9956	433.9314	2.551143	2.857236	0.306093	0.306
D	433.9246	410.0647	2.85728	3.023533	0.166253	0.166
D	410.0586	396.9019	3.023578	3.123805	0.100227	0.100
D	396.8962	388.8016	3.12385	3.188887	0.065037	0.065
D	388.7961	383.4367	3.188932	3.233505	0.044573	0.045
D	383.4313	379.6842	3.23355	3.265462	0.031912	0.032
D	379.6842	374.911	3.265462	3.289072	0.02361	0.024
D	376.9587	374.911	3.289072	3.307037	0.017965	0.018

Table 59. Spectral sequence of Carbon-IV (Shaded yellow are calculated and hypothesized.)

The differences between Hydrogen and Deuterium in the calculation of the neutron-wave are as follows:

Neutron-wave

Hydrogen	Deuterium	Hydrogen	Deuterium	Difference		Difference
Wavelength	Wavelength	λ in eV	λ in eV	Energy eV	Neutron nodes	between nodes
121.566824	121.53379	10.1988704	10.2016426	0.0027722	0.0033475	0.00057530
102.57222	102.542741	12.0875251	12.091	0.0033475	0.0035322	0.00018470
97.25367	97.235063	12.7485604	12.752	0.0035322	0.0035506	0.00001840
94.9743	94.9484754	13.0545241	13.0580747	0.0035506	0.0035963	0.00004570
93.78034	93.7548369	13.2207271	13.2243234	0.0035963	0.0036252	0.00002890
93.07482	93.049497	13.3209421	13.3245673	0.0036252	0.0036814	0.00005620
92.62256	92.5970945	13.3859859	13.3896673	0.0036814		

Hydrogen	Deuterium	Hydrogen	Deuterium	Difference		Difference
Wavelength	Wavelength	λ in eV	λ in eV	Energy eV	Neutron nodes	between nodes
656.2711	656.09245	1.88922579	1.88974021	0.00051442	0.00070	0.00018386
486.1287	485.99564	2.55044454	2.55114282	0.00069828	0.00080	0.00010193
434.0462	433.92464	2.85648	2.85728021	0.00080021	0.00085	0.00005031
410.174	410.05862	3.02272764	3.02357816	0.00085052	0.00087	0.00002296
397.0072	396.89619	3.12297683	3.12385031	0.00087348	0.00089	0.00001833
388.9049	388.79614	3.18803977	3.18893158	0.00089181	0.00090	0.00001130
383.5384	383.43128	3.23264708	3.23355019	0.00090311		

Table 60. Calculation of calculation of the strength of a neutron wave in Deuterium.

The energy of the neutron-wave is small. Because of this, the NIST spectral line measurements would have to be accurate to several decimal places and not as I have calculated for estimations as was done here. Also, I may have chosen the wrong fine structure line in estimating values. A comparison of the Neutron-wave and how the proton-wave changes:

Neutron	Neutron	Hydrogen						
Deuterium Lyman	Deuterium Balmer	Antinodes	Comparison of proton-wave antinode energies					
		10.198	He I	Li II	Li I	Be I	B II	B I
0.0005753		1.889	1.868999					1.856514
0.0001847	0.0001839	0.661	0.65505	0.66166	0.68760	0.84898	0.65721	0.649788
0.0000184	0.0001019	0.306	0.30373	0.30992	0.31578	0.38257	0.35711	
0.0000457	0.0000503	0.1662	0.16520	0.16719	0.17057	0.20094	0.14876	
0.0000289	0.0000230	0.1002	0.09971		0.10250	0.11833	0.12033	0.100909
0.0000562	0.0000183	0.0651	0.06476		0.06627			0.076687
	0.0000113	0.0449	0.04442		0.04549			0.037967
		0.0318	0.03179		0.03235			0.032037
		0.0237						0.023734

Table 61. Table showing effect of Neutron-wave on proton-wave energies.

3. CONCLUSIONS

Obviously, the point of this new model of the atom was not to invalidate statistical mechanics developed by Boltzman, Einstein, Boze, et al. The point is to model a single atom according to the easily verifiable spectral data in order to progress beyond current QM theory and improve upon and unify the statistical methods developed for Quantum Mechanics and provide a way forward to a unified field theory. My intent was not only to describe a new model that answers how, but to really understand why nature behaves as it does. With that in mind, I will here present hypotheses and models based on Sollism Theory with the intent of understanding how the underlying mechanisms of the atom work relying heavily on logic and abstract reasoning with a view to uniting the macro world of causality and invariance with the micro world where these two classical ideas were discarded from the inception of Quantum Mechanics because the atom itself was not well understood when QM was developed from 1913 to 1932. Although the above spectral data points to a new model that must be taken at face value from the incontestable spectral data, the following deductions arise from theoretical applications of this new model which may or may not be proven with further research, but are important to a complete model.

3.1 Deduction of Electron Transitions under this new model

The spectrum described in this research shows that electrons all reside in spherical shells within the nodes of a nucleon wave. We see how Bohr concluded that there were quantum jumps as the proton-wave only allows certain transitions for the electron. We should however consider how the energy of the proton-wave can remain stable while the energy of the electron loses and gains energy.

Max Planck had a stroke of genius when he pictured oscillators on the surface of a body[lxxi]. Einstein used this description of oscillators in 1905[lxxii]. As we know, Planck and Einstein were describing subatomic particles as oscillators. Particles radiate at the oscillation frequency according to classical mechanics. The electron-particle in the node oscillates at the frequency that it transitioned to the node. It does not radiate the energy that it is holding in its electron-wave while oscillating in the node, however, this does not break the laws of electrodynamics as the proton-wave reflects the energy back to the electron under the Second Law of Thermodynamics i.e. that energy is not transferred from low energy systems to high energy systems. The proton-wave with its high energy can be described as a system of high energy where the energy is equivalent to the mass it surrounds. The electron-wave always contains lesser energy than the proton-wave. Thus, the electron-particle retains its energy as a frequency of kinetic energy i.e. oscillation.

We can even assume that the electron does not need to be immobile in its shell, but can vibrate and change position in the shell without radiating by a second mechanism other than under the Second Law of Thermodynamics. The reason the electron does not radiate as it vibrates in a single node is that the energy of the oscillation shoots off the electron and collides with the proton-wave and then the proton-wave deflects the energy back to the electron-wave. And the electron-particle absorbs the energy and oscillates and has kinetic energy so it collides with the proton-wave which deflects the energy back to the electron-wave so the oscillation continues in a reversible cycle. This creates continuous oscillation without radiation while the electron resides within a node. We can see from the spectrum that the proton-wave does not readily absorb energy because the energy of the antinodes is the same across all series, so that it is reasonable that the proton would return the energy to the electron. The reason that the proton-wave does not absorb the energy of the electron-wave is due to the Second Law of Thermodynamics in that heat i.e. energy moves from a more excited system to a less excited or cooler system. The proton-wave has an energy proportionate to its mass. The electron-wave has an energy proportionate to its mass. Excess energy in the electron-wave does not readily transfer to the larger energy system of the proton-wave. Therefore, the electron-wave must transfer the energy to other electrons that have less energy in their electron-waves. If an electron-wave is not in contact with another electron-wave of lesser energy and the electron is holding a full quantum of energy, the electron-wave must radiate the energy as light. (We note that the proton-wave may absorb energy and radiate as is known to happen in particle accelerators[lxxiii]. Proton-wave absorption will be treated later.)

When the electron absorbs energy from the environment, the electron-wave absorbs the light, and the particle oscillates at the new frequency as shown in the tables above. We deviate from current theory in that in this new model all oscillations without emission or absorption occur in a reversible cycle between the electron-wave and proton-wave.

First let's treat the electron as if it gains or loses energy before the transition. An electron with energy E is oscillating with frequency v in node a which is a node other than ground states equal to zero excess energy, and has kinetic energy v_{kin} where $v_{kin} = E$. After it loses or gains energy, the electron now has energy $E_1 = b$ where b is the energy of the adjoining antinode. The electron then transitions through antinode b and while inside antinode b, it has potential energy v_{pot}, because the energy is stored in the electron-wave while transitioning.

$$v_{kin}(E_1 a) = v_{pot}(E_1 b)$$

After the energy loss or gain, the motion through the antinode does not change the electron energy, therefore, there is no absorption or emission of radiation while the electron is moving through the antinode having lost the energy to overcome the Coulomb force of the nucleus or gained enough energy to overcome the Coulomb force to an outer shell. After emerging from the antinode, the electron oscillates at the energy that it initially had plus the energy that it lost or gained before transition. When the electron is oscillating in the node, it is in a reversible cycle with the proton-wave P due to the Law of Atomic Equilibrium explained below. The kinetic energy of the electron transfers to the proton-wave and then is transferred back to the electron from the proton-wave.

$$v_{kin}(E_1a) \rightarrow v_{pot}(P) \rightarrow v_{kin}(E_1a)$$

The preceding relation may not be a precise statement, since the potential energy may not be absorbed by the proton-wave but may only be deflected by it, since the proton-wave maintains its antinode energy despite the energy of the electron. The reversible cycle may be a perfect deflection of energy where there is no transfer of energy to the proton-wave, such that, the kinetic energy of the electron is deflected back from the proton-wave to potential energy in the electron-wave, such that the electron changes the potential energy to kinetic energy in a reversible cycle.

$$v_{kin}(E_1a) \rightarrow v_{pot}(E_1a) \rightarrow v_{kin}(E_1a)$$

If while the electron is oscillating in an outer node, instead of a deflection, the proton-wave is causing the electron to create a reflection, then the oscillation has reflective symmetry and no energy is ever transferred from the electron to the proton-wave.

Also, the current model explains that the electron can only transition outward and inward to its ground state which is described as n=1, n=2, etc. These whole numbers represent the consecutive spheres in space where the nodes occur. Under this new model described by the spectrum, the electron is restricted by the antinodes of the proton-wave. Therefore in zero ground state atoms when the electron is in ground state, the electron mass to electron-wave energy is in natural state equilibrium i.e. there is no excess energy in the electron-wave so that the excess energy in the electron-wave is equal to zero. Therefore, when energy is absorbed to transition from ground state, the electron is only allowed to transition to the nodes described by the surrounding antinodes. For instance, there are two antinodes on either side of the electron in Balmer ground state. There is the 10.2 eV inner antinode or the 1.89eV outer antinode. If the electron gains 10.2 eV of energy, it may transition either to Lyman ground state, or it may transition outward through several antinodes. This is when the Coulomb force between the proton and electron makes the inward transition more likely, then the electron is more likely to transition to Lyman ground. This means that the electron has 10.2 eV in Lyman ground state when it transitions inward toward the proton. Therefore, a Lyman electron has more energy than the other series as it takes more energy to transition to the Lyman series. Once in Lyman ground state, the electron is holding too much energy for ground state and must transition back out through the 10.2 eV antinode again, because it has 10.2 eV of energy. The electron did not lose the 10.2 eV of energy in the transition to Lyman ground state and so it oscillates back outward as it cannot stay in Lyman ground state because it has enough energy to penetrate the outer antinode. Later, when it emits the 10.2 eV as radiated light, it falls back into ground state at zero energy. To transition outward again, it must gain energy. To recap the transition, a Balmer ground state electron absorbs 10.2 eV, transitions inward through 10.2 eV antinode. Transitions outward with no emission or absorption through the 10.2 eV antinode. Then later may emit 10.2 eV and transition back to Lyman ground state where it will have zero kinetic energy. No particle ever has zero energy because each particle has a standing partial quanta wave of energy. That wave holds the particle together and if that wave is gone, the particle does not exist.

The change to the atomic model under Sollism Theory is so fundamental that it is worth restating known phenomena. Therefore, whether an electron transitions outward or inward depends on the overall state of the system of atoms. If all atoms of a local system are in a state of increasing energy, the electron absorbs energy as the system gains energy or heats up. As the system cools, the electron emits energy. Similarly, an electron in an atom that is surrounded by less energetic atoms will emit energy, and an electron in an atom that is surrounded by more energetic atoms will absorb energy. This is in accordance with the known laws of thermodynamics. Atoms behave in accordance with universal laws.

Under Sollism theory, the electron may oscillate in a node without losing energy or radiating due to the proton-wave in which the electron resides. An anomaly in our current science is that we have no explanation for the continuous background radiation of the emission spectrum. If all light comes from electrons radiating at discrete frequencies, where does the rest of the light come from? The current model does not explain how we can see the absorption spectrum. Without the continuous spectrum, the entire sky would be Fraunhofer lines or missing colors, meaning how could the sun produce all colors in the rainbow? We all acknowledge that lead atoms absorb x-rays of .01 to 10nm wavelength so that lead is a good radiation barrier, but the atomic spectrum of the element lead is absorbing wavelengths in the range of 106 to 1049nm[lxxiv]. So we should try to build a model that will allow for a continuous spectrum and not disrupt the atomic spectra of elements. We begin with the assumption that all particles are surrounded by standing waves and that the standing wave absorbs and emits the energy.

We know through Compton Scattering that an electron is able to absorb a portion of a gamma ray and ionize and can thereafter be described as a free electron resulting in an "increase in the wavelength of the scattered beam"[lxxv].

The absorption spectrum of light is viewed across a continuous spectrum. Where Ek is the kinetic energy of the electron, meaning the frequency of oscillation of the electron, where E is any form of energy or amount of energy and Ex is excessive energy gain than required for a full transition to an allowed spectral node, then in the case of light absorbed where λ is the wave length so that λE is a wavelength of any energy:

$$Ek + \lambda E =+ \lambda(Ek) - \lambda(Ex)$$

For each different amount of light energy absorbed (absorption shown by the + sign), the amount of -Ex (excess energy) is not absorbed but scattered back into the spectrum of continuous light. Wavelength λ absorbed takes a discrete value corresponding to the node of the vibrational energy state of an allowable electron transition (up to its ionization energy such as in Compton scattering, but much of the environmental energy will not be gamma radiation) so that the absorption spectrum remains discrete. And Ex takes any value depending upon the original frequency of light impacting the electron less than the energy absorbed by the electron for an allowed transition. This excess energy is scattered and creates the continuous spectrum, in which, only the amount needed for a full quantum has been absorbed. This does not interfere with our perception of the absorption spectrum. And Ex takes infinite values depending upon the incident light impacting the electron. Therefore, the electron can utilize any amount of incident energy present in the environment. Therefore, a star that contains mostly hydrogen can radiate a continuous spectrum of light.

Only the amount of energy for one transition to an allowed node is absorbed from any amount of energy and the remaining energy is scattered. When the incident light is of an energy exactly described by the energy of the electron transition, then the absorption is equal to Ek, then there is no scattering but complete absorption as in the current model.

The excess scattered light will be of a different frequency because it was only partially absorbed. This scattered excess light will not interfere with the absorption spectrum, because it is not the frequency of the absorbed light.

The spectral lines represent the majority of transitions taking place in many millions of atoms as a system, so this would not disrupt the spectrum. The continuous background of the spectrum tells us that most radiation is random, not discrete. And most allowable electron transitions are not happening often enough to create spectral lines. Only the most traversed antinodes create spectral lines and this mostly occurs in or near the ground state lines. This does not mean other transitions are not happening, but the other transitions do not happen frequently enough to create an intensity of emission or a noticeable absorption. In the same way, other non-discrete emissions and absorptions are happening and these are so random and at such a variety of wavelengths as to not create emission or absorption lines. However, the proton-wave restricts the allowable energies for transitions causing the electrons to fall into a pattern of absorption and emission that when multiplied by the atoms in the system create the atomic spectrum.

The emission spectrum is equal to the absorption spectrum, because the proton-wave antinodes restrict the absorption of the electron to the energy necessary to transition through the antinodes. Because the absorption spectrum causes the electron to have discrete amounts of energy when the electron is in an excited state, then when the electron loses the energy to return to ground state zero, it can only lose the energy that it has gained, and it has to lose enough energy to penetrate through the next energy antinode and can lose no further energy until it can transition through a lower energy antinode, so that the emission spectrum is discrete automatically.

3.2 Zeeman and Stark Effect

Despite the spectra showing that the proton-wave maintains consistent energy in its antinodes, we know from the Zeeman[lxxvi] and Stark[lxxvii] effect that the proton-wave can be seen to be in different energy states between different atoms of the same element.

Under this new model of the atom, this is caused when the proton-wave is gaining energy by the proton transitioning in other proton-waves. The proton-wave for the same element can have a larger atom due to energy absorption. The proton-wave energy absorption causes some proton-waves to have an overall expansion, thus shifting the spectral lines creating a spectrum where some nodes and antinodes of the proton-wave in some atoms are expanded thereby showing a fine structure line, while other proton-waves in other atoms show the normal spectral line. We always view the spectrum of, at least, many millions of atoms. Where either energy or magnetism is affecting some atoms and not others, the fine structure line will appear in some atoms and not in others. One line will appear in the normal position in some atoms. The expanded proton-wave line will appear as one line in other atoms that have more energy in their proton-wave. The overall effect is to lessen the intensity of the normal spectral line and increase the intensity of the fine structure line depending upon the number of atoms affected by electromagnetism which causes a response in energy variations of the atomic proton-wave.

Therefore, the cause of the Zeeman and Stark effect are that the protons themselves sit in each other's waves. As the proton-wave is infinite according to the Balmer-Rydberg formula, this must be so. The protons appear to have discrete energy levels when they move in Nuclear Magnetic Resonance (NMR)[lxxviii]. Since the proton-wave appears stable in Hydrogen I because it is sustaining antinodes of determined energy levels, and does not change with the energy levels Lyman, Balmer, etc. series of the electron, the proton must be transitioning through the proton-wave of an adjoining proton to cause fine structure. Hydrogen is diatomic and therefore a pair of Hydrogen atoms sit closely in each other's proton-waves. When magnetism pulls the protons from each other, it causes them to shift in their mutual positions of overlapping proton-waves. The protons themselves would sit in nodes surrounded by the weaker energy antinodes of their mutual proton- waves than those that the electrons sit in. So a shift would cause only a slight energy difference in the infrared. When the proton is sitting in the proton-wave of another proton, where the source of magnetism moves the respective protons further apart from each other, one proton will shift in the other proton's wave. Where the proton sits in the proton-wave of another proton will depend on the chemistry of the molecule. If protons are oriented in the same environment and chemically equivalent, the transition levels will be the same. So the signal will be the same for all transitions that are equally spaced from each other, because they will sit in the same nodes of their respective protons.

Take the Balmer series. A shift of the proton A sitting in proton B's wave where proton A was sitting in the node described in the Balmer equation by substituting 90 and 91 for the lines between the energy, we get:

Balmer Lines	energy in eV	m Balmer Equation
364.7401186	3.39925395	90
364.7361798	3.39929066	91
Difference	0.000036710	Antinode

If the proton shifted through this antinode by magnetism, the values of the Hydrogen atoms that had shifted would show a .000036 eV shift in their frequency while other protons not receiving enough energy to shift in their proton-waves would not change the electron frequency of their atoms. Therefore, the overall effect in the spectrum would be less intensity in the normal frequency of the node and more intensity with more electrons experiencing the shift in the proton-waves as a .000036 eV shift in frequency. We can see this effect at the 656.27248nm line of Hydrogen when it has fine structure equal to 656.28518nm. However, this is probably oversimplified and the transition is of higher energy and distributed across the entire proton-wave.

Because each proton sits in the standing wave of other protons, a portion of the energy of the weaker outer antinodes of the partial quantum standing wave of the protons are encircling other protons in molecules. In other words, the two Hydrogen atoms making up diatomic Hydrogen are sitting in each other's standing wave, therefore, the two separate atoms are encircled by a portion of each other's outer standing waves which creates energy exterior to and encircling both atoms. Since the proton-wave extends its energy to an infinite extent, all protons sit in each other's proton-waves. This describes partially superimposed spheres of energy binding protons in molecules. This explains Van Der Waal forces[lxxix].

So, in this model of the atom, the fine structure lines can be traced to an off-set of the proton-wave caused by a proton transition enlarging some atoms as in the Helium I shown in Table 10 above where the two proton-waves in the nucleus are slightly offset from one another.

3.3 Heat

Let's imagine an electron being fired at Hydrogen from a LEED gun infinitely far away from the atom at 5 eV. If the electron impacts the Hydrogen ion, the electron will seek a level of ground state. It will pass through antinodes emitting as it approaches ground state, but it cannot come to rest in the 10.2 eV Lyman ground state because it does not have enough energy to transition through all the antinodes to Lyman ground state. So the 5 eV electron must come to rest in Balmer ground state. To arrive at Balmer ground state, the electron will have emitted 3.4 eV of energy as it passed all the antinodes to arrive at Balmer ground state because all antinodes to Lyman state are 13.6 and subtract the 10.2 Lyman antinode and you have 3.4 eV energy of all the antinodes to Balmer ground state. The electron will have 1.6 eV of energy in Balmer ground state, but it cannot transition through the 1.89 eV antinode so it must lose 1.6 eV of energy without radiating 1.6 eV of light which is not a full quantum and come to rest in Balmer ground state. We must stand by the fact that non-full quantum light is not radiated, so we have to ask what happens to the excess energy since ground state must be zero in Neutral Hydrogen. We have seen that partial quanta may form a standing wave if that partial quanta is surrounding a particle. However, as we know from the spectrum of the proton in Neutral Hydrogen, the proton-wave has a strong tendency to maintain its energy as the antinodes do not change with changes in the energy of the electron.

In this case, the excess energy in the above scenario cannot be radiated. We know that heat travels in two ways: one is through radiation at the speed of light in full quanta, and the other is through contact by convection and conduction at less than light speed. We can see now that we can account for excess energy in the electron-wave that is not radiated, through the definition of heat by convection and conduction as contact energy transfer of electron-waves that release partial quantum energy. When the energy is less than E_k i.e. the exact kinetic energy necessary for the node of the proton-wave which energy is equal to a full quantum, it is partial quanta energy and is transferred through contact with particle-waves as heat. In this way, we allow for atomic emission of excess energy that does not interfere with the radiation spectrum of light as there is no radiation absorption or emission, but a flow of heat energy by contact. Because both convection and conduction do not travel at light speed, they cannot be radiated light. Because the standing partial quantum wave around particles is energy that does not travel at light speed, we can imagine other energy that does not travel at light speed. An electron can oscillate a full quantum and produce radiated light, or an electron can pass energy by electron-wave contact and transfer heat.

The electron can convert energy of heat, electricity, and light, by absorbing a quantum from either, the full quantum and re-emitting it with the frequency of its oscillation in the node, or transferring excess partial quantum energy by electron-wave contact. Heat is partial quanta contact energy. Kinetic energy is the particles manner of showing it contains excess energy. The particle oscillates when it contains excess energy in its wave. However, the oscillation does not contain the energy, but is a byproduct of the fact that excess energy is present. The energy is this wave energy. Wave energy is measured here in electron volts and causes a particle to oscillate at the voltage of excess energy that the particle holds in its wave.

We infer that the electron wave is storing the energy for its transitions through the proton-wave as the spectrum tells us that the proton-wave stays the same energy. However, from studying a free electron in a vacuum tube, we infer that the electron tends to radiate excess energy at multiple frequencies when it is not close to the proton, so when the electron is outside of the higher energy proton-wave antinodes, the electron freely releases energy to tend to its natural electron-wave lowest energy state. These releases of energy are still discrete but not as distinguishable as when the electron is in the higher antinodes of the proton-wave. In the same way, an electron in a cloud chamber appears to be moving continuously, but it is moving on the lower energy antinodes of the proton.

Thus energy is released from an atom in only certain quanta due to a physical reason. Under the Law of Atomic Equilibrium, a particle must retain its energy field to mass ratio to remain stable and cannot emit energy in excess of this ratio. Any emission of energy greater than the energy needed for the node is reflected back from the proton energy wave and given back to the particle.

An electron in the Lyman series may gain 10.20 eV of energy and transition, then gain another 1.89eV of energy and transition. This line would appear to be in the Balmer series, but the electron would be in the Lyman excitation state. The electron would emit a Lyman line when the electron fell back to ground state.

Another scenario would be an elastic collision of an electron from a LEED gun. Let's imagine an electron fired at 50 eV from a LEED gun at a Hydrogen ion atom from an infinite distance. The electron would lose 13.6 eV of energy as it approached Lyman ground state, but as it still retains 36.4 eV of energy, it will have an elastic collision with the proton-wave and ionize back out gaining 13.6 eV of energy so that the energy is again 50 eV after the collision.

It is given that the laws of conservation of energy and matter exist in the atom. We will assume that an equilibrium exists between a proton's mass and its energy in its particle wave. We can infer that the proton particle and proton-wave tend to preserve an energy to mass ratio. The mass of a proton particle is generally fixed and the energy of the proton-wave is generally fixed in the spectrum. We infer that the wave of every particle can store energy, but that its natural ground state is the amount of energy to mass ratio that keeps the particle stable described by **e=m**. The Law of Atomic Equilibrium is that all particles tend to natural ground state of their energy wave to particle mass ratio. The Law of Energy is that unradiated energy tends to accumulate in a spherical shape of coherent standing waves in space that creates particles. The universe naturally clumps energy around particles. Where **p** is a particle: **p = m + e**, or **m + e = 1**. Unbroken (in latin: Sollus).

Electrical current is full quanta light energy that travels by electron-wave contact. Electrical current is contact-energy. Heat energy is partial quanta energy that travels through contact of the electron-waves. Radiation of heat and radiation of light are both the same thing, that is, they are full quanta energy that is radiated at the speed of light i.e. they are both light and what Max Planck discovered is exactly what Einstein discovered, in that, heat radiation is light radiation. In heat, the electron releases excess energy in two separate processes. It releases full quanta as light and it releases partial quanta as slow-speed contact energy. The energy is the same, but released in two forms.

If all electron-waves in an atom are aligned so that they are all in contact and all electron-waves of their respective molecules are aligned so that they are in contact, electricity or electrical current can flow freely without resistance as in a superconductor. When all electron-waves are not in contact, the electrical current flows by electron-wave contact and the more irregular the path of contact, the more resistance. When the energy of heat or electricity are equal to the frequency of the vibrating electron in its node, the electron converts the heat or electricity to full quanta light radiation corresponding to the electron's frequency. The radiation of light from heat and the radiation of light from electricity are the same energy being released in the same way by electrons are full quanta light radiation. Therefore, the light from heat radiation, electrical radiation in a light bulb, and radiation from the sun are indistinguishable. Of course, we can manipulate the frequency, but it is the same full quanta energy that originated as excess energy in the energy wave of a particle.

The partial quanta waves belonging to particles have zero velocity relative to their particle. And the wave-particle energy-mass relation is stable around protons and electrons. Because of this protons and electrons are long-lived particles. Energy holds matter together. All mass is surrounded by energy. For these two particles i.e. the proton and electron, there is no associated velocity for the particle-waves. This appears to be a universal law of the atom—a law of equilibrium **e - m = 0**. On the other hand, not all partial quanta waves form around particles in a stable energy to mass relation which creates particles with short half-lives. Where partial quanta energy is absorbed in the particle-wave in excess of **e=m**, the partial quanta can move through contact in the form of heat. Where **e - m > 0**, work is done i.e. energy transfer.

The Universal Law of Equilibrium says that every stable particle resists changes to its particle-mass to energy-wave ratio. For every stable particle, the ratio of particle-mass to energy-wave is equal. Conversely, for every unstable particle, the ratio of particle-mass to energy-wave is an inequality. When **e=m**, the particle is stable.

Potential energy is the measure of excess energy above natural equilibrium state that is stored in the electron-waves, proton-waves, and neutron-waves of the atoms describing the mass that has potential energy.

It should be noted that this does not interfere with the laws of the statistical mechanics of Maxwell, Bozeman, et al. The reason is that an increase in energy stored in the particle causes the particle to vibrate at a higher frequency so this automatically increases the kinetic energy of the system. This never destroys the absorption or emission spectrum as this contact energy is always less than needed for an allowed transition so there is no light radiation or light absorption by partial quantum excess energy transferred through electron-wave contact.

3.4 Electricity

This model of the atom can describe the cause of electricity thus: The Zeeman and Stark Effect are the transition of protons which cause electricity i.e. magnetic changes cause electricity. In induction of electricity, a non-neutral charge pulls on the protons in matter. The protons must shift through the antinodes of other protons. When the proton-particles shift, the proton-waves must absorb energy to shift through the antinode. The proton-wave absorbs energy from the environment in the amount necessary for transition through the proton-wave. This, however, is not the ground state for the proton. When the magnetic force is no longer applied to the proton-particle, it emits the energy gained and shifts back to its ground state. This energy emission is in the form of full quantum energy that is emitted through electron-wave contact.

Protons have proton-waves that extend to infinity as is deduced above from the Balmer-Rydberg equation. The problem of the proton-waves exchanging energy is that the web of infinite proton-waves would cause energy transfer without resistance if energy flowed through the proton-waves. However, this is solved if energy is transferred through the electron-waves. Therefore, the excess energy in the proton-wave can be passed to the electron-waves, because the electron-waves are systems of lower energy than the proton-waves.

So proton-waves shift toward the opposite magnetic pole of the source charge. This causes the proton-wave to gain energy in the amount of the antinode or antinodes that the proton is forced to transition through. This causes the antinodes of the proton-wave to enlarge. This causes the electrons to appear to transition at slightly different antinode levels causing fine structure.

Protons can emit excess energy through light radiation in a particle accelerator. Electron-waves are most likely finite as seen from the fact that there is resistance to electrical current meaning that electron-waves are not always in contact with other electron-waves. And heat and current only flows when atoms are close enough together that the electron-waves are in contact. Therefore, the electron-waves must not always be in contact. The electron-wave is finite.

The proton must release the excess energy that was absorbed by the forced transition by magnetism. For protons, as with electrons, the ground state is defined by the element and its molecular structure. When protons are in the state of matter defined as a solid, the protons and electrons are so closely bound that radiating full quanta as light is constantly reabsorbed by near electrons. Electricity could be thought of as tiny increments of light radiation and absorption in a solid, but it is more likely that the energy emitted when the proton transitions back to ground state is released to the electrons that reside in the proton-wave. These electrons then release the energy through electron contact.

We know that protons as well as electrons are shifted by a magnetic force. Therefore, electrons are also transitioned to a higher energy state by the magnet and when the magnetic force is removed the electrons also settle back down to ground state releasing their energy.

We know that lightning has a different form than sunlight. Therefore, energy is not being emitted in lightning or sparks as pure radiated light. The current of electricity that we see is rather energy that has not been radiated by the electron, but is transferred by electron-wave contact by contact transference of energy.

The proton transitions through the antinode of another proton. The antinode is of a particular energy as is shown in the spectrum. In order to transition through an antinode, the proton must absorb energy from the environment even if the proton is being forced through the antinode by magnetism. Therefore, the energy wave of the proton must gain energy and the proton-wave antinodes increase in energy. Therefore, the electron transitioning in the proton-wave will have different transition energies as the proton has more energy in its proton-wave. The protons sit in the lower antinodes of other protons so the transition energies are in the infrared as is known from Nuclear Magnetic Resonance (NMR). When the magnet is taken away or fluxes away from the proton, the proton-particle returns to ground state and its proton-wave emits the excess energy to the electron which releases it through contact with other electron-waves.

A magnet pulls the proton to a different node. Electrical energy is pulled from the energy in the surrounding environment. Protons are forced to transition. They absorb energy from the environment. The proton-wave engorges by the amount of energy of the antinode of proton transition. This changes the electron transition lines slightly in the affected atoms. The magnet is taken away. The protons transition to ground state and emit energy in the form of non-radiated contact transference. As long as the magnet continues to return to pull the proton from ground state to a higher energy level, the electrical current continues as long as there is a magnetic flux.

It may appear in experiment that electrons cause the flow of electricity, but electrons flow through a cathode ray tube in order to try to correct the imbalance of charge. The spectrum tells us it is the Stark and Zeeman Effect or the magnetic effect that causes electricity. That effect is related to the transition of the protons as is seen in the spectrum where the lines are shifted to the extent of the proton transitions.

In a super conductor, the material must be cold so that the electrons move to lower ground states of less energy away from the nucleus and the atoms become enlarged. The electron-waves then can all become in contact with each other and the "contact light" that is electricity can flow freely.

Induction can be illustrated by dragging a finger across a keyboard and observing the hammers lift inside the piano. In the same way, a magnet pulls apart the protons and electrons shifting them to higher more energetic nodes. Energy must be absorbed during the transitions of protons and electrons through the antinodes in an equal amount to the antinode energy. In induction, a magnet is like the finger dragged across the keyboard lifting the particles to higher states of energy. As the magnet moves away, the particles drop down to lower energy states and release the energy as contact-energy electrical current.

The Maxwell equations describe the radiation of full quantum light. However, there are two fields. One describes radiated full quantum light and one describes the proton-wave radiating through space. The magnetic lines of force existing around bodies are the weaker antinodes of the proton-waves appearing around matter in a substance where the radius of the magnetic poles in the matter is greater than in neutral matter causing a magnetic imbalance such that it distorts the proton-waves. We see that just as the proton-wave becomes coherent around the nucleus, the lower antinodes extending beyond matter become coherent around matter.

3.5 Light

The photon of a radiating electromagnetic wave—light—is best understood as a cycle of a propagating or traveling standing wave with nodes and antinodes where one full cycle is described from node to node. A full quantum is a full photon which is described by one full cycle so that $e\lambda=hc$ meaning one wavelength represents one photon. It can readily be seen that this describes light as a radiating wave that behaves as a particle. This describes a model that more closely resembles the observed phenomena and in a simple way unites the wave theory of light with the photon particle theory of electromagnetic waves. Although contrary to current particle physics theory, this also correlates nicely with the above atomic theory of particles being surrounded by standing waves of energy.

Michael Faraday taught us that light can be affected by magnetism[lxxx]. Faraday's polarization experiment is pictured in Figure 1.

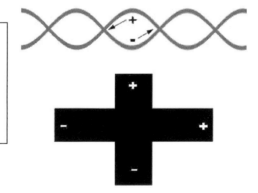

Polarization of light. A bar magnet perpendicular to the light wave has no effect. Poles presenting parallel to the light wave temporarily move the dual charges of the photon.

Figure 1. Picturing the polarization of light in the Faraday experiment.

In Faraday's experiment, the polarization of light caused by the mirror created a plane of charge perpendicular to the direction the light was traveling so that a magnet attracting the light from a perpendicular direction had no effect on the polarization. However, a magnet attracting the charges away from the perpendicular re-polarized the plane of charges in the light to an orientation parallel to the direction of the light. In this alternative view, the EM waves could have charge although the charge would have to always be in a neutral configuration.

The propagation of light in Maxwell's equations describe an electromagnetic field based on the inverse square law. The Maxwell equations are automatically quantized by the inverse square law which describes spheres in space similar to the proton-wave pattern. The quantized energy is between the spheres of the traveling standing wave of EM radiation. Where the radius describes lengths equal to the wavelengths of electromagnetic waves, the spheres describe the nodes and the difference between spheres describes the energy of the wave. In this way, Planck's constant is inherent in Maxwell's equations. Where the unit for r is the wavelength of light, the difference between wavelengths represents the energy of the light. This applies both to electric current and radiated light that both travel at the speed c.

In the neutral atom, the proton's partial quantum light energy of the standing wave behaves like the Neutral Hydrogen atom and each proton-wave of the nucleus is polarized by the electron so that one proton-wave attracts one electron-particle. Or more precisely, each electron attracts the positive poles of a single proton standing wave. In single electron atoms, the light of each proton standing wave becomes polarized so that all the electric poles of the light are aligned so that the positive poles of the light are attracted to the negative electron and the standing wave becomes coherent.

Radiated full quantum light therefore is a traveling standing wave with each photon describing dipoles in its electric field which field is perpendicular to its magnetic field. Each spherical photon contains multitudinous charges which split with distance from the source. In a single wave of radiated light, the dipole of one photon is transposed with respect to the adjoining photon as like charges repel and unlike attract so that light has a neutral charge. Michael Faraday's experiment showed that a magnet can temporarily rotate the poles of the photon to align in the same direction. This phenomenon describes what is happening in the single electron atom.

As the two charges of the photon spread across space in either a radiated light wave or a partial quantum standing particle-wave, the charges lose energy as described by the inverse square law. This can be understood as the charges splitting or dividing over space with distance such that the overall charge remains neutral and the charge weakens with distance.

In quantum entanglement, a photon spherical wave is split between unlike charges.

The mathematical formula for all atoms may thereby be described by how the number of electrons rotates the magnetic poles of the antinodes of the proton standing waves that make up the nucleus.

To visualize how the electrons are traveling on different polarized proton-wave configurations in the same atom, we may take James Clerk Maxwell literally that the electric field is linear and described as a plane and the magnetic field is rotating around the plane which is "plane-polarized light."[lxxxi] So by defining radiated light by Maxwell's wave theory and Einstein's particle theory, we get a traveling wave that has the configuration of a standing wave, that is, it has nodes and antinodes and the energy is carried between the nodes. Therefore, if the electric field is a plane within the antinodes of the proton-wave and the magnetic field is not another plane, but fills the space above and below the electric plane, then we can picture the electrons polarizing the plane to themselves. In this way, the electrons in the same atom could travel across proton-waves of different magnitudes of energy.

In Bose-Einstein Condensate, we see the coherent waves of the atom shown in phase[lxxxii]. Also, we can see how a super atom forms, because the colder an atom becomes, the further out its electrons transition from the proton as Sollism Theory predicts many ground states per element. The spectrum is caused by introducing heat or light energy into the element and most of the spectral lines are in the higher ground states. However, a true Hydrogen ground state near absolute zero would be very far from the Lyman series ultraviolet ground state. The lesser energy antinodes reside further away from the nucleus and each has its own ground state. As the electrons lose energy, the atom becomes larger as the electrons move to lower and lower ground states.

Quantum entanglement is the result of the wave nature of photons in a traveling standing wave. The wave front could be split, but continue to have the properties of the wave.

4. Discussion

I realize that this is a Copernican wrench into a 100-year old system that has been refined, but the spectral evidence appears to allow no other interpretation than a standing wave of energy surrounding each particle in the nucleus. This model has benefits for a better understanding of electromagnetism, motion, heat, chemistry, biology, and other sciences, as well as clarifying much of the experimental results of the last century. The energy of the proton-wave into infinite space will eventually have repercussions in astronomy and explain dark matter and dark energy. The portion of the proton-wave in the nucleus should eventually explain the nuclear weak force. Because this theory proves that both the macro and micro worlds obey the same laws, it can lead to an Atomic Theory of Gravity which includes Newton's outer inverse squared energy, Einstein's inner relativity energy, and the greater spheres of energy exterior to all matter.

It is my belief that this model of the atom will be a first step in clarifying many of the big questions left unanswered by quantum mechanics. I know many experiments that could be cited that would reinforce this view of the atom, however, there are too many to cite in this first article as the concept here is revolutionary and will be controversial for some time to come. Some of my conclusions may need to be refined, but the basic data from the spectral evidence cannot be controverted. I believe this is the beginning of a new era in atomic theory and that many new equations can be found from examining the spectra from this standpoint. Many mathematical relationships still need to be discovered from this basic data.

The current Quantum model explains that the Hydrogen lines are there because they have to be there although they are contrary to the laws of electrodynamics. However, when we introduce causality, it unifies the universe. Unfortunately, science influences philosophy, and I am fully aware of the philosophical implications, and frankly, I prefer an indeterministic philosophy of the universe with some of its more interesting and fantastical notions, but philosophy cannot dictate science. However, no model of the universe has yet been the underlying reality. The goal of a scientific model is to determine measurement and predict nature, so the model must be deterministic to do this, even if, in reality, the universe is indeterministic. Science is still young. Every theory will eventually be revised and become more precise until one day perhaps thousands of years from now, there will be no more anomalies to explain.

Having said that, let us take one more leap of logic and ask what is being compressed by the energy when a particle is created? The only answer that arises is space. Therefore, we may conclude that the only fundamental things in the universe are energy, space, and charge.

4. Materials and Methods

The majority of the research was done directly from the NIST tables published online.

5. Conclusions

This article is the tip of the iceberg and much more research needs to be done in order to find more accurate equations for atomic behavior based upon this model.

Funding: This research received no external funding.

Conflicts of Interest: The author declares no conflict of interest.

References

E. Rutherford, F.R.S., "The Scattering of α and β Particles by Matter and the Structure of the Atom", Philosophical Magazine, Series 6, vol. 21, May 1911, p. 669-688.

Davisson, C. J., "Are Electrons Waves?", Franklin Institute Journal 205, 597 (1928).

Max Planck, "On the Theory of the Energy Distribution Law of the Normal Spectrum," Verh. Dtsch. Phys. Ges. Berlin 2, 237 (1900).

Niels Bohr, "On the Theory of Atoms and Molecules," Philosophical Magazine, Series 6, Vol. 26, July 1913, p.1-25.

NIST "NIST Atomic Spectra Database Lines Data," National Institute of Standards and Technology. Available online: www.physics.nist.gov [accessed April 2017].

Size of the Hydrogen atom given by "Hydrogen Atom," Wikipedia.

Einstein, A. (1907) Theorie der Strahlung und die Theorie der Spezifischen Wärme. Annalen

der Physik, 4, 180-190.

Debye. P. (1912) Zur Theorie der spezifischen Wärme. Annalen der Physik, 4, 789-839.

https://doi.org/10.1002/andp.19123441404

NIST Hydrogen Ionization Energy. Available online: http://physics.nist.gov/cgi-bin/ASD/ie.pl [accessed April 2017].gives an ionization energy of 13.5984 and NIST Constants gives an Energy of 13.60569.

NIST Hydrogen Ionization Energy. Available online: http://physics.nist.gov/cgi-bin/ASD/ie.pl [accessed April 2017].gives an ionization energy of 13.5984

Albert Einstein, "Does the inertia of a body depend on its energy content?" Annalen

der Physik, 1905.

R.Adler, B. Casey, et al., "Vacuum catastrophe: An elementary exposition of the cosmological constant problem," Dept. of Physics, San Fran. State U, 94132, 7 July 1994.

NIST "NIST Atomic Spectra Database Lines Data," National Institute of Standards and Technology, Available online: www.physics,nist.gov [accessed April 2017].

NIST "Strong Lines of Helium," National Institute of Standards and Technology, Available online: www.physics,nist.gov.

NIST "NIST Atomic Spectra Database Lines Data," National Institute of Standards and Technology, Available online: www.physics,nist.gov [accessed April 2017].

NIST "NIST Atomic Spectra Database Lines Data," National Institute of Standards and Technology, Available online: www.physics,nist.gov [accessed April 2017].

Max Planck, "On the Theory of the Energy Distribution Law of the Normal Spectrum," Verh. Dtsch. Phys. Ges. Berlin 2, 237 (1900).

Albert Einstein, "Einstein's Proposal of the Photon Concept—a Translation of the Annalen der Physik Paper of 1905," American Journal of Physics, Vol. 33, No. 5, May 1965.

"Particle accelerator," Wikipedia. [accessed April 2017].

NIST "Strong Lines of Lead," National Institute of Standards and Technology. Available online: www.physics,nist.gov [accessed April 2017]..

Compton, Arthur H., "A Quantum Theory of the Scattering of X-rays by Light Elements" Phys. Rev. 21, 483, (1 May 1923).

Zeeman, P. (1897). "On the influence of magnetism on the nature of the light emitted by a substance". Philosophical Magazine. 5th series. 43: 226–239.

J. Stark, Beobachtungen über den Effekt des elektrischen Feldes auf Spektrallinien I. Quereffekt (Observations of the effect of the electric field on spectral lines I. Transverse effect), Annalen der Physik, vol. 43, pp. 965–983 (1914). Published earlier (1913) in Sitzungsberichten der Kgl. Preuss. Akad. d. Wiss.

"Nuclear Magnetic Resonance Spectroscopy," Wikipedia [accessed April 2017].

van der Waals; J. D. (1873). Over de continuiteit van den gas- en vloeistoftoestand (On the Continuity of the Gaseous and Liquid States) (doctoral dissertation). Universiteit Leiden.

Michael Faraday, "On the magnetization of light and the illumination of magnet lines of force," Philosophical Transactions (1846) Vol. 136-137.

James Clerk Maxwell, "On Physical Lines of Force," Part IV. "The Theory of Molecular Vortices Applied to the Action of Magnetism on Polarized Light," Philosophical Magazine, 1861.

M.H. Anderson, et al, "Observation of Bose-Einstein Condensation in a Dilute Atomic Vapor," Science Vol. 269, pp. 198-201, 14 Jul 1995.

© 2019 by Janeen A. Hunt. Submitted for possible open access publication under the terms and conditions of the Creative Commons Attribution (CC BY) license (http://creativecommons.org/licenses/by/4.0/).

References:

1. S Lakshmibala, *"Heisenberg, Matrix Mechanics and the Uncertainty Principle"*, Resonance, Journal of Science Education, Volume 9, Number 8, August 2004.
2. Carl Rod Nave, *Hyperphysics-Quantum Physics*, Department of Physics and Astronomy, Georgia State University, CD 2005.
3. Dr. Kenjiro Takada, Emeritus professor of Kyushu University, *Microscopic World 1,2&3*, Internet seminar, http://www2.kutl.kyushu-u.ac.jp/seminar/MicroWorld1_E/MicroWorld_1_E.html
4. E. Rutherford, F.R.S., "The Scattering of α and β Particles by Matter and the Structure of the Atom", *Philosophical Magazine,* Series 6, vol. 21, May 1911, p. 669-688.
5. Ernest Rutherford, "The Structure of the Atom", *Philosophical Magazine* Series 6, Volume 27, March 1914, p. 488 – 498.
6. Niels Bohr, "On the Constitution of Atoms and Molecules", *Philosophical Magazine,* Series 6, Volume 26, July 1913, p. 1-25.
7. Gerald Edward Brown and A. D. Jackson, *The Nucleon-Nucleon Interaction*, (1976) North-Holland Publishing, Amsterdam.
8. W. Pauli, "On the Connexion between the Completion of Electron Groups in an Atom with the Complex Structure of Spectra", *Z. Physik* 31, 765ff (1925).
9. R. Machleidt and I. Slaus, *The nucleon-nucleon interaction*, J. Phys. G **27** (2001) R69 (topical review).
10. Kenneth S. Krane, "Introductory Nuclear Physics", Wiley & Sons (1988).
11. N.A. Orr, F.M. Marqu'es y, "Clustering and Correlations at the Neutron Dripline", *Laboratoire de Physique Corpusculaire*, arXiv:nucl-ex/0303005 v1 18 Mar 2003, IN2P3-CNRS, ISMRA et Universit'e de Caen, Boulevard Mar'echal Juin, 14050 Caen cedex, France.
12. Louis de Broglie, "Waves and Quanta.", Nature. Vol. 112 (1923): 540.
13. M R Andrews *et al.* 1997 "Observation of interference between two Bose condensates" Science 275:637.
14. Davisson, C. J., "Are Electrons Waves?", *Franklin Institute Journal* 205, 597 (1928)
15. Finning FW, Holt Fr, "Possible emission of the dineutron in fission." *Nature.* 1950 May 6;165(4201):722.
16. Ieki K, Galonsky A, Sackett D, Kruse JJ, Lynch WG, Morrissey DJ, Orr NA, Sherrill BM, Winger JA, Deak F, Horvath A, Kiss A, Seres Z, Kolata JJ, Warner RE, Humphrey DL., "Is there a bound dineutron in 11Li?", *Phys Rev C Nucl Phys.* 1996 Oct;54(4):1589-1591.
17. J. Bardeen, L. N. Cooper, and J. R. Schrieffer, "Theory of Superconductivity", *Phys. Rev.* 108 (5), 1175 (1957).
18. Carl Rod Naves analysis of the Davisson-Germer Experiment at http://hyperphysics.phy-astr.gsu.edu/hbase/quantum/davger2.html.
19. Richard K. Gehrenbeck, "Physics Today," 31 (1), 34 - 41 (1978), reprinted in the AVS publication "History of Vacuum Science and Technology" (ISBN: 0-88318-437-0).
20. Dr. H. Geiger and E. Marsden, "The Laws of Deflexion of α Particles through Large Angles", *Philosophical Magazine,* Series 6, Volume 25, Number 148, April 1913.
21. Rohlf, James William, "Modern Physics from a to Z0", Wiley, 1994.

22. Porterfield, William W.; Kruse, Walter, "Loschmidt and the Discovery of the Small" *J. Chem. Educ.* 1995 72 870.
23. Niel's Bohr Library, Microfilm BSC 6. N. Bohr Scientific Correspondence through 1922, R , Section 3. Correspondence between N. Bohr and E. Rutherford (96/241 in 1912-22).
24. Fisica Moderna, A História do Modelo de Bohr, http://www.if.ufrgs.br/tex/fis142/fismod/mod06/m_s04.html.
25. Hideki Yukawa, "Meson theory in its developments", Nobel Lecture, December 12, 1949, http://nobelprize.org/physics/laureates/1949/yukawa-lecture.pdf
26. Janine Raven, "The Electron Centennial Page-Probing the Proton", 1997 http://www.dpgraph.com/janine/partons.html.
27. "Radiation — Waves and Quanta (1) Note of Louis de Broglie", presented by Jean Perrin. (Translated from *Comptes rendus*, Vol. 177, 1923, pp. 507-510) http://www.davis-inc.com/physics/broglie/broglie.shtml#N_1_.
28. "Mesons in the 2005 Review of Particle Physics" http://pdg.lbl.gov/2005/listings/mxxxcomb.html.
29. Table of isotope binding energies using data from the Atomic Mass Data Center. http://www.einstein-online.info/en/spotlights/binding_energy/binding_energy/index.txt
30. Prof. James Schombert, 21st Century Science, Lecture 15, Copenhagen Interpretation, University of Oregon, http://abyss.uoregon.edu/~js/21st_century_science/lectures/lec15.html.
31. BG Englert, *Phys. Rev. Lett.*, Vol. 77, 2154, 1996.
32. Brown, Jonathan, The Physical Science Encyclopedia, New York, Cornell University Press, 1980.
33. Hugo Pfoertner, Densest Packings of n Equal Spheres in a Sphere of Radius 1, http://www.randomwalk.de/sphere/insphere/spisbest.txt
34. The ZEUS Experiment, http://www-zeus.desy.de.
35. Albert Einstein, 'The Born-Einstein Letters' Max Born, translated by Irene Born, Macmillan 1971.
36. Richard Feynman, quoted by Mermin, D, Physics Today, 30(4):38, 1985.
37. David Cassidy, "Quantum Mechanics 1925-1927, Heisenberg's breakthrough", American Institute of Physics and David Cassidy, http://www.aip.org/history/heisenberg/p07b.htm.
38. David Cassidy, "Quantum Mechanics 1925-1927, Heisenberg's breakthrough", American Institute of Physics and David Cassidy, http://www.aip.org/history/heisenberg/p08.htm.
39. W. Heisenberg, IAEA Bulletin special supplement (1968), p. 45.
40. Azim O. Barut, Biography of Alfred Landé, http://www.physik.uni-frankfurt.de/paf/paf38.html
41. Irving Langmuir, Journal of the American Chemical Society, 41, 868 (1919).
42. Charles R. Bury, "Langmuir's Theory Of The Arrangement Of Electrons In Atoms And Molecules," *Journal of the American Chemical Society,* Vol. 43, p. 1602-1609 (1921).
43. Niels Bohr, "Atomic Structure", *Nature,* March 24, 1921
44. Edmund C. Stoner, "The Distribution of Electrons among Atomic Levels", *Philosophical Magazine*, October 1924, Series 6, Volume 48, No. 286 p. 719 – 736.
45. http://library.thinkquest.org/19662/low/eng/improved-bohr.html accessed 2006.

46. Edmund Blair Bolles, "Einstein Defiant: Genius versus Genius in the Quantum Revolution" (2004) Joseph Henry Press, p. 239, http://www.nap.edu/openbook/0309089980/html/239.html.
47. Carl Rod Nave, *Hyperphysics-Quantum Physics*, Department of Physics and Astronomy, Georgia State University, CD 2005. http://202.113.227.137/songz/index/hyper/hbase/nuclear/nucuni.html
48. M. Planck, *Deutsch. Phys. Gesell. Verh.* 13.13, Feb. 15, 1911.
49. Davisson CJ, Germer LH., "Reflection of Electrons by a Crystal of Nickel." Proc Natl Acad Sci U S A. 1928 Apr; 14(4): 317-322. http://www.pubmedcentral.gov/picrender.fcgi?tool=pmcentrez&blobtype=pdf&artid=1085484
50. C. J. Davisson and L. H. Germer "Reflection And Refraction Of Electrons By A Crystal Of Nickel," *Proc Natl Acad Sci U S A.* 1928 Aug; 14(8): 619-627. http://www.pubmedcentral.gov/picrender.fcgi?tool=pmcentrez&blobtype=pdf&artid=1085652.
51. Arne Hessenbruch, "Discovering the Nanoscale", D. Baird, A. Nordmann & J. Schummer (eds.), Amsterdam: IOS Press, 2004ISBN: 1-58603-467-7.
52. Quoted by Aage Petersen in "The philosophy of Niels Bohr", Bulletin of the Atomic Scientists (September 1963).
53. Erwin Schrödinger - Biography, Nobel Prize.org http://nobelprize.org/physics/laureates/1933/schrodinger-bio.html
54. Schrödinger Erwin, The Interpretation of Quantum Mechanics. Ox Bow Press, Woodbridge, CN, (1995).
55. Asimov, Isaac *Understanding Physics*, Book I, Chap.1, 1993.
56. Michael Baum, NIST Press Release, "From Supernova to Smoke Ring: Recent Experiments Underscore Weirdness of the Bose-Einstein Condensate" March 12, 2001 http://www.nist.gov/public_affairs/releases/tn6240.htm.
57. Nikolay Noskov, "Brilliance and poverty of quantum mechanics", Translated from Russian by Jury Sarychev "Science of Kazakhstan", 1 (85), January 1-15, 1997.
58. A.A. Sokolov, I.M. Ternov. "Quantum mechanics and nuclear physics." M.: *Prosveshchenie*, 1970, p. 39-40.
59. Catharine H. Colwell, PhysicsLAB Online, Copyright © 1997-2006, http://dev.physicslab.org/Document.aspx?doctype=3&filename=AtomicNuclear_DavissonGermer.xml.
60. A. Einstein, "Physikalische Gesellschaft Zuerich" (On the quantum theory of radiation), Mitteilungen 18, 47 (1916); the same paper appeared also in Physik. Zeitschr. 18, 121 (1917).
61. C.Davisson, L.H.Germer, "Diffraction of Electrons by a Crystal of Nickel", *Phys. Rev.*, 30, 705 (1927).
62. Albert Einstein as quoted in Ideas and Opinions, Albert Einstein, Crown Publishers Inc. 1954, 1982 p. 270, 334.
63. L.J. Wang, A. Kuzmich, A. Dogariu, Nature 406 (2000).277-279.
64. Measurement of the Single-Photon Tunneling Time" by A. M. Steinberg, P. G. Kwiat, and R. Y. Chiao, Physical Review Letters, Vol. 71, page 708; 1993.
65. Excitations in Inverted Two-Level Media' by R. Y. Chiao, A. E. Kozhekin, and G. Kurizki, Physical Review Letters, Vol. 77, page 1254; 1996.

66. M. Mojahedi, E. Schamiloglu, F. Hegeler and K. J. Malloy, "Time-Domain Detection of Superluminal Group Velocity for Single Microwave Pulses," Physical Review E, Vol. 62, pp. 5758-5766, October 2000.
67. R. B. Lindsay, "Note On 'Pendulum' Orbits In Atomic Models." *Proc N. A. S.*, Vol. 13, 1927, p.413.
68. Mike Sutton, "Getting the numbers right – the lonely struggle of Rydberg" Chemistry World, Vol. 1, No. 7, July 2004. [Note: this article errs in saying Balmer used Planck's constant before it was invented. This error is due to Balmer using "h" for his constant of ~365nm.]
69. J R Rydberg, 'On the Structure of the Line Spectra of the Chemical Elements" Phil Mag 5th series, vol 29 (1890), 331–337.
70. Joyce Calarco, "An Historical Overview of the Discovery of the X-Ray", Yale-New Haven Teachers Institute, 2005. http://www.yale.edu/ynhti/curriculum/units/1983/7/83.07.01.x.html.
71. Richard P. Feynman, "QED, The Strange Theory of Light and Matter", Princeton University Press, 1985. ISBN 0-691-08388-6.
72. Karl von Meyenn and Engelbert Schucking, "Wolfgang Pauli" Physics Today, February 2001 Volume 54, Number 2.
73. Alberto P. Guimaraes, "From Lodestone to Supermagnets", Wiley-VCH Verlag BmbH & Co., 2005, p.123, ISBN: 3-527-40557-7.
74. Werner Heisenberg, "Encounters with Einstein and Other Essays on People, Places, and Particles", Princeton University Press, 1983.
75. Andrew Duffy, "Magnetism on the Atomic Level", Boston University, http://webphysics.davidson.edu/physlet_resources/bu_semester2/c16_atomic.html
76. Douglas G. Giancoli, "Physics, Fourth Edition", 1995, Prentice-Hall, ISBN 0-13-102153-2, p. 814.
77. A. H. Compton, Phys. Rev. Ser. 2, 21, 483 (1923).
78. Michael Eckert, "Werner Heisenberg: controversial scientist", Physics World, December 2001. http://physicsweb.org/articles/world/14/12/8/1
79. J. Smith, "Generation of Electrons, Electromagnetic Radiation and Neutrons; Absorption, Fluorescence and Detection" JVS 8/23/97revision of 4/2/83 ms, University of Chicago, 1998. http://cars9.uchicago.edu/JoeSmith/radiation.htm
80. "Introduction to Quantum Mechanics" The University of Nottingham, p.16 http://www.maths.nottingham.ac.uk/personal/scc/qti/coursenotes.pdf
81. Arthur I. Miller, "Science: A thing of beauty", New Scientist magazine, 2537, 04 February 2006, page 50.
82. Rhodes, Richard, "The Making of the Atomic Bomb", 1986, Touchstone Publishers, p.23-28.
83. Rutherford, *Proc. Roy. Soc.,* A, vol. 97, p. 374 (1920).
84. J. Chadwick, FRS, "The Existence of a Neutron", *Proc. Roy. Soc.,* A, 136, p. 692-708.
85. Alex Keller, "The Infancy of Atomic Physics: Hercules in his Cradle", Oxford University Press, 1983, p.160.
86. Raji Heyrovska, "Compton wavelength, Bohr radius, Balmer's formula and g-factors." Cornell University Library. http://arxiv.org/ftp/physics/papers/0401/0401050.pdf

87. F. A. Muller, "The Equivalence Myth of Quantum Mechanics", *Stud. Hist. Phil. Mod. Phys.*, Vol. 28, No. 1, pp. 35-61, 1997.
http://www.phys.uu.nl/~wwwgrnsl/muller/eqmythparti.pdf
88. Gilbert N. Lewis, "The Atom and the Molecule", Journal of the American Chemical Society, Volume 38, 1916, pages 762-786.
89. Faye, Jan, "Copenhagen Interpretation of Quantum Mechanics", The Stanford Encyclopedia of Philosophy (Summer 2002 Edition), Edward N. Zalta (ed.), http://plato.stanford.edu/archives/sum2002/entries/qm-copenhagen/.
90. Bohr, Niels Henrik David, "Essays, 1958-1962, on atomic physics and human knowledge.", New York, Interscience Publishers, 1963.
91. Werner Heisenberg, "Physics and Beyond", New York, Harper and Row, 1971.
92. Edward F. Redish, "The Spinning Electron", University of Maryland
http://www.physics.umd.edu/rgroups/ripe/perg/abp/TPProbs/Problems/R/R17.htm
93. A. Pais, in Physics Today (December 1989) M.J. Klein, in Physics in the Making (North-Holland, Amsterdam, 1989)
http://www.lorentz.leidenuniv.nl/history/spin/spin.html
94. Bohr, Niels Henrik David, "Atomic theory and the description of nature", New York, AMS Press, 1934, 1978.
95. Werner Heisenberg, "Natural Law and the Structure of Matter", Bideford Gazette Lts, Rebel Press, 1970.
96. Pais, "Einstein and the quantum theory", Rev. Mod. Phys. 51, 863 (1979).
97. Max Born, "Natural Philosophy of Cause and Chance", Dover Publications Inc., NY, 1949, 1964.

Footnotes

[i] Darling, David, "Max Planck and the origins of quantum theory" Encyclopedias, http://www.daviddarling.info/encyclopedia/Q/quantum_theory_origins.html retrieved Feb. 24, 2107.

[ii] Asimov, Isaac, "Understanding Physics" Vol. 2., "Quanta" p.129. 1966, 1993 Barnes and Noble.

[iii] Darling, David, "Max Planck and the origins of quantum theory" Encyclopedias, http://www.daviddarling.info/encyclopedia/Q/quantum_theory_origins.html retrieved Feb. 24, 2107.

[iv] Asimov, Isaac, "Understanding Physics" Vol. 1., "The Search for Knowledge" p.5. 1966, 1993 Barnes and Noble.

[v] Louisa Gilder, The Age of Entanglement, Chapter 2, ISBN-13: 978-1400044177.

[vi] Eckert, Michael, ""How Sommerfeld extended Bohr's model of the atom (1913-1916), Equopean Physcial Journal H, December 2013 DOI: 10:1140/epjh/e2013-40052-4

[vii] "Q: What exactly is the vacuum catastrophe and what effects does this have upon our understanding of the universe?," June 16, 2011 by The Physicist, http://www.askamathematician.com/2011/06/q-what-exactly-is-the-vacuum-catastrophe-and-what-effects-does-this-have-upon-our-understanding-of-the-universe/

[viii] R.Adler, B. Casey, et al., "Vacuum catastrophe: An elementary exposition of the cosmological constant problem," Dept. of Physics, San Fran. State U, 94132, 7 July 1994.

[ix] Albert Einstein to Max Born, January 27, 1920, in The Collected Paper of Albert Einstein, vol. 9, tr. A. Hentschel (Princeton: Princton University Press, 2004), Document 284, p. 237.

[x] Louisa Gilder, The Age of Entanglement, ISBN-13: 978-1400044177.

[xi] K. von Meier (ed.), Wolfgang Pauli, Scientific Correspondence with Bohr, Einstein, Heisenberg, a.o., Volume IV, Part I: 1950-1952 (Springer, Berlin and Heidelberg, 1996).

[xii] C.P. Enz, B. Glaus, and G. Oberkofler (eds), Wolfgang Pauli und sein Wirken an der ETH Zurich (vdf Hochschulverlag ETH, Zurich, 1997).

[xiii] Murray Gell-man, 1967 Nobel Prize Speech.

[xiv] Crease, Robert, and Goldhaber, Alfred, "The Quantum Moment: How Planck, Bohr, Einstein, and Heisenberg Taught Us to Love Uncertainty," W.& W. Norton, 2014, pp. 56-57.

[xv] Adam Becker, "What is Real? The Unfinished Quest for the Meaning of Quantum Physics" Hardcover – March 20, 2018, Chapter 3.

[xvi] Cushing, James T., "Bohm's theory: Common Sense dismissed," Studies in History and Philosophy of Science Part A, Vol. 24, Issue 5, December 1993, Pp. 815-842.

[xvii] Rosinger, Elemer E., "What is wrong with von Neumann's theorem on 'no hidden variables'", Quantum Physics, arXiv:quant-ph/0408191.

[xviii] Becker, Adam, "What is Real?", published by Basic Books, First Edition, p. 90, 2018.

[xix] Becker, Adam, "What is Real?", published by Basic Books, First Edition, p. 146, 2018.

[xx] Becker, Adam, "What is Real?", published by Basic Books, First Edition, p. 238, 2018.

[xxi] Jeremy Bernstein, "Hitler's Uranium Club: The Secret Recordings at Farm Hall," Heisenberg said, "It would have been so beautiful if we had won [WWII]." Springer, P. 43 (2103).

[xxii] Baggott, J, The Quantum Story-A History in 40 Moments, p. 105, Oxford University Press, 2011.

[xxiii] Baggott, J, The Quantum Story-A History in 40 Moments, p. 80, Oxford University Press, 2011.

[xxiv] Baggott, J, The Quantum Story-A History in 40 Moments, p. 105-111, Oxford University Press, 2011.

[xxv] Niels Bohr, Collected Works, vol. 6. Ed. Jorgen Kalckar (New York: North-Holland, 1985), p.52.

[xxvi] Stoner, Edmund, Philosophical Magazine, October 1924.

[xxvii] Kossel, W. (1916). "Über Molekülbildung als Frage des Atombaus" [On the formation of molecules as a question of atomic structure]. Annalen der Physik (in German). 354 (3): 229–362.

[xxviii] N. Bohr, "Der Bau der Atome und die physikalischen und chemischen Eigenschaften der Elemente", Zeitschrift fer Physik, vol. 9, pp. 1-67, 1922. http://dx.doi.org/10.1007/BF01326955

[xxix] Annalen der Physik (ser. 4), 22, 180–190, 800 https://einsteinpapers.press.princeton.edu/vol2-trans/228

[xxx] Asimov, Isaac, "Understanding Physics" Vol. 1., p.5, 235. 1966, 1993 Barnes and Noble.

[xxxi] Yukawa, Hideki, "Meson theory in its developments", Nobel Lecture, December 12, 1949.

[xxxii] Asimov, Isaac, "Understanding Physics" Vol. 2., "Quanta" p. 129-131. 1966, 1993 Barnes and Noble.

[xxxiii] Barnes, Joshua E., "Spectra in the Lab", 2005. https://www.ifa.hawaii.edu/users/barnes/ASTR110L_F05/spectralab.html

[xxxiv] "Physics World, "'Quantum microscope' peers into the hydrogen atom", May 23, 2013. http://physicsworld.com/cws/article/news/2013/may/23/quantum-microscope-peers-into-the-hydrogen-atom. Based on the article: A.S.Stodoina, et al., "Hydrogen Atoms under Magnification: Direct Observation of the Nodal Structure of Stark States", Phys. Rev. Lett., 110, 213001—Published 20 May 2013.

[xxxv] A. S. Stodolna, et al., "Hydrogen Atoms under Magnification: Direct Observation of the Nodal Structure of Stark States", Phys. Rev. Lett. 110, 213001 – Published 20 May 2013

[xxxvi] Gilder, Louisa, "The Age of Entanglement," Chap 6.

[xxxvii] Asimov, Isaac, "Understanding Physics" Vol. 2., "Quanta" p. 131. 1966.

[xxxviii] Baggott, J, The Quantum Story-A History in 40 Moments, p. 202, Oxford University Press, 2011.

[xxxix] Baggott, J, The Quantum Story-A History in 40 Moments, p. 290, Oxford University Press, 2011.

[xl] Einstein's Proposal of the Photon Concept—a Translation of the Annalen der Physik Paper of 1905, A. B. Arons and M. B. Peppard, Am. J. Phys. 33, 367 (1965); doi: 10.1119/1.1971542.

[xli] Einstein's Proposal of the Photon Concept—a Translation of the Annalen der Physik Paper of 1905, A. B. Arons and M. B. Peppard, Am. J. Phys. 33, 367 (1965); doi: 10.1119/1.1971542.

[xlii] Richard Feynman video, "Part 3-The Electrons and ther interactions." The Douglas Robb Memorial Lectures, University of Auckland, New Zealand, 1979.

[xliii] R. M. Macfarlan, "Optical Spectral Linewidths in Solids,"Lasers, Spectroscopy and New Ideas, Volume 54 of the series Springer Series in Optical Sciences pp 205-223.

[xliv] Asimov, Isaac, "Understanding Physics" Vol. 2., "Quanta" p. 131. 1966, 1993 Barnes and Noble.

[xlv] W. Heisenberg, "Uber quantenthoretische Umdentung kinematischer und mechanischer Beziehungen", Z. Phys. 33. 879-893 (1925).

[xlvi] S. Weinberg, Dreams of a Final Theory (New York, Pantheon, 1992).

[xlvii] Aitchison, MacManus, Snyder, "Understanding Heisenberg's ''magical' paper of July 1925: a new look at the calculational details," Feb 1, 2008.

[xlviii] Kalckar, Jorgen (ed.), Niels Bohr Collected Works. Volume 6, pp iii-xxvi, 3-495. Foundations of Quantum Physics 1 (1926-1932). Elsevier, `1985.

[xlix] Gilder, Louisa, The Age of Entanglement, Chapter 6, Light Waves and Matter Waves, 2008,

[l] Min Wang, et al., Neuronal basis of age-related working memory decline, Nature, 2011; [DOI: 10.1038/nature10243].

[l] Einstein, Albert (1905-06-30). "Zur Elektrodynamik bewegter Körper" (PDF). Annalen der Physik. 17 (10): 891–921. Bibcode:1905AnP...322..891E. doi:10.1002/andp.19053221004. Retrieved 2017-01-15. See also a digitized version at Wikilivres:Zur Elektrodynamik bewegter Körper.

[li] A. Einstein: Autobiographical Notes translated and edited by Paul Arthur Schilpp, Open Court Publishing Company, Peru, IL. And the Estate of Albert Einstein.

[liii] Becker, Adam, "What is Real?", published by Basic Books, First Edition, p. 113, 2018.

[liv] Baggott, J, The Quantum Story-A History in 40 Moments, p. 64, Oxford University Press, 2011.

[lv] E. Rutherford, F.R.S., "The Scattering of α and β Particles by Matter and the Structure of the Atom", Philosophical Magazine, Series 6, vol. 21, May 1911, p. 669-688.

[lvi] Davisson, C. J., "Are Electrons Waves?", Franklin Institute Journal 205, 597 (1928).

[lvii] Max Planck, "On the Theory of the Energy Distribution Law of the Normal Spectrum," Verh. Dtsch. Phys. Ges. Berlin 2, 237 (1900).

[lviii] Niels Bohr, "On the Theory of Atoms and Molecules," Philosophical Magazine, Series 6, Vol. 26, July 1913, p.1-25.

[lix] NIST "NIST Atomic Spectra Database Lines Data," National Institute of Standards and Technology. Available online: www.physics.nist.gov [accessed April 2017].

[lx] Size of the Hydrogen atom given by "Hydrogen Atom," Wikipedia.

[lxi] Einstein, A. (1907) Theorie der Strahlung und die Theorie der Spezifischen Wärme. Annalen der Physik, 4, 180-190.

[lxii] Debye. P. (1912) Zur Theorie der spezifischen Wärme. Annalen der Physik, 4, 789-839. https://doi.org/10.1002/andp.19123441404

[lxiii] NIST Hydrogen Ionization Energy. Available online: http://physics.nist.gov/cgi-bin/ASD/ie.pl [accessed April 2017].gives an ionization energy of 13.5984 and NIST Constants gives an Energy of 13.60569.

[lxiv] NIST Hydrogen Ionization Energy. Available online: http://physics.nist.gov/cgi-bin/ASD/ie.pl [accessed April 2017].gives an ionization energy of 13.5984

[lxv] Albert Einstein, "Does the inertia of a body depend on its energy content?" Annalen der Physik, 1905.

[lxvi] R.Adler, B. Casey, et al., "Vacuum catastrophe: An elementary exposition of the cosmological constant problem," Dept. of Physics, San Fran. State U, 94132, 7 July 1994.

[lxvii] NIST "NIST Atomic Spectra Database Lines Data," National Institute of Standards and Technology, Available online: www.physics,nist.gov [accessed April 2017].

[lxviii] NIST "Strong Lines of Helium," National Institute of Standards and Technology, Available online: www.physics,nist.gov.

[lxix] NIST "NIST Atomic Spectra Database Lines Data," National Institute of Standards and Technology, Available online: www.physics,nist.gov [accessed April 2017].

[lxx] NIST "NIST Atomic Spectra Database Lines Data," National Institute of Standards and Technology, Available online: www.physics,nist.gov [accessed April 2017].

[lxxi] Max Planck, "On the Theory of the Energy Distribution Law of the Normal Spectrum," Verh. Dtsch. Phys. Ges. Berlin 2, 237 (1900).

[lxxii] Albert Einstein, "Einstein's Proposal of the Photon Concept—a Translation of the Annalen der Physik Paper of 1905," American Journal of Physics, Vol. 33, No. 5, May 1965.

[lxxiii] "Particle accelerator," Wikipedia. [accessed April 2017].

[lxxiv] NIST "Strong Lines of Lead," National Institute of Standards and Technology. Available online: www.physics,nist.gov [accessed April 2017].

[lxxv] Compton, Arthur H., "A Quantum Theory of the Scattering of X-rays by Light Elements" Phys. Rev. 21, 483, (1 May 1923).

[lxxvi] Zeeman, P. (1897). "On the influence of magnetism on the nature of the light emitted by a substance". Philosophical Magazine. 5th series. 43: 226–239.

[lxxvii] J. Stark, Beobachtungen über den Effekt des elektrischen Feldes auf Spektrallinien I. Quereffekt (Observations of the effect of the electric field on spectral lines I. Transverse effect), Annalen der Physik, vol. 43, pp. 965–983 (1914). Published earlier (1913) in Sitzungsberichten der Kgl. Preuss. Akad. d. Wiss.

[lxxviii] "Nuclear Magnetic Resonance Spectroscopy," Wikipedia [accessed April 2017].

[lxxix] van der Waals; J. D. (1873). Over de continuiteit van den gas- en vloeistoftoestand (On the Continuity of the Gaseous and Liquid States) (doctoral dissertation). Universiteit Leiden.

[lxxx] Michael Faraday, "On the magnetization of light and the illumination of magnet lines of force," Philosophical Transactions (1846) Vol. 136-137.

[lxxxi] James Clerk Maxwell, "On Physical Lines of Force," Part IV. "The Theory of Molecular Vortices Applied to the Action of Magnetism on Polarized Light," Philosophical Magazine, 1861.

[lxxxii] M.H. Anderson, et al, "Observation of Bose-Einstein Condensation in a Dilute Atomic Vapor," Science Vol. 269, pp. 198-201, 14 Jul 1995.

Reprinted article is copyrighted under © 2019 by Janeen A. Hunt. Submitted for possible open access publication under the terms and conditions of the Creative Commons Attribution (CC BY) license (http://creativecommons.org/licenses/by/4.0/).

Made in the USA
Middletown, DE
13 January 2024